Medical Parasitology

Khaled ElShewy

Medical Parasitology

A Body System Approach

Springer

Khaled ElShewy
Medical Parasitology
Faculty of Medicine, Suez Canal University
Ismailia, Egypt

ISBN 978-3-031-64738-3 ISBN 978-3-031-64736-9 (eBook)
https://doi.org/10.1007/978-3-031-64736-9

This Springer imprint is published by the registered company Springer Nature Switzerland AG
The registered company address is: Gewerbestrasse 11, 6330 Cham, Switzerland

Preface

This book provides a concise, clear, and affordable overview of the medically important parasites. The book is straightforwardly written for medical students and physicians to equip them for approaching more simplified literature. The book differs from the traditional parasitology systematic books. It is written following the integrated modular system being introduced in many countries. Parasites are grouped according to the body systems they infect. To avoid repetition, parasites with multisystem effects are grouped in the system they mainly infect, and they have been referred to in other systems. The book considers the various stages of the parasite's lifecycle, modes of transmission, epidemiology, and diagnosis of the parasitic disease. So as to enhance the student's ability to integrate between basic and clinical subjects, the text focuses on the pathogenesis and the subsequent clinical presentation of the commonly seen parasitic infections in humans. Although anti-parasitic treatment is an ever-changing area, it has been brought up-to-date in this text just to help practitioners. Each section is followed with an endnote highlighting key points and with some review questions, to facilitate reviewing. The drawings, tables, and colored diagrams involved in this book are totally original and unique aiming at simplifying the pathogenesis and structural morphology. Structural morphology and dosages of anti-parasitic drugs are written in small fonts for those who need further details. Hopefully, readers will find this book useful; comments and suggestions for improving the quality of the book will be most welcome.

Cairo, Egypt — Khaled ElShewy

Contents

1 Introduction to Medical Parasitology

Test your learning and check your understanding of this book's contents: use the "Springer Nature Flashcards" app to access questions using https://sn.pub/oyFLIx. To use the app, please follow the instructions in chapter 1.

1. Go to https://flashcards.springernature.com/login
2. Create a user account by entering your e-mail address and assigning a password.
3. Use the following link to access your SN Flashcards set: ▶ https://sn.pub/oyFLIx

If the link is missing or does not work, please send an e-mail with the subject "SN Flashcards" and the book title to customerservice@springernature.com.

Parasites are the most adverse of all microorganisms. They range from unicellular amebae to multicellular meter-long worms. Studying parasites is very important because of their medical, veterinary, and economic importance. Parasitic diseases represent major health problems, especially among underdeveloped countries and among immunocompromised in developed countries. A high morbidity and mortality have been reported in these countries resulting from parasitic infections. The main task of studying parasites is to control and hence prevent parasites from causing diseases. For this purpose, we have to know what a parasite is and how it can survive, evolve, and cause disease.

The following topics are necessary to start studying medical parasitology:

A. Definition and terms used in medical parasitology.
B. Biology of parasites.
C. Host–parasite interaction.
D. Host defense against the parasite.
E. Classification of medically important parasites.
F. Diagnosis of parasitic disease.

Definitions and Terms Used in Parasitology

What Is a Parasite?

The parasite is a live organism living in or on another organism (host).

Types of Parasites

Parasites could be:

1. **Obligate parasite**: a parasite that cannot survive outside its specific host.
2. **Facultative parasite:** a parasite that can live outside its host as a free-living, and under certain conditions, it can parasitize it.
3. **Endoparasite:** a parasite living inside its host.

K. ElShewy, *Medical Parasitology*, https://doi.org/10.1007/978-3-031-64736-9_1

4. **Ectoparasite:** a parasite that could infest the surface of its host.
5. **Pseudo parasites**: nonparasite objects detected in the human body that resemble parasites, e.g., pollens, plant fibers, molds, and fungi.
6. **Opportunistic parasite:** a nonpathogenic or low virulent parasite that could infect immunocompromised individuals and cause fulminate to fatal infections.
7. **Zoonotic parasite:** a parasite that is normally transmitted between animals and humans.
   - **Parasite isolate:** a particular parasitic sample taken from a field or a community and maintained in the lab.
   - **A strain of parasite:** denotes a group of parasites that share a common characteristic, e.g., the same geographic area, genes, or drug resistance.

Nomenclature of Parasite

The parasite is named using a binomial system; it includes both the genus of the parasite (first name and should start in a capital letter) and its species (second name, starting in a small letter). Both names should be written in italics, e.g., *Schistosoma mansoni*.

What Is a Host?

The host could be a vertebrate or invertebrate animal that harbors the adult parasite or one of its developmental stages during the life cycle. The place where the parasite resides inside its host is termed **habitat,** where the parasite lives and exhibits its adverse effects.

Types of Hosts

1. **Final** (definitive/main): the host that harbors the adult or sexually differentiated form of the parasite (in which sexual reproduction takes place).
2. **Intermediate**: the host that harbors the immature (developmental) or sexually undifferentiated stages of the parasite.
3. **Vector:** an arthropod acting as an intermediate or final host.
4. **Normal host** (natural host): the host in which natural infection causes rapid and complete development of the parasite.
5. **Abnormal host:** it is the host in which the parasite cannot develop as usual; instead, the parasite suffers a retarded or incomplete development.
6. **Reservoir host**: a host that serves as a continuous and sustained source of infection and reinfection of humans.
7. **Dead-end host:** a host that could not transmit the parasite to another host.
8. **Paratenic host:** a host serving as a vehicle to the parasite or one of its developing stages without any developmental changes.

Biology of the Parasite

To survive and reproduce, many parasites evolve through different morphologic stages, environments, and hosts. The sequence of these stages and environments is called the **life cycle** of the parasite (Fig. 1.1).

Parasite could have one or more of the following life cycles:

1. **Direct** (simple life cycle): when the parasite directly infects the same host species without an intermediate host, e.g., from human to human.
2. **Indirect** (complex life cycle): when the parasite infects more than one host species, e.g., human–animal–human.
3. **Free Living:** some parasites live freely in the environment; under certain circumstances, they can infect humans and cause serious diseases.

A detailed understanding of the parasite life cycle facilitates the development of strategies for the management (diagnosis, treatment, prevention, and control) of parasitic infections. This can be achieved through knowing the following from the life cycle:

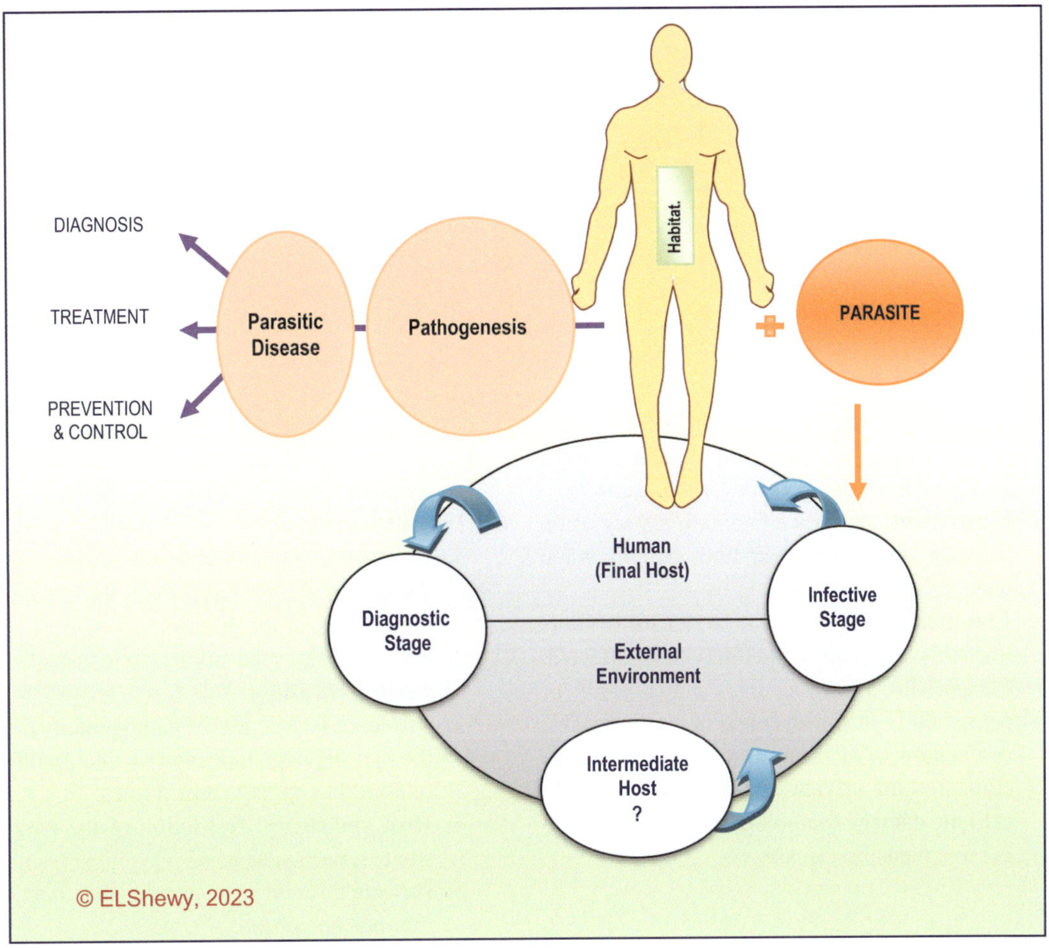

Fig. 1.1 General life cycle of the parasite

- The target **host** of the parasite.
- The **environmental** conditions necessary to complete the parasite's life cycle.
- **Infective stage (I.S):** a developmental stage that is capable of starting and causing the parasitic infection in the host.
- **Mode of infection:** the method(s) by which the infective stage enters the host.
- **Diagnostic stage (D.S):** the stage(s) that could be detected in the host and/or discharged out of him, and its presence denotes parasitic infection.

Reproduction of Parasites in the Host

Generally, helminths do not multiply inside their hosts; instead, they send their eggs or larvae outside the host to complete the life cycle in other hosts. The protozoan parasite, on the other hand, reproduces mainly in their host using one or more than one of the following:

1. **Sexual reproduction (gametogony, sporogony):** occurs in the final host. It is mediated through male and female gametes that join to form a zygote, and it ends with the formation of either sporozoites or oocysts.
2. **Asexual reproduction (schizogony):** occurs in the intermediate host. It comprises multiple asexual mitotic divisions of the parasite; it ends with the formation of many merozoites.
3. **Endodyogony:** asexual reproduction with a multiple division of the nuclear material inside the same parasitic cell.

4. **Binary fission:** transverse or longitudinal division of the parasite into two.

Host–Parasite Interaction

Parasites are frequently associated together or with their hosts for common and shared benefits (protection, nutrition, and dispersion). The pattern of such association could be a symbiotic, a mutualism, a commensalism, or a parasitism:

- **Symbiosis:** it is an obligate association between two organisms that both benefit, e.g., bacterial endosymbionts living inside *Giardia lamblia*.
- **Mutualism:** includes a facultative association between two organisms; both benefit when they are associated.
- **Commensalism:** only one organism benefits from this association; the other is not affected.
- **Parasitism:** a parasite lives either partly or permanently at the expense of its host. This association is of prime importance from the clinical point of view, as it always results in extreme damage (pathological changes), leading to a parasitic disease.

Pathological Changes Caused by Parasites Are Due to the Following Mechanisms

1. Competition with the host for essential nourishments, e.g., vitamins and minerals.
2. Mechanical injury, e.g., large-sized parasites causing pressure atrophy, obstruction of an organ, and/or laceration caused by parasite hooks or lancets.
3. Multiplication of parasite in its host (multiplying parasites are more dangerous than non-multiplying ones).
4. Alteration of the host's homeostasis through metabolites or by-products excreted or secreted by the parasite.
5. Immune reactions stimulated by the parasite could be injurious, e.g., allergy and atopy.
6. Suppression of the host immunity, e.g., immunocompromising parasites.

Factors Affecting the Severity of Parasitic Disease

(a) **Host factors**:
  1. Immune status of the host.
  2. The racial status of the hosts determines the susceptibility to specific species of parasites.
  3. Genetic makeup of the host. (Some hosts are genetically immune to certain parasites.)
  4. Host behavior and economic status; standards of living, poverty, health education, availability of medical care, and healthy water supply.

(b) **Parasite factors:**
  1. Intensity of parasitic infection: generally, the higher the parasitic load, the greater the severity of the disease.
  2. Pathogenicity (the ability to cause disease) of parasites.
  3. Virulence (the degree of pathogenicity) of the parasite: e.g., indicated by fatality rate or ability to damage host tissues.
  4. **Host specificity:** restriction of the parasite to a particular host.
  5. Parasite's ability to evade the host's immune response.

  During parasitic and other infectious diseases, two expressions are commonly used:

  - **Prepatent period:** the time taken from the start of infection until the appearance of the parasite inside or outside the host.
  - **Incubation period:** the time between infection and appearance of the clinical manifestations.

Nomenclature of Parasitic Disease

The resulting parasitic disease is named by omitting the last one or two letters and adding the suffix **asis**, **iasis,** or **iosis** to the stem of the parasite name, e.g., ***Schistosoma***, Schistosom**iasis**.

Host Defense against the Parasite

Host defense against parasitic infection involves both natural (innate) defense mechanisms directed against any parasites and adaptive (acquired) defense mechanisms mainly directed against and stimulated by specific parasites. On the other hand, parasites have evolved several strategies to survive or evade these defense mechanisms.

Natural Host Defenses against Parasites

1. Intact skin, salivary enzymes, tears, and sweat.
2. Stomach acidity (unfavorable for parasites).
3. Genetic makeup of host (some hosts are resistant to specific parasites).
4. Maternal immunity (mother to infant).
5. Wandering phagocytic cells.
6. Complements.

Acquired Defense Mechanisms Include

1. Humoral antibodies production (IgG, M, D, E, A).
2. Cell-mediated (cytotoxic cells).
3. Complements formation.
4. Immune mediators (cytokines, interleukins (ILs), TNF …).

Parasite Evasion Mechanisms Include

1. Intracellular location of the parasite (to hide from direct contact with immune mediators), e.g., *Leishmania* and *Plasmodium*.
2. Intraluminal location of parasites (away from antibodies), e.g., intestinal worms.
3. Antigenic variation and continuous changing of the outer structures of the parasite (hard to be identified), e.g., *Trypanosoma and Schistosoma.*
4. Location inside cyst forms (protective shell), e.g., *Entameba histolytica*.
5. Exhaustion of the immune system of the host through stimulation of producing a huge number of nonspecific antibodies, e.g., *Leishmania donovani* and *Toxoplasma*.

Opportunistic Parasites

These are parasites that normally cause no or mild symptoms when infecting immunocompetent individuals, while in immunocompromised individuals, they are aggressive and even life-threatening (**see Appendix C**).

Causes of Immunocompromisation

1. HIV infection.
2. Anticancer drugs and radiotherapy.
3. Antirejection drugs that are administered during organ transplants.
4. Chronic debilitating disease, e.g., uncontrolled diabetes, renal and liver cell failure.
5. High doses of corticosteroids.
6. Malignancies and bone marrow depression.
7. Severe malnutrition.
8. Chronic parasitic diseases, e.g., trypanosomiasis, schistosomiasis, leishmaniasis, and malaria.

Classification of the Medically Important Parasites

Parasites of medical importance to humans range from single unicellular protozoan parasites to multicellular helminths (worms). In addition to protozoa and helminths, certain arthropods are considered important ectoparasites of enormous medical importance. The following is a simplified overview of the classification of the most important parasites affecting humans (Fig. 1.2).

Protozoa

- Protozoa are unicellular organisms capable of performing all life activities (feeding, reproduction, excretion). Most protozoa have trophozoite active form and inactive quiescent cyst form.
- Most of the protozoa have a direct life cycle; some others have a complex cycle that needs one or more host species to develop.
- Protozoa reproduce through different reproduction mechanisms: sexual (male and female

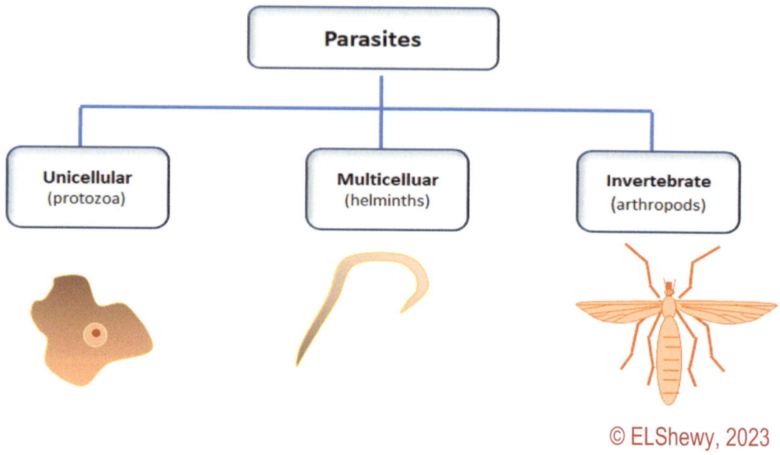

Fig. 1.2 General classification of parasites

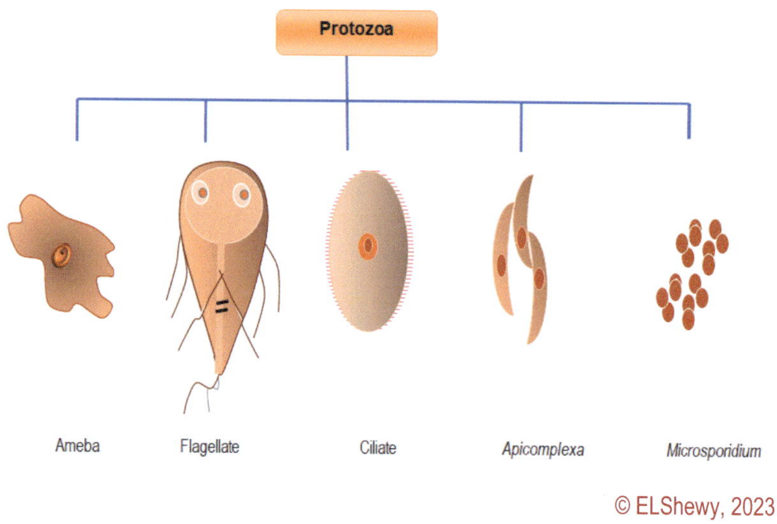

Fig. 1.3 General classification of the protozoa

gametes forming a zygote), asexual (mitotic division), and binary fission.

- Most of the protozoa are highly pathogenic to humans and should be carefully.
- differentiated from the morphologically identical nonpathogenic species.
- Protozoan infection may be carried from one host to another by direct contact, contaminated food, blood, or through an arthropod vector.

According to their mode of movement, protozoan parasites with remarkable medical and clinical importance are grouped into four major groups (Phyla) (Fig. 1.3):

1. ***Sarco-mastigophora***:
   This includes two subgroups of parasites:
   (a) *Sarcodina* (amebae): it includes parasites that move with **pseudopods**, and they are mostly nonpathogenic. Some of these parasites are free living, and some others are pathogenic, e.g., *Entameba histolytica.*
   (b) *Mastigophora* (flagellates): it includes parasites that move with **flagella;** they inhabit the intestine, blood, and the oral and urogenital cavities, e.g., *Giardia.*
2. ***Apicomplexa*:**
   It includes parasites characterized by having a sporozoite with an **apical complex** and a complex life cycle (formerly called *sporozoa*). They usually reproduce by asexual and sexual reproduction in the same or in alternating hosts. They could affect blood, intestine, and other human tissues, e.g., Plasmodium.

3. ***Ciliophora:***
   It includes parasites that move by **cilia**. The most important human protozoan parasite in this group is *Balantidium coli*.
4. ***Microspora (Microsporidia):***
   This group has no mode of movement; it includes intracellular obligatory parasites. They are the smallest of all protozoa. They primarily infect immunocompromised individuals. They have little or no host specificity

Helminths

- Helminths (worms) are multicellular (metazoan) parasites, which range from a few millimeters to several meters long.
- Helminths could infect the human body through the mouth (ingested with contaminated food or drinks) or the skin (piercing the skin or injected by insect bite).
- Helminths are characterized by having variant developmental stages throughout their life, e.g., eggs (ova), larvae, and cercariae; these stages could be helpful in the diagnosis of helminth infection.
- Morphologically, eggs of helminths could possess some structures that help in the diagnosis of the parasite species, e.g., spine, knobs, and/or an operculum.
- The development of the parasite inside the eggs is characteristic and helpful for its diagnosis, e.g., embryonated egg, immature egg, egg with segmented ovum, and so on.
- Larvae are also essential for the identification of certain helminth species, e.g., filariform larvae of hookworm and *Strongyloides*.
- Most of the helminths have a simple life cycle without an intermediate host. However, some others have a complex life cycle with alternating hosts, including snails or murine animals.

Medically important helminths are grouped in two main *Phyla* (Fig. 1.4):

1. **Platyhelminthes** (flatworms):
   Platyhelminthes are characterized by a flat, bilaterally symmetric body. They are mostly hermaphroditic (containing male and female sex organs), and they include:
   - **Trematodes** (flukes): leaf-like elongated worms with cup-shaped muscular suckers for attachment. They are further subgrouped into blood, intestinal, liver, and lung trematodes, e.g., *Schistosoma mansoni*.
   - **Cestodes** (tapeworms): elongated ribbon-like segmented bodies with scolex (head) with specialized organs for attachment (bothria, suckers, hooks); neck; immature;

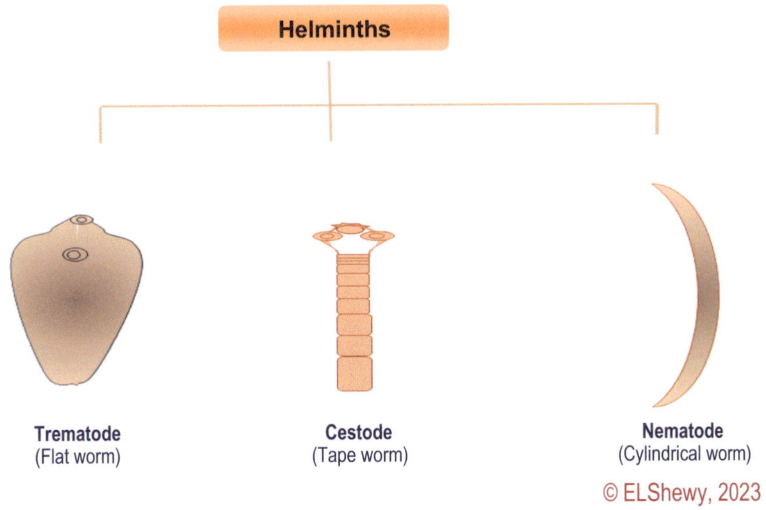

Fig. 1.4 General classification of the helminths

mature and gravid (full of eggs) segments. Cestodes are hermaphroditic, e.g., *Taenia*.

2. **Aschelminths** (roundworms):

   These include rounded worms or **nematodes**. They are elongated cylindrical worms. They have muscular bodies with pointed ends and with no means of attachment. They are mostly heterophroditic (separate sex). Some nematodes lay eggs (oviparous), and some others lay larvae (larviparous). They mainly inhabit the human intestine, e.g., *Ascaris lumbricoides,* and some of them could infect different human tissues, e.g., lymphatic, skin, and subcutaneous tissues, e.g., *Wuchereria bancrofti.*

Arthropods

(See Chapters 18, 19, and 20)

Diagnosis of the Parasitic Diseases

Diagnosis of a parasitic disease follows the same general outlines as any other infectious disease. The following are general outlines for diagnosis of a suspected parasitic infection:

1. Personal history (age, residency, food habits, previous diseases, drug history …).
2. Travel history.
3. Occupational history (food handlers, farmers, health providers …).
4. Complaints of the patient.
5. Clinical examinations.
6. Investigations:
   (a) **Laboratory investigations**.
   (b) Imaging (X-ray, computerized axial tomography (CAT), magnetic resonance imaging (MRI), and ultrasonography).
   (c) Exploratory procedures (e.g., surgical exploration, biopsy, endoscopy).

Laboratory Investigations

Importance of laboratory investigations:

- Physical examination alone is not enough to diagnose most of the parasitic diseases.
- Laboratory investigations are needed to establish an accurate diagnosis and differentiation between pathogenic and nonpathogenic species.
- Laboratory investigations help physicians to select the proper treatment, thus saving time and costs.
- They help to follow the prognosis of the disease and the effect of treatment.

Common Biological Samples Used in Diagnosis of Parasites

- Fresh or preserved stool (feces).
- Urine.
- Blood (whole blood or serum) and lymph.
- CSF and sputum.
- Fluid aspirates (bronchoalveolar and duodenal).
- Tissue biopsies (skin, liver, spleen, mucosa, lymph node).

There are numerous laboratory techniques for each parasite; these can be divided into:

1. **Direct:** it detects the parasite itself, e.g., microscopy, culture, and animal inoculation.
2. **Indirect:** it detects markers denoting the parasitic infection and not necessarily the parasite itself.
3. **Molecular:** Polymerase chain reaction (PCR) and proteomic assays.
4. **Special diagnostic procedures:** endoscopy, colonoscopy and entero-test (Fig. 1.5).

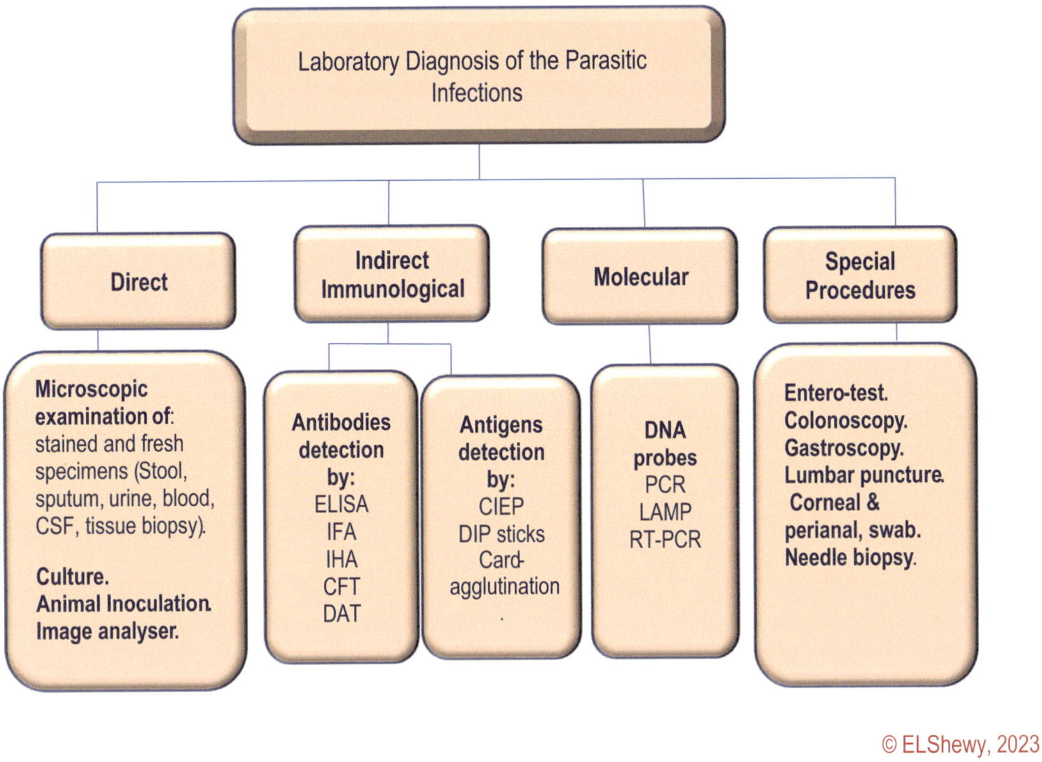

Fig. 1.5 Scheme of laboratory diagnostic procedures used in parasitic infections

Direct Diagnostic Methods

A. **Microscopic Examination:**

This method depends on direct visualization of the parasite or one of its developmental stages (diagnostic stage) in different biological samples.

Biological samples might need concentration using different techniques (sedimentation or floatation) to be examined directly as a wet mount or after fixation and staining.

The following is a simplified listing **(no technical details)** of some microscopic examinations commonly used to diagnose parasitic infections:

(a) **Stool Examination:**

Also called ova and parasites (O&P) examination. It is mostly used for the detection of gastrointestinal parasites; it includes:

- **Macroscopic** examination (detection of physical characteristics of the specimen; color, odor, appearance, and presence of blood, mucus, worms).
- **Microscopic** examination (revealing the diagnostic stages), it includes:

  Wet mount smear (fresh fecal samples).

  Permanent stained (preserved and stained fecal samples): Polyvinyl alcohol (**PVA**), merthiolate iodine formalin (**MIF**), and **10% formalin** are common preservatives used. Trichrome, iodine, and iron and hematoxylin (I&H) are commonly used stains.
- To raise the level of detection, stool examination could be done after the concentration of samples using different floatation (with zinc sulfate or Sheather's solution) or sedimentation techniques (with formalin–ethyl acetate).

  - Examination of fresh unpreserved stool is useful for detection of the fragile trophozoites, demonstration of motile stages of the parasite, and the detection of physical characteristics of the stool sample.
  - **Harada Mori filter paper**, **Charcoal plate,** and **Baermann** techniques are specific techniques commonly used to detect and separate certain helminthic larvae in stool samples.

(b) **Urine Examination:**

Generally, urine is examined for the detection of parasites affecting the genitourinary tract. This includes microscopic examination of a fresh urine sample or concentrated samples using sedimentation or filtration techniques (centrifugation, nucleopore).

(c) **Blood Examination:**

Microscopic examination of blood films to detect blood parasites includes:

i. Thin blood film (TBF):
  - Needs fixation and staining of the film by expert personnel.
  - Accurately determines the species and morphological details of the parasite.

ii. Thick blood film (THBF):
  - Can be used in field works for mass screening by nonexpert personnel.
  - Does not determine the species of the parasite.
  - Does not need fixation or staining.

B. **Culture:**
  - The culture of some biological samples is commonly used. It allows the parasite to multiply and increases the possibility of its detection.
  - Many artificial culture media are used for this purpose. Each parasite or a group of parasites has a particular culture medium with specific ingredients.
  - After the development of the parasite in a culture medium, a sample is taken and processed for microscopic identification, as mentioned earlier.
  - Drawbacks of culture techniques include:
  - It needs technical expertise, and it takes a long time.
  - Difficult to perform under field conditions.
  - Old, preserved, or refrigerated samples are not suitable for culture.

C. **Animal Inoculation:**

As in culture, biological samples could be used to inoculate (intra-peritoneal, intranasal, and subcutaneous) some experimental animals to increase the possibility of detection. Experimental animals (mice, guinea pigs, hamsters) simulate natural hosts; after inoculation of the sample and establishment of experimental infection, specimens are taken and examined as mentioned above. The drawbacks are the same as in culture.

Indirect Diagnostic Methods

These methods do not detect the parasite itself; instead, they detect the parasite antigens or the antibodies raised against it by the host immune system. These diagnostic techniques are called immunological assays; the following are the most commonly used:

A. **Antibodies detection assays (serological assays):**

(Known antigens tested against unknown antibodies in the patient's serum)
  - Agglutination tests: direct agglutination test (**DAT**).
  - Card agglutination tests (**CAT**).
  - Enzyme-linked immunosorbent assay (**ELISA**).
  - Immunofluorescent (**IF**): direct (**DIF**) and indirect (**IFA**).
  - Intradermal skin tests (**ID**): hypersensitivity reaction against injected known parasite antigen(s).

Types of antiparasitic antibodies:
Detection of specific IgM antibodies denotes a recent parasitic infection, while, IgG denotes an old infection, and detection of both of them means a recent on top of an old infection. Meanwhile, IgA is a secretory antibody against surface parasitic infection, e.g., parasites infecting the mucosal surfaces.

B. **Antigens detection assays:**
   (Known antibodies tested against unknown antigens in the biological samples)
   - Counter-immuno-electrophoresis (CIEP).
   - Dipstick assay and coproantigens detection tests.

Types of parasitic antigens:
Secretory antigens, somatic structural antigens, and excretory antigens are common parasitic antigens detected in biological samples.

Advantages of using immunological assays:

- Good for screening a large number of samples in a short time.
- They are valuable in the absence or reduced load of the parasites in the samples.
- High sensitivity and specificity.

Disadvantages of using immunological assays:

- Not all parasites have immunoassay tests.
- Need for equipped labs and expert personnel.
- Cross-reactions between many parasites.
- Detection of antibodies does not tell us if it is a recent or an old infection.

N.B. The principles and uses of the most common immunodiagnostic tests are mentioned in **Appendix (E).**

Molecular Methods

Parasitic cells or their free circulating DNAs can be shed into blood and/or excreta. Detection of such DNA allows the diagnosis of parasitic infections without the whole parasite being present. Polymerase chain reaction (PCR)-based methods are now available for detection and amplification of such DNA molecules. There are several different PCR; the most commonly used are conventional PCR, multiplex PCR, and real-time PCR.

Advantages of Using Molecular Tests

- Detection of asymptomatic infections with low parasitic burden.
- Monitor immunosuppressed patients for probable parasitic infections.
- Screening important parasitic infections in asymptomatic donors.
- Multiplex PCR allows the detection of many parasitic species within a single sample.
- Highly sensitive.
- PCR sequencing distinguishes different parasitic species.
- PCR sequencing helps to identify novel parasitic species.
- Molecular techniques recognize better the epidemiology of parasitic diseases.

However, these techniques need special technical expertise and expenses, which are usually lacking in the field setting in endemic areas. Molecular techniques are mainly confined to research labs.

Special Diagnostic Procedures

As mentioned above, some procedures are necessary for obtaining biological samples to be used for further investigations. The following are the most common procedures used for these purposes:

- **Entero-capsule test:** a gelatine capsule with string is swallowed to get samples from the upper GIT.
- **Endoscopy:** to take specimens from upper GIT.

- **Bronchoalveolar lavage:** to get bronchial secretion.
- **Aspiration and tissue biopsy:** to get a specimen from the spleen, liver, bone marrow, and lymph node.
- **Colonoscopy:** to take specimens from lower GIT.
- **Swabs:** corneal, anal, vaginal, mucosal, and perianal.
- **Lumbar puncture:** to obtain CSF from the spinal column.
- **Skin biopsy:** it includes needle biopsy, punch biopsy, and scratch biopsy.
- **Adhesive tap:** to take a perianal swab.
- **Xenodiagnosis**: the use of insect vectors to diagnose parasitic infection.

Review Questions

1. **Thin blood film (THBF) is better than thick blood film (TBF) because it**
   (a) does not need staining
   (b) is rapid
   (c) keeps RBCs intact
   (d) is cheaper
2. **Indirect diagnostic procedures for parasitic infections include**
   (a) microscopic examination
   (b) animal inoculation
   (c) antibodies detection tests
   (d) culture
3. **Examination of a fresh stool sample is better than a preserved sample because it**
   (a) detects the morphology of the parasite
   (b) helps identification of the parasite's motility
   (c) increase the possibility of detection of a small number of parasites
   (d) is cheaper
4. **Detection of the parasite's antigens is more reliable than the detection of its antibodies because it**
   (a) is cheaper
   (b) denotes a past infection
   (c) is an easy procedure
   (d) denotes an actual infection
5. **Molecular diagnosis of parasitic diseases is better than traditional methods because it**
   (a) needs no technical expertise
   (b) is highly sensitive
   (c) can be done under field conditions
   (d) takes short time
6. **Microspora are characterized by**
   (a) having cilia for movement
   (b) being intracellular parasites
   (c) being large in sizes
   (d) infecting RBCs
7. **Nematodes are characterized by**
   (a) being hermaphrodite
   (b) infecting only the small intestine
   (c) having no attachment organs
   (d) depositing eggs only
8. **Symbiotic association between two organisms means**
   (a) an obligate association and only one organism benefits
   (b) a facultative association and both organisms benefit
   (c) an obligate association and both organisms benefit
   (d) a facultative association and only one organism benefits
9. **Dead-end host means a host that**
   (a) is going to die at the end
   (b) could transmit infection to the same host
   (c) could not transmit the parasite to another host
   (d) could act as a reservoir for the parasite

2 Nematode Helminths of the Small Intestine

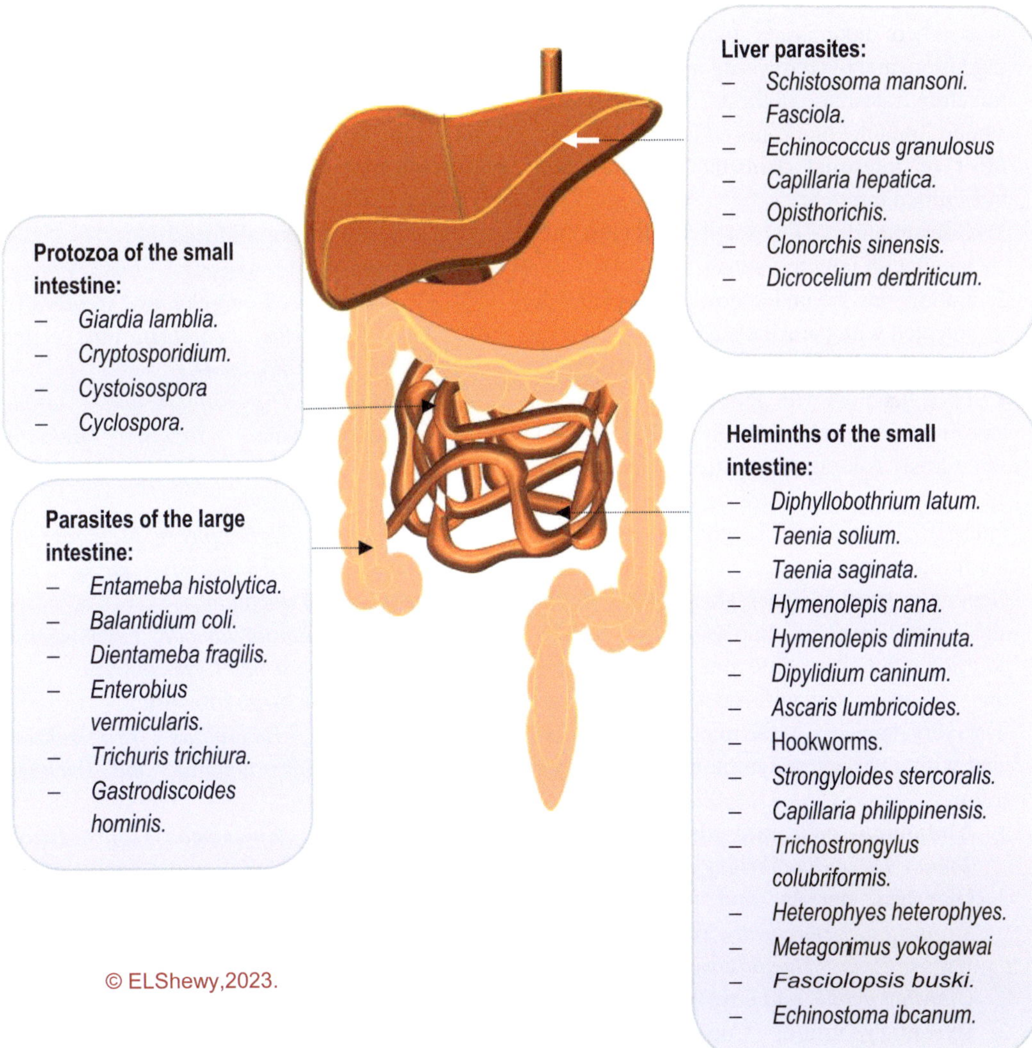

Intestinal Parasites

K. ElShewy, *Medical Parasitology*, https://doi.org/10.1007/978-3-031-64736-9_2

> Test your learning and check your understanding of this book's contents: use the "Springer Nature Flashcards" app to access questions using ▶ https://sn.pub/oyFLIx. To use the app, please follow the instructions in Chap. 1.

Introductory

- Intestinal parasites are those that primarily infect the small intestine, colon, and the rectum.
- Intestinal parasitic infections are most common where poor water quality, inadequate sanitation, and overcrowding are present.
- Infection rates are highest in children and among immunocompromised individuals.
- Most of intestinal parasitic infections are transmitted through:
  1. Consumption of foods and drinks contaminated with infective stages.
  2. Eating raw or undercooked meat or fish infected with parasitic stages.
  3. Flies.
  4. Food handlers with poor hygiene.
- Autoinfection (infection from and to the same host) is a characteristic of some parasites infecting the intestine. This includes internal, external, and retrograde autoinfection.
- Generally speaking, parasites infecting the small intestine cause diarrhea and/or malabsorption, while those infecting the large intestine are associated with constipation and/or dysentery. Parasites of the intestine are associated with a variety of symptoms and signs as follows:
  1. **Abdominal colic and discomfort:** e.g., *Taenia, Diphyllobothrium, andAscaris.*
  2. **Diarrhea, nausea, and vomiting:** e.g., *Strongyloides stercoralis, Trichostrongylus, Hymenolepis, Dipylidium, Heterophyes, Giardia lamblia, and Cryptosporidium.*
  3. **Dysentery and/or Constipation:** e.g., *Entamoeba histolytica, Balantidium coli, and Trichuris trichiura.*
  4. **Pruritus ani:** e.g., *Enterobius vermicularis.*
  5. **Anemia and malnutrition:** e.g., Hookworms, *Diphyllobothrium latum, Trichuris trichiura, and Ascaris lumbricoides.*
  6. **Indigestion, fatty dyspepsia, and distention:** e.g., *Giardia lamblia, Cryptosporidium, and Heterophyes.*
  7. **Malabsorption:** e.g., *Giardia lamblia, Cryptosporidium, Capillaria philippinensis, Cyclospora, and Cystoisospora.*
  8. **Unexplained weight loss and physical retardation:** Caused by most of the chronic intestinal worm infection.

Together with some protozoa, all three forms of helminths could infect human intestine, i.e., trematodes, cestodes, and nematodes. The following are the **commonest** parasites infecting human intestine:

A. **Parasites of the small intestine:**
   - **Nematodes:** *Ascaris lumbricoides,* Hookworms, *Strongyloides stercoralis, Trichostrongylus colubriformis, and Capillaria philippinensis.*
   - **Cestodes:** *Diphyllobothrium latum, Taenia solium, Taenia saginata, Hymenolepis nana & diminuta, and Dipylidium caninum.*
   - **Trematodes***: Heterophyes, Metagonimus yokogawai, Fasciolopsis buski, and Echinostoma ilocanum.*
   - **Protozoa:** *Giardia lamblia, Cryptosporidium, Cyclospora, and Cystoisospora.*

B. **Parasites of the large intestine:**
   - **Helminths:** *Enterobius vermicularis, Gastrodiscoides hominis, and Trichuris trichiura.*
   - **Protozoa:** *Entamoeba histolytica, Balantidium coli, and Dientamoeba fragilis.*

The majority of human parasites are those infecting human intestine. So, in this chapter and in the succeeding chapters (3, 4, 5), we will discuss the helminths (nematodes, cestodes, and trematodes) and the protozoan parasites infecting the small intestine. In Chaps. 6 and 7, we will discuss protozoa and helminth parasites of the large intestine.

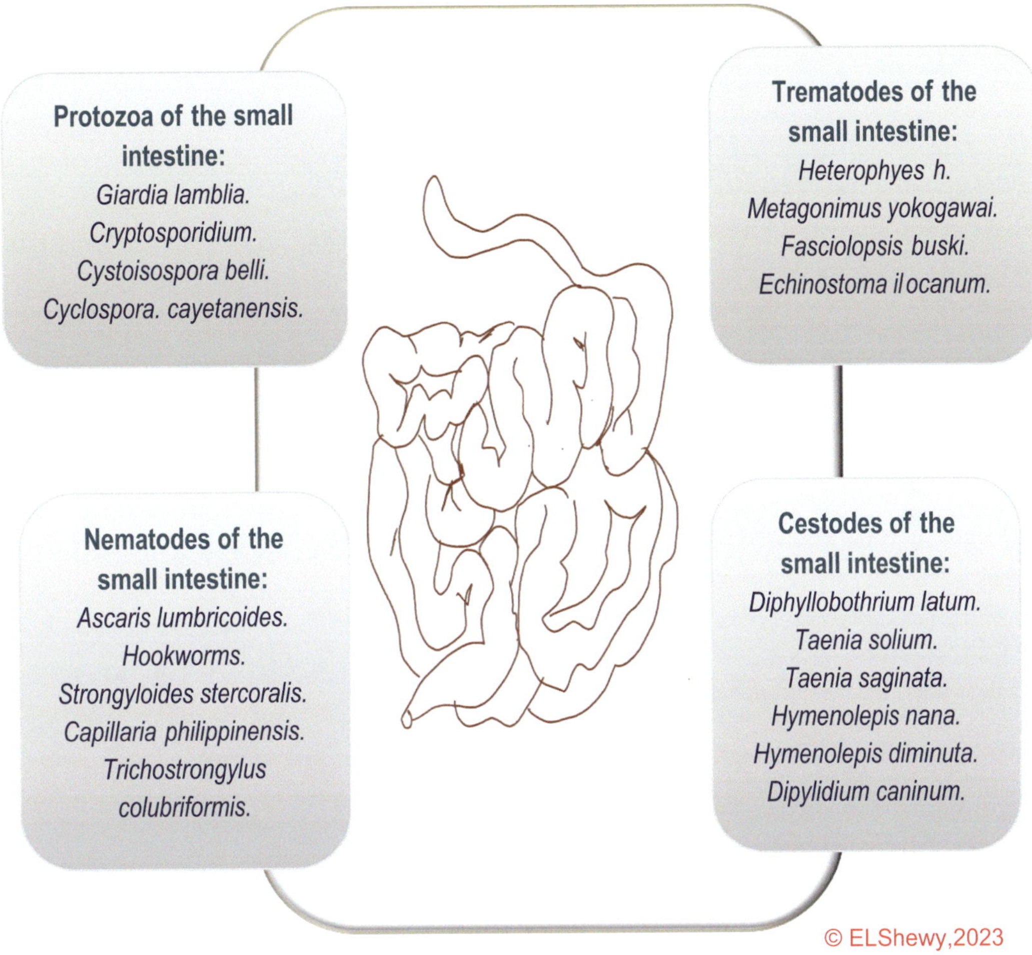

Parasites of the Small Intestine

Ascaris lumbricoides (*A. lumbricoides*)

(Round Worm, Abdominal Snake)

A. lumbricoides is the largest of the intestinal nematodes. Infection with *Ascaris* causes severe intestinal and extraintestinal complications, especially among young children.

Geographical Distribution

A. lumbricoides prevails throughout the temperate and tropical areas. Victims are young individuals, those living in rural areas, slums, crowded camps, and in poor sanitary conditions.

Parasite Morphology

The following are the morphological criteria of *Ascaris* adult male and female worms with their characteristic eggs (Fig. 2.1):

- Adult worm (♀).
  - It has a rounded cylindrical body with tapered ends.
  - 15–30 cm in length and 5–6 mm in width.
- Adult worm (♂).
  - It has a cylindrical body with tapered anterior ends.
  - 10–15 cm in length and 5–6 mm in width.
  - Pot hook posterior end with two spicules.
- Eggs (four types of eggs):
  1. Normal fertile egg:

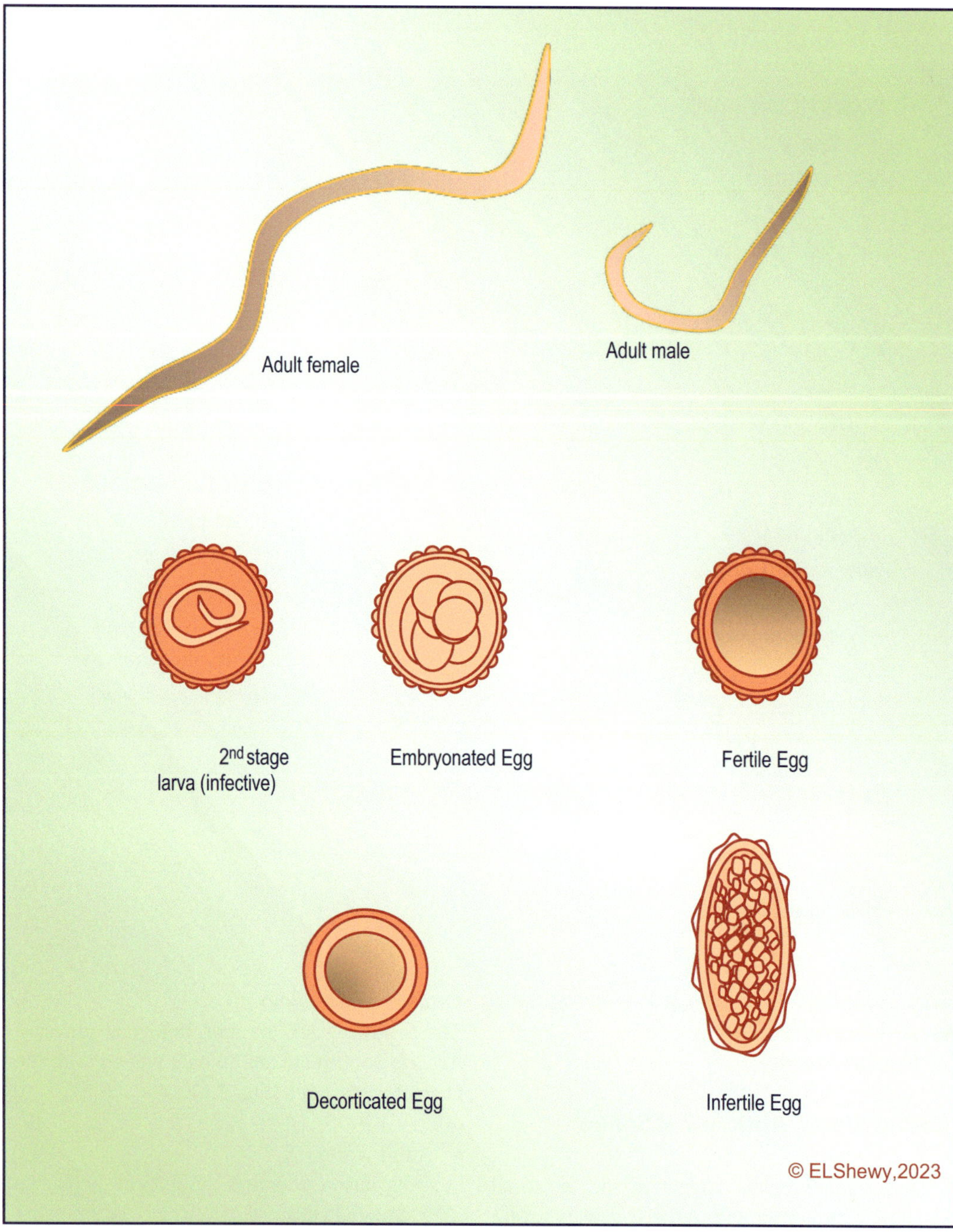

Fig. 2.1 *Ascaris lumbricoides*: adult and egg morphology

- It is oval with a thick smooth shell and a mamillated outer covering.
- 60 x 45 μm.
- Brown yellowish color.

2. Embryonated egg:
   - It has the same criteria as the fertile egg but with an embryonic mass of a segmented ovum or infective larva. It is a long-lived egg (up to many years).

3. Decorticated fertile egg:
   - The outer mamillated coat is lost.
   - The inner shell is present and may stain with bile.
4. Infertile egg:
   - It is larger and narrower than the fertile one.
   - It has very thin outer and inner coats, and it contains many retractile granules.

Transmission

A. lumbricoides is transmitted through:

1. Oral ingestion of food (green leafy vegetables) and beverages contaminated with the infective stage (IS) (embryonated egg with second stage larva).
2. Mechanical transmission of the infective stage through insects (house flies, cockroaches, …).
3. Use of human manure as a fertilizer.
4. Contamination of the municipal water with sewage.

Habitat

Adult worm inhabits the small intestines of human, particularly the upper parts of the small intestine.

Life Cycle

- Female *Ascaris* lays eggs (about 200,000 eggs/day).
- Eggs pass with the patient's stool to the external environment to be fully matured and develop into larvae. After the first molt, second larval stage is formed, and the eggs become infective.
- Ingestion of these eggs is followed by hatching and release of their larvae.
- Larvae penetrate intestinal lymphatics to reach the liver, heart, and finally, the lung capillaries. After 4 days (2 weeks from ingestion), bursting of the alveolar capillaries occurs and larvae molt two times and creep on the mucosal lining of the trachea, pharynx, and then esophagus to reach the small intestine. In the small intestine, they undergo the fourth molt to develop into adult worms (Fig. 2.2).

Final hosts:	Humans.
Infective stage:	Embryonated egg with the second stage larva.
Diagnostic stage:	*Ascaris* eggs; sometimes adult worms could be detected in the patient's stool.
Intermediate host:	None.

N.B.

- *A. lumbricoides* undergoes four molts during its life cycle; one in the external environment, two in the lung, and the fourth in the intestine.
- *Ascaris's* life cycle takes 2–3 months from ingestion to the appearance of its eggs in the stool.
- The life span of the adult worm may reach 1 to 2 years.

Pathogenesis

The following pathological changes are demonstrated during *Ascaris* infection:

1. Migrating larvae are trapped in the alveolar capillaries leading to minor hemorrhage and exudates. Exudates initiate an allergic reaction with eosinophilia and pneumonitis (**parasitic pneumonia**).
2. Adult worms do not attach to the intestinal wall; instead, they keep position through continuous body contraction. They may block the intestinal lumen, especially in heavy infection.
3. Worms, during their movement in the intestine, could migrate to the appendix or biliary duct, causing their obstruction or they could extend out of the intestine to the esophagus. Adult worms rarely come out during vomiting through the mouth or nostrils.
4. Nutritional deficiencies usually occur among heavily untreated cases.
5. Excretory and secretory by-products of the adult worms may lead to allergic manifestations and may affect the central nervous system.

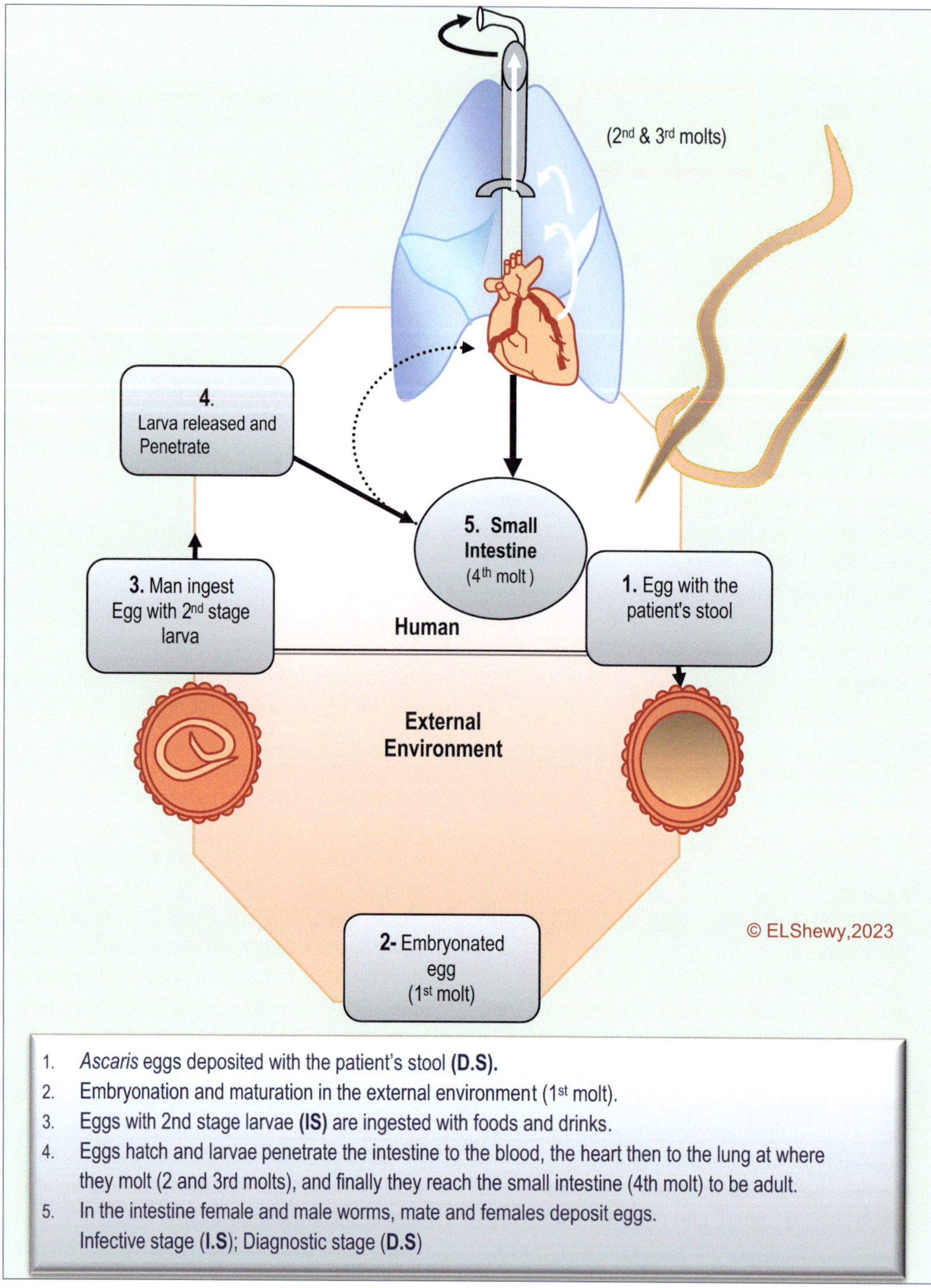

Fig. 2.2 *Ascaris lumbricoides*: life cycle

Clinical Picture

1. Pneumonitis (Loffler's syndrome or parasitic pneumonia), 4 days to 2 weeks after ingestion of the infective stages. It is presented with a wheezy chest, asthmatic manifestations, and blood-tinged sputum that could contain larvae.
2. Abdominal colic, vomiting, bowel disturbance, and hunger pain.
3. Abdominal tenderness, malnutrition, cachexia, sometimes signs of appendicitis or intestinal obstruction, and gall bladder problems.
4. Insomnia and nervous disturbance.
5. Chronic infection in children is associated with retarded mental and physical growth.

Diagnostic Procedures

Diagnosis of *Ascaris* infection includes the following:

- **Direct:**
  - Microscopic stool examination using floatation concentration procedures for detection of the characteristic eggs (fertilized and others).
  - Infertile eggs do not concentrate well with the floatation procedure. So, formalin ethyl acetate (FEA) sedimentation concentration is recommended.
- **Indirect Immunological:**
  Serological tests, e.g., ELISA and IFA, are used for detection of the specific antibodies.
- **Other Investigations:**
  Blood picture, plain and barium X-ray abdomen, and cholangiogram.
  N.B. Microscopic examination of the sputum could reveal the presence of *Ascaris* larvae with many eosinophils.

Treatment

In mixed infection with other parasites, *Ascaris* should be treated first to obviate the untoward activity. The following drugs are commonly used in ascariasis:

1. **Mebendazole** (Vermox):
   - 100 mg orally /2 times daily /3 days for both adults and children.
   - **Side effects (SE):** diarrhea and abdominal pain.
2. **Levamizol hydrochloride** (Ketrax):
   - One single dose of 2 mg/kg/ body weight (b.w).
   - One tablet for preschool age, 2 tablets for 6–12 years, and 3 for adults.
3. **Pyrantel pamoate** (Antiminth):
   - 11 mg/kg/ b.w (maximum 1 gm).
   - **SE**: headache, dizziness, and fever.
4. **Piperazine citrate**:
   - 75 mg/kg /once daily/2 days.
   - Contraindicated in renal and convulsive patients.

N.B. In intestinal obstruction due to ascariasis, the following should be tried:

- Nasogastric suction until no vomiting.
- Nasogastric dose of piperazine 150 mg/kg b.w after 1–2 h.
- If vomiting recurs, suction with another piperazine dose is repeated 3 times, if failed surgical intervention is necessary.

Prevention and Control

1. Following the preventive rules for all helminthic infections (cutting nails, washing hands, good personal hygiene, clean foods, drinks, healthy food handlers, and health education).
2. Targeted treatment of the infected cases.
3. Proper sewage disposal and abstinence from using human manure as fertilizer.
4. Combating flies and other insects.

Hookworms

Ancylostoma duodenale (*A. duodenale*) and *Necator americanus* (*N. americanus*)

Hookworms are small intestinal nematodes infecting both humans and animals. Infection with human species usually causes hookworm disease characterized by anemia and gastrointestinal disturbance, while nonhuman species results in cutaneous lesions.

Geographical Distribution

There are two species of human hookworms:

1. *A. duodenale* prevails in Europe and in the areas of the Mediterranean (Old World).
2. *N. americanus* prevails in the western hemisphere (New World).

Parasite Morphology

The following are the morphological criteria of *A. duodenale.* (Fig. 2.3)

- Adult worm (♀).
  - Small worm with a pointed posterior end, and its size is about 10–13 x 0.6 mm.
  - It has a buccal capsule with ventral teeth and cutting plates.
- Adult worm (♂).
  - It is smaller than the female with a curved posterior end, and 2 copulatory spicules.
  - 8–10 x 0.45 mm.
  - The buccal capsule has ventral teeth.
  - Bursa of the male at the posterior end with 13 rays, tripartite, and with a caudal spine.

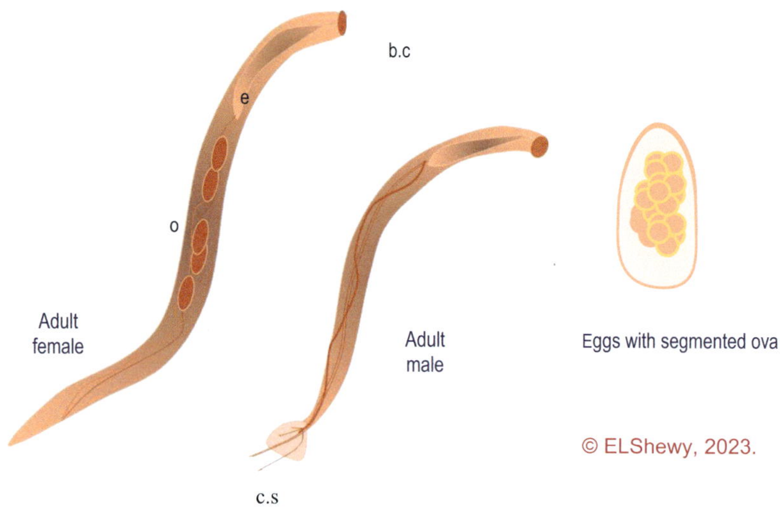

Fig. 2.3 *Ancylostoma duodenale*: morphology

b.c: buccal capsule; e: esophagus; o: ovary; c.b: copulatory bursa; c.s: copulatory spicules.

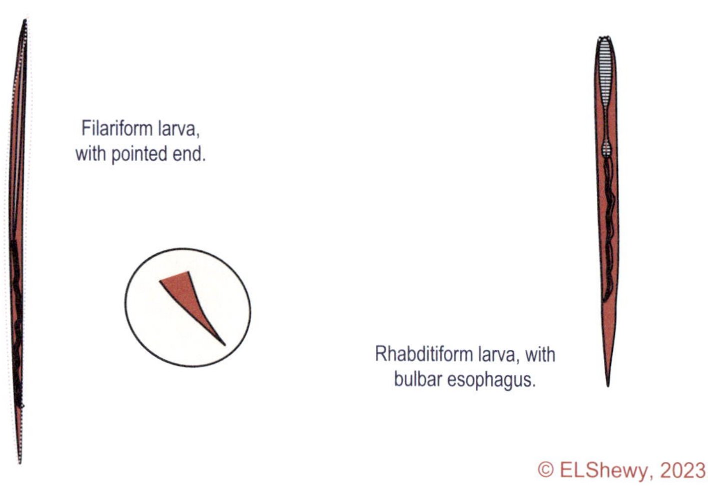

- Larvae:
  - Rhabditiform larva: 250–500 μ x16 μ. It has a bulbar esophagus.
  - Filariform larva: 500–700 μ x 20 μ. It has a pointed tail and a cylindrical esophagus.
- Egg:
  - Oval with bluntly rounded ends and a thin shell.
  - 60 x 40 μm.
  - Transparent in color.
  - It contains a segmented ovum (4 or more cells stage).

N.B.

- Eggs of both worms (*duodenale* and *americanus*) are indistinguishable. However, *N. americanus* adult worms differ from that of *A. duodenale* in the following: Female *N. americanus* has a size of 9–11 x 0.35 mm, with a pointed posterior end, and 2 productive sets,
- The male *N. americanus* has a size of 5–9 x 0.3 mm, curved posterior end, with 2 copulatory spicules.
- The head of *N. americanus* curves dorsally, and its buccal capsule has dorsal cutting plates.
- The male bursa of *N. americanus* has 14 rays, bipartite, with no caudal spine.

Transmission

Hookworm is transmitted through penetration of the human-bared skin by the filariform larva.

Habitat

Adult worms inhibit the small intestine of humans.

Life Cycle

- Female hookworms lay fertilized eggs to pass with stool.
- In a good external environment, hatching occurs and larvae exit (rhabditiform larvae).
- Rhabditiform larvae feed and undergo 2 molts to be filariform larvae (IS) that can live outside, for 2 weeks before finding a host.
- Filariform larvae penetrate the bared skin of humans (attracted by body heat).
- Larvae migrate through the small blood vessels to the heart and lung alveoli, where the third molt occurs, followed by the moving of the larvae to the trachea, larynx, and pharynx to be swallowed to the esophagus and finally settle in the intestine.
- In the small intestine, the fourth molt occurs into adult worms. (Fig. 2.4).

Final hosts:	Humans.
Infective stage:	Filariform larva.
Diagnostic stage:	Eggs and/or larvae in stool or in soil.
Intermediate hosts:	None.

N.B. *A. duodenale* lives 1–5 years; while *N. americanus* lives up to 18 years.

Pathogenesis

The attachment of hookworm to the intestinal mucosa causes minimal atrophy with flattening of the intestinal mucosa. Migrating larvae in the lung causes pneumonitis. However, the main pathological changes are hematologic as follows:

1. Each worm can suck around 0.3 ml of blood/day. *A. duodenale* causes more blood loss than *N. americanus*.
2. Hemoglobin level decreases to 4 gm/dl, producing hypochromic microcytic anemia.
3. Hyperplasic bone marrow is usually remarkable in the case of chronic ancylostomiasis with erythroid and myeloid hyperplasia of the spleen.

Clinical Picture

Hookworm infection, especially among young patients is characterized by:

1. An allergic skin reaction (like ground itch), due to penetration of the skin by larvae.
2. Pneumonitis (parasitic pneumonia) with cough, fever, and dyspnea, particularly, if the number of migrating larvae in the lung tissue is large.
3. Flatulence, epigastric pain, nausea, vomiting, diarrhea start on the day 30th– 45th day.
4. Eggs demonstrated in the patient's stool after 2 months from infection.

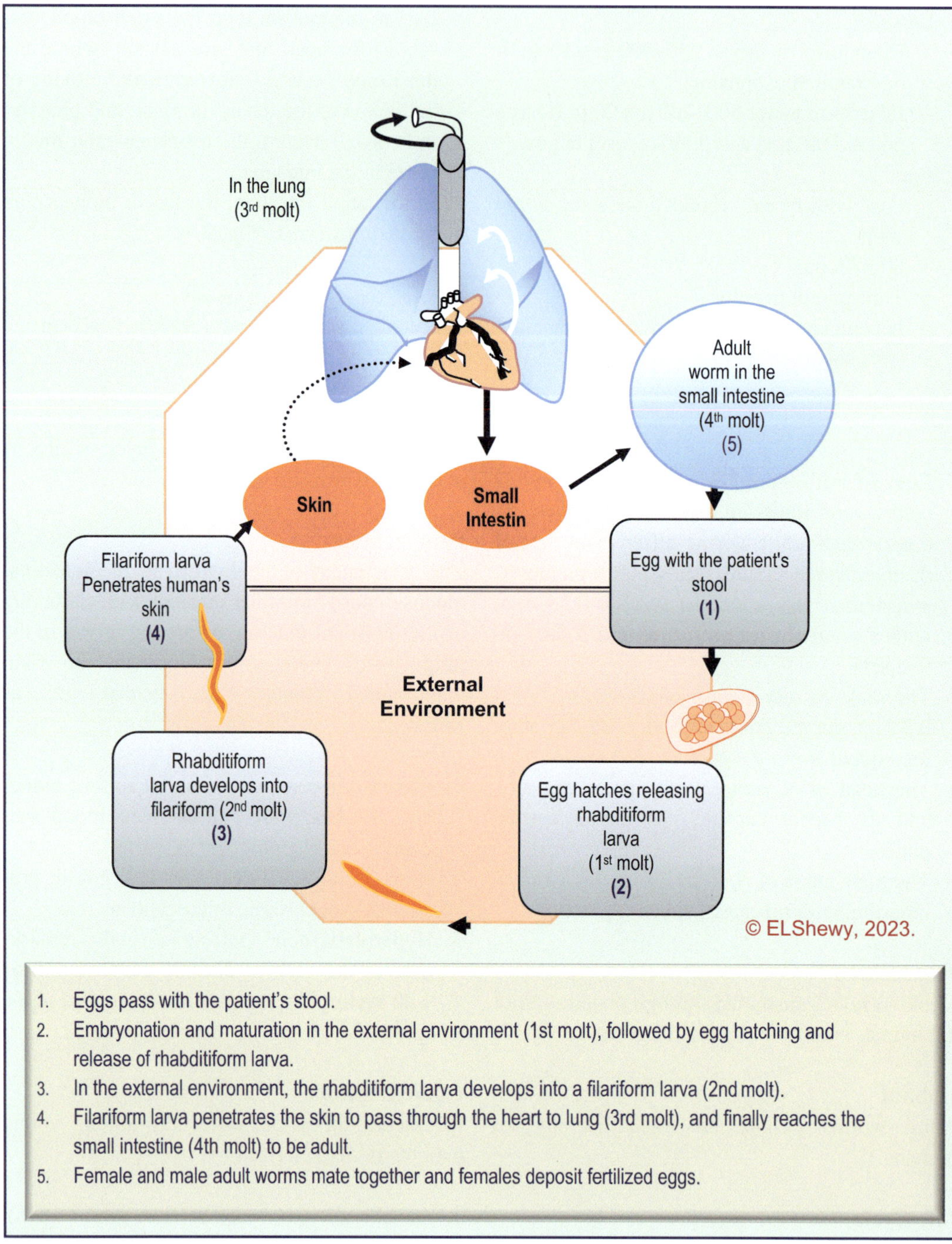

Fig. 2.4 Hookworm: life cycle

5. Mild chronic blood loss leads to symptoms of anemia (microcytic hypochromic anemia), e.g., pallor, easy fatigability, sweating, and lack of concentration. These manifestations range from mild to severe depending on the heaviness of infection.
6. Hypoproteinemia with facial and peripheral edema especially in marked blood loss.

7. **Pica** (habitual ingestion of nonfood substances) is demonstrated in some cases.

Diagnostic Procedures

Diagnosis of hookworm infection includes the following:

- **Direct:**
  - Microscopic stool examination using floatation concentration procedures for detection of the characteristic eggs (oval, thin shell with segmented ova).
  - Stool examination for detection of the characteristic rhabditiform and/or filariform larvae. The following are commonly used techniques:
    (a) Baermann technique.
    (b) Charcoal plate culture.
    (c) Harada Mori test.
- **Indirect Immunological:**
  - Serological tests are used for detection of the specific antibodies, e.g., ELISA, IHA, and IFA.
- **Other Investigations:**
  - Blood picture.

N.B. Microscopic examination of the sputum could reveal the presence of migrating larvae with many eosinophils.

Treatment

The following drugs are commonly used for ancylostomiasis:

1. **Mebendazole**:
   - 100 mg orally/2 times daily / 3 days, for both adults and children.
   - It is good for both *A. duodenale* and *N. americanus.*
   - **SE**: diarrhea and abdominal pain.
2. **Pyrantel pamoate**:
   - 11 mg/ kg/ body weight (maximum 1 gm).
   - For both *A. duodenale* and *N. americanus.*
   - **SE**: headache, dizziness, and fever.
3. **Albendazole:** (preferred by WHO).
   - 400 mg daily/3 days or in a single oral dose for adults.
   - 200 mg as a single oral dose for children.

N.B. The concomitant anemia should be treated by giving ferrous sulfate 200 mg x 3 times/ day/3 months.

Prevention and Control

1. Following the same preventive rules for all helminthic infections including health education and proper sanitation (proper sewage disposal).
2. Walking barefoot should be prohibited.
3. Treatment of the infected cases.
4. Protection of the exposed skin.
5. Abstinence from using human feces as fertilizer.

- **Parasitology:** Hookworm inhabits the small intestine of humans; it is characterized by having hooks and lancets. Infection occurs through filariform larvae by piercing the human skin . Larvae have heart-to-lung circulation with 4 molts. Each adult worm can suck 1/3 ml of blood/daily.
- **Clinical:** Microcytic hypochromic anemia with hyperplasia of the bone marrow, together with pneumonitis and asthmatic attacks. Additionally, a hunger pain, malnutrition, and stunted growth.

Strongyloides stercoralis (*S. stercoralis*)

(Thread Worm)

S. stercoralis is a nematode worm commonly seen among immunocompromised individuals (**opportunistic parasite**). It is responsible for a severe form of enteritis and occasionally, a disseminated fatal hyperinfection.

Geographical Distribution

S. stercoralis has the same distribution as hookworm infection with an increased incidence among immunocompromised individuals.

Parasite Morphology

Eggs of *S. stercoralis* hatch in the host intestine, so they are rarely detected in the stool. Rhabditiform and filariform larvae of *S. stercoralis* are commonly detected in the stool and should be carefully differentiated from those of hookworm (Figs. 2.5 and 2.6).

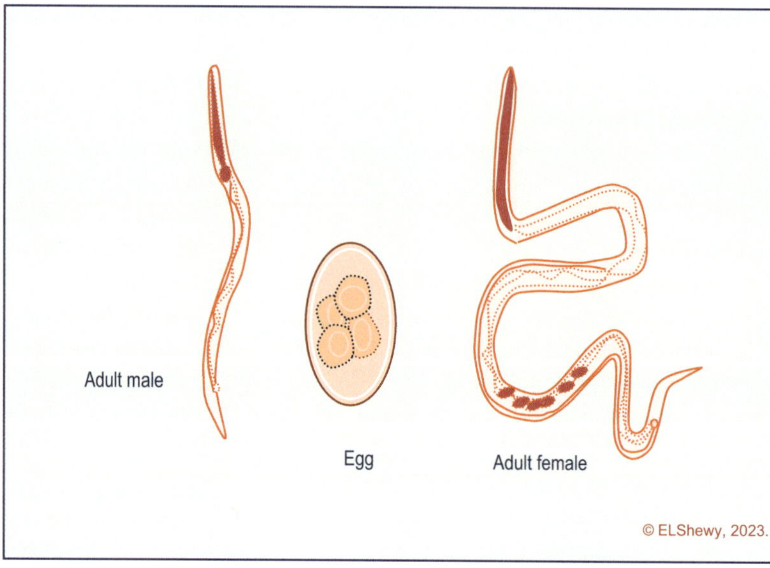

Fig. 2.5 *Strongyloides stercoralis:* morphology

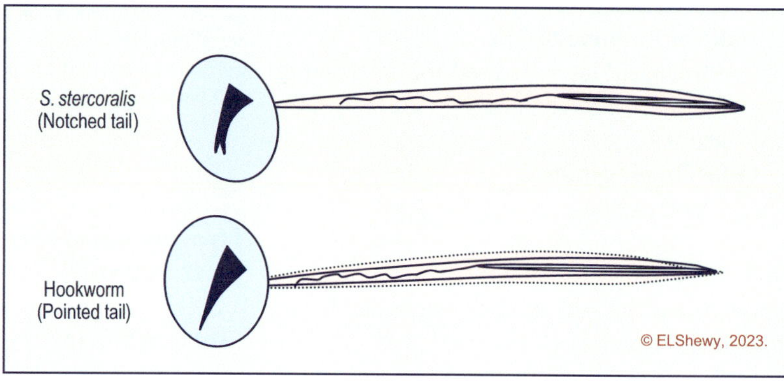

Fig. 2.6 Posterior ends of filariform larvae of *S. stercoralis* and hookworm

- Adult worm (♀):
  - It has a cylindrical, thread-like transparent body.
  - 2 mm x 0.03 mm
  - Its posterior end is pointed, while the anterior one is rounded.
  - It has a short buccal cavity with 4 lips.
  - It has 2 uterine branches laying eggs in the intestine.
- Adult worm (♂).
  - Little is known about it; the female is considered parthenogenic (self-fertilizing).
- Larvae:
  - Rhabditiform larva: 250 μ x16 μ, with bulbar esophagus, its buccal cavity is relatively short (1/3 of the length).
  - Filariform larva: 600 μ x 20 μ**,** straight esophagus extends to ½ of the body**.** The tail is **notched.**
- Egg:
  - Is rarely seen in the stool.
  - It looks like the hookworm egg, but with a uniform size and shape.
  - 60 x 40 μm.
  - It contains a fully segmented ovum.

Transmission

S. stercoralis is transmitted through:

1. Penetration of the host skin by the infective filariform larvae.
2. Internal autoinfection especially among immunocompromised individuals.

Habitat

Adult worm of *S. stercoralis* inhabits the mucosal lining of the small intestine of humans.

Life Cycle

S. stercoralis has 2 life cycles: (Fig. 2.7).

1. **Parasitic life cycle (Direct)**
   - Female *S. stercoralis* lays fertilized eggs in the mucosa of the small intestine. Eggs hatch rapidly to release rhabditiform larvae.
   - Rhabditiform larvae either pass with the patient's stool and directly develop into filariform larvae that penetrate another host skin to reach circulation or they do not pass with the stool and directly transform into filariform larvae that penetrate the intestinal mucosa of the same host to reach circulation.
   - In the circulation, filariform larvae pass from the heart to the lung and through the mucosal lining of the trachea, pharynx, and

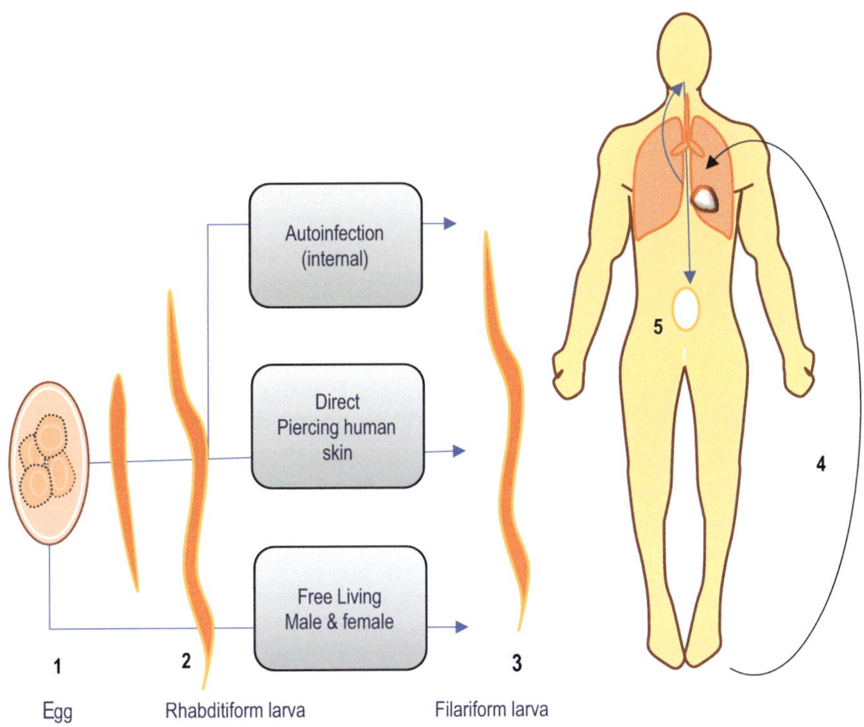

1. Egg hatches inside the human body and releases rhabditiform larva **(D.S)**
2. Rhabditiform larva transforms to filariform larva (I.S)
3. Filariform larva causes the human infection through three routes: internal retrograde autoinfection, direct and free-living cycle with piercing bared human skin.
4. After the heart to lung circulation, filariform larva settles in the small intestine to be adult male and female.
5. The female deposits eggs in the intestine and repeats the cycle.

Fig. 2.7 *Strongyloides stercoralis*: life cycle

then the esophagus to reach the small intestine to mature into adult males and females.

2. **Free living:**
   - Under a suitable external environment; rhabditiform larvae in the stool develop into adult worms that live freely in soil for one or two generations. They then produce free-living rhabditiform larvae that molt twice and develop into filariform larvae. These larvae could pierce bared human skin and start infection.

   To summarize: The filariform larva is the infective stage for *S. stercoralis* and infection is mediated through the following 3 routes:

   1. **External parasitic infection** through the rhabditiform larvae, which transform into the infective filariform larvae.
   2. **External free-living** rhabditiform larvae transforming into adults, then rhabditiform and infective filariform larvae.
   3. **Internal infection (auto or retro infection)** before they pass outside, rhabditiform larvae transform into infective filariform ones that penetrate intestinal mucosa or perianal skin to go to the circulation.

Final hosts:	Humans.
Infective stage:	Filariform larva.
Diagnostic stage:	Rhabditiform and/or filariform larvae in the stool or the soil.
Intermediate hosts:	None.

Pathogenesis

1. Penetration of the skin with filariform larvae results in skin lesions that range from minor to occasionally chronic cutaneous larval migration **(larva currens).**
2. Migrating larvae in the lung cause minor to patchy pneumonitis with larvae in the sputum **(parasitic pneumonia).**
3. Adult worms could be found in any part of the GIT especially, among immunocompromised patients. They usually cause extensive ulcerations and sloughing of the mucosa.

Clinical Picture

Clinical manifestations of strongyloidiasis differ according to the intensity of infection as well as the immune status of the host as follows:

1. **Mild to moderate infection with competent immune response:**
   - Skin lesions (ground itch) due to penetration of the skin with filariform larva.
   - Pneumonitis (fever, cough, and dyspnea), although it is milder than ascariasis.
   - Moderate to severe diarrhea.
   - Malabsorption and steatorrhea.
2. **Heavy infection with *S. stercoralis* (hyperinfection syndrome).**
   - Severe G.I.T symptoms (colic, diarrhea, and steatorrhea).
   - Malnutrition.
   - Ulceration of the small intestine and even the large intestine could be demonstrated (causing dysenteric manifestations).
   - Fever, dyspnea, wheezing, hemoptysis, cough, and general weakness.
3. **Immunocompromised patients (disseminated strongyloidiasis).**
   - Severe pulmonary manifestations (fever, cough, and dyspnea).
   - Severe GIT manifestations (diarrhea, indigestion, ulceration, and bleeding).
   - Dissemination of the infection to other organs, e.g., liver, pancreas, heart, kidneys, and CNS.
   - Death.

Diagnostic Procedures

Diagnosis of *S. stercoralis* infection includes the following:

- **Direct:**
  - Repeated microscopic stool examination to detect the characteristic larvae.
  - Detection of the characteristic rhabditiform and/or filariform larvae is done with the aid of.
    (a) Baermann technique.

    (b) Charcoal plate, agar plate, and filter paper cultures.
    (c) Harada Mori test.

- **Indirect Immunological:**
  - Serological tests (ELISA) are used for the detection of the specific antibodies against *S. stercoralis*. However, there are remarkable cross-reactions with other helminthic infections, e.g., *Ascaris* and filarial worms.
- **Other Investigations:**
  - String-capsule (**Entero-test**) and duodenal aspiration are specially performed for the detection of the characteristic eggs.
  - Sputum examination in disseminated infection and real-time PCR.
  - X-ray of the abdomen, with severe strongyloidiasis, shows loss of mucosal pattern and tubular narrowing.

N.B.

- Differentiation between larvae of hookworm, *S. stercoralis* and *Trichostrongylus* is very important as they could be detected in soil (geohelminths). (Fig. 2.8).
- Demonstration of the *S. stercoralis* eggs in feces, although very rare, should be differentiated from the eggs of hookworm.

Treatment

The following drugs are commonly used for strongyloidiasis. In immuno compromised patients, the treatment of strongyloidiasis must be prolonged.

1. **Thiabendazole** (drug of choice):
   - 25 mg/kg/body weight (b.w) / 2 times daily, for 2 days.
   - **SE:** nausea, vomiting, and vertigo.
2. **Albendazole:**
   - 16 mg /kg /b. w/once daily, for 3 days.
   - The 3 days course should be repeated after 2 weeks.
   - In hyperinfection syndrome, 400 mg daily for 15 days is recommended.
3. **Ivermectin:**
   - 100–200 μg /kg/ b.w, once daily, for 2 days.
   - Cure rate of about 88–100%.

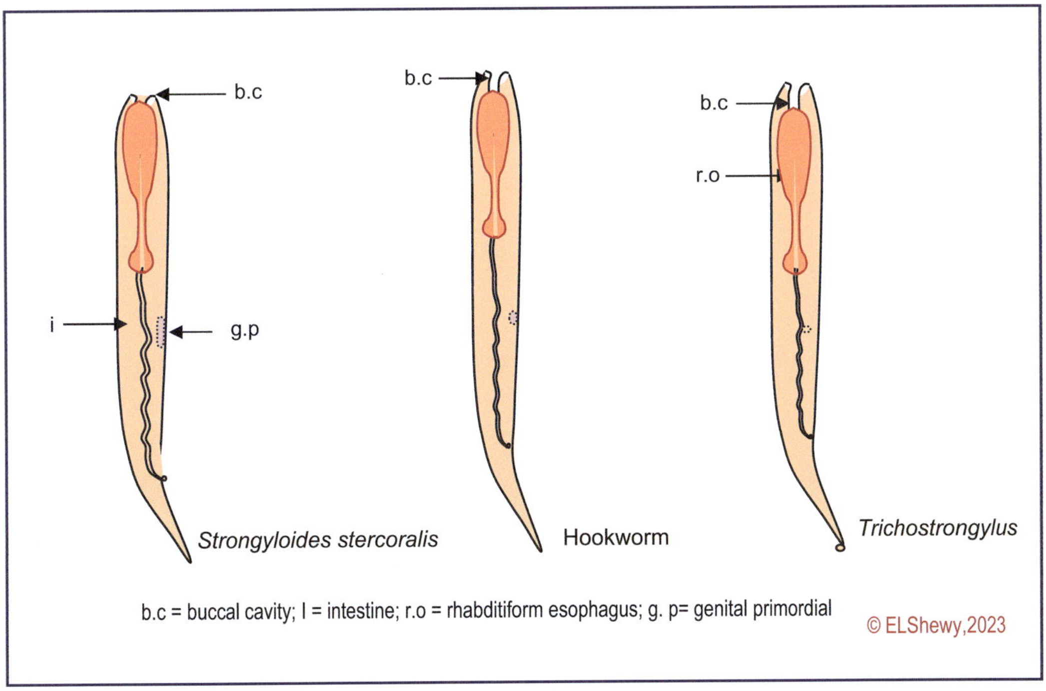

Fig. 2.8 Rhabditiform larvae detected in stool

Prevention and Control

1. Following the same prevention rules for all helminthic infections with health education and proper sanitation.
2. Treatment of the infected cases.
3. Protection of the exposed skin.
4. Abstinence from using human feces as fertilizer.
5. Special care is needed for immunocompromised patients as strongyloidiasis is a common opportunistic infection among them.

- **Parasitology:** *S. Stercoralis* is a very small nematode worm inhabiting the small intestine, and it is characterized by deposition of larvae. It has a free and parasitic life cycle. Infection occurs by many routes, including internal or retro infection.
- **Clinical:** Symptoms and signs differ according to the intensity of infection as well as the host's immune status. Fulminating and disseminating infection among immunocompromised individuals.

Trichostrongylus colubriformis

(*T. colubriformis*)

Trichostrongylus colubriformis is an intestinal nematode that mainly parasitizes herbivorous animals. It occasionally infects humans **(zoonotic),** and it is related to hookworms but with some morphological and life cycle differences.

Geographical Distribution

T. colubriformis infection is reported in Armenia, the Middle East, Japan, Korea, China, and Russia.

Parasite Morphology

Adults male and female of *T. colubriformis* look like hookworms, but with slender hair-like bodies and attenuated anterior ends. (Fig. 2.9).

- Larvae:
  - Rhabditiform larva is similar to hookworm rhabditiform larva but with a peripheral knob at the posterior end.
- Egg:
  - It looks like the hookworm egg, with a more pointed end.
  - 90 x 45 μm.
  - It has a very thin shell.
  - It contains a fully segmented ovum (32-cell stage).

Transmission

T. colubriformis is transmitted through occasional **ingestion** of the infective larva with human foods or drinks.

Habitat

T. colubriformis inhabits the small intestine of herbivores (cattle, sheep, goats, and donkeys) and occasionally humans.

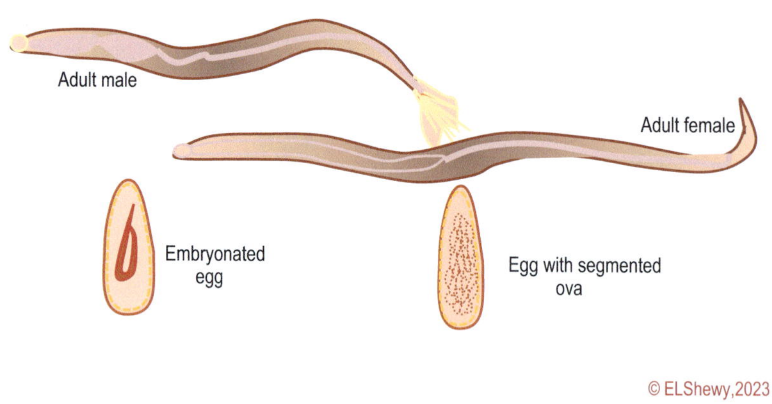

Fig. 2.9 *Trichostrongylus colubriformis:* morphology

Life Cycle

The life cycle of *T. colubriformis* is postulated in (Fig. 2.10).

Final hosts:	Humans.
Infective stage:	Filariform larva.
Diagnostic stages:	Eggs and/or larvae in the stool.
Intermediate host:	None.
Reservoir hosts:	Cattle, goats, and sheep.

Pathogenesis

Larvae of *T. colubriformis* **do not undergo** heart-to-lung circulation. They attach themselves to the intestinal mucosa and grow into adults ingesting blood and causing chronic blood loss with eosinophilia.

Clinical Picture

Light infection passed unnoticed, in heavy infection, the patient suffers anemic manifestations (pallor, easy fatigability, sweating, and lack of concentration), and a marked emaciation.

Diagnostic Procedures

Diagnosis of *T. colubriformis* infection is mainly done through microscopic stool examination using direct and floatation concentration techniques. Larvae of hookworms, *S. stercoralis* and *T. colubriformis* should be differentiated (Fig. 2.8).

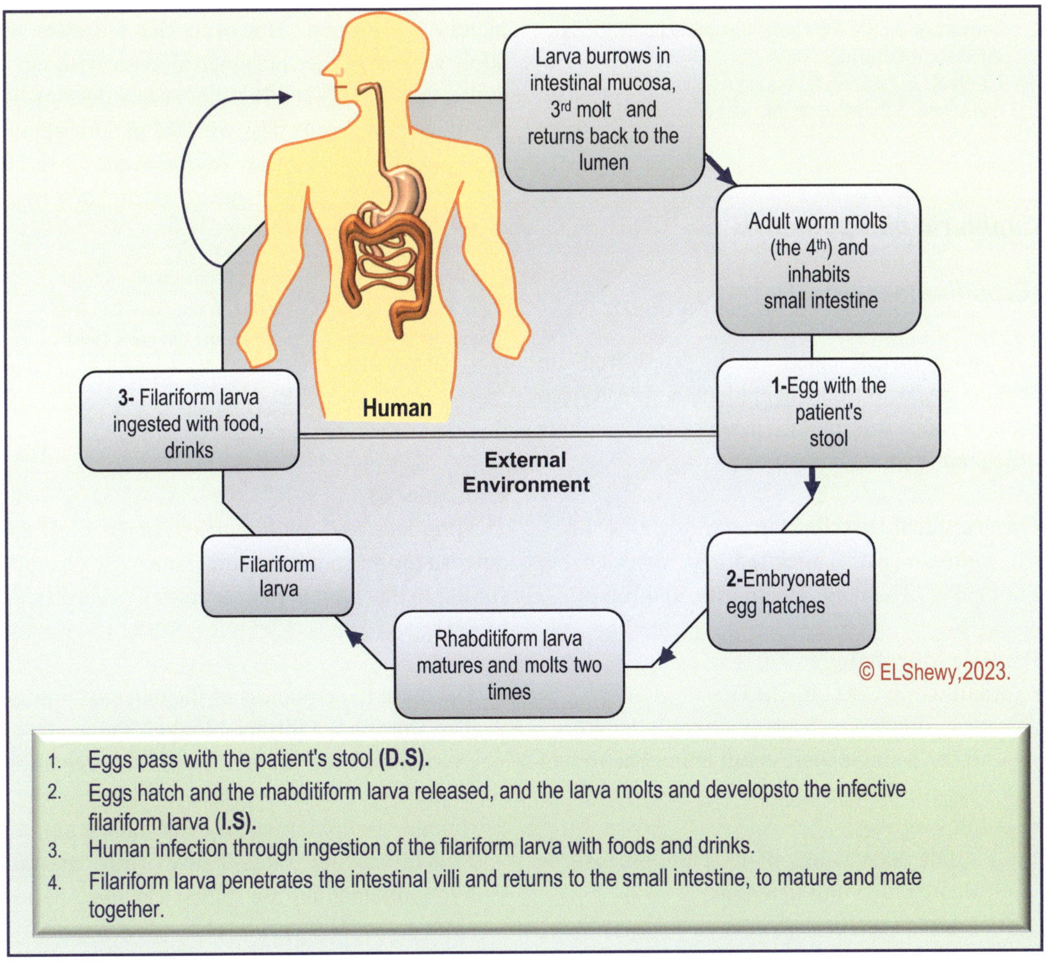

Fig. 2.10 *Trichostrongylus colubriformis:* life cycle

Treatment

Mebendazole:

- 100 mg orally / 2 times /3 days for both adults and children.
- It is the drug of choice.
- **SE**: diarrhea and abdominal pain.

Prevention and Control

1. Following the same prevention rules for all helminthic infections.
2. Health education and proper food and drink sanitation.
3. Treatment of the infected cases.
4. Abstinence from using human feces as fertilizer.

- **Parasitology:** *T. Colubriformis* is a small zoonotic nematode which mainly infects herbivores. It looks like hookworms, and occasionally, it infects humans on eating foodstuffs contaminated with the infective filariform larvae.
- **Clinical:** Light infection is asymptomatic, heavy one causes a marked anemia and emaciation.

Capillaria philippinensis

(*C. philippinensis*)

C. philippinensis was first discovered in Philippines, with frequent epidemics spread from time to time. It is mainly a **zoonotic** nematode infrequently infect human.

Geographical Distribution

C. philippinensis infection is reported in Philippines, Thailand, Japan, Iran, and Egypt.

Parasite Morphology

- Adult worm (♀):(Fig. 2.11).
  - It is slender in shape, 4–6 mm in length.
  - It has a single uterus containing one row of eggs or larvae.
- Adult worm (♂).
  - Adult male worm of C. philippinensis is slender, 4–5 mm in length.
  - It has a sheathed spicule.
- Egg:
  - It is barrel-shaped with bipolar plugs (do not protrude), 50 x 25 μm.
  - It has a thick shell with a pitted outer cover.
  - It contains a segmented ovum.

Transmission

C. philippinensis is transmitted through eating undercooked fresh and brackish water fish containing the larval stages.

Habitat

C. philippinensis inhabits the small intestine (jejunum) of fish-eating birds and humans. Occasionally, it could be detected in the colon, stomach, or even in the esophagus.

Life Cycle

The life cycle of *C. philippinensis* is not completely understood. However, the presence of adult worms in the intestinal mucosa with eggs and/or larva suggests that the worms develop in the mucosa and eggs pass with the patient's stool to be taken by brackish or freshwater fish in which larval stages have been detected. (Fig. 2.12).

Final hosts:	Birds, occasionally, humans.
Infective stage:	Larvae in undercooked fish.
Diagnostic stage:	Eggs in the patient's stool.
Intermediate host:	Fish.
Reservoir hosts:	Birds.

Pathogenesis

The presence of *C. philippinensis* in the intestinal mucosa results in nonspecific pathologic changes similar to that seen in malabsorption syndrome as follows:

1. Decrease the thickness of the mucosal lining.
2. The intestinal villi are blunted, flattened, or even completely obliterated, with deepening of the crypts of Lieberkühn.
3. Submucosal inflammatory cell infiltration.
4. Protein-wasting enteropathy with marked hypoproteinemia, low blood calcium, potassium and cholesterol levels.
5. Mild eosinophilia.

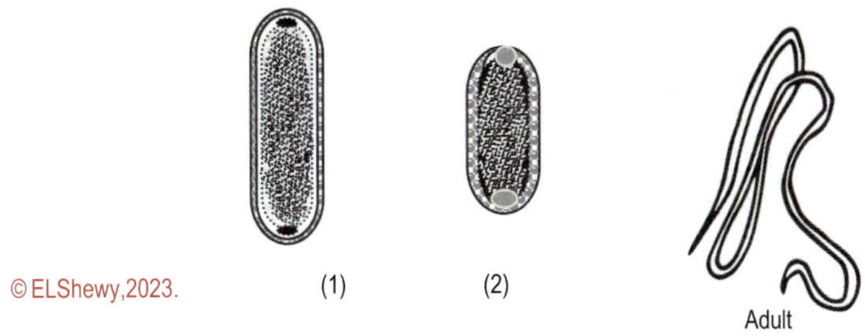

Fig. 2.11 Eggs of *C. hepatica* (1), *C. philippinensis* (2) and adult worm of *Capillaria*

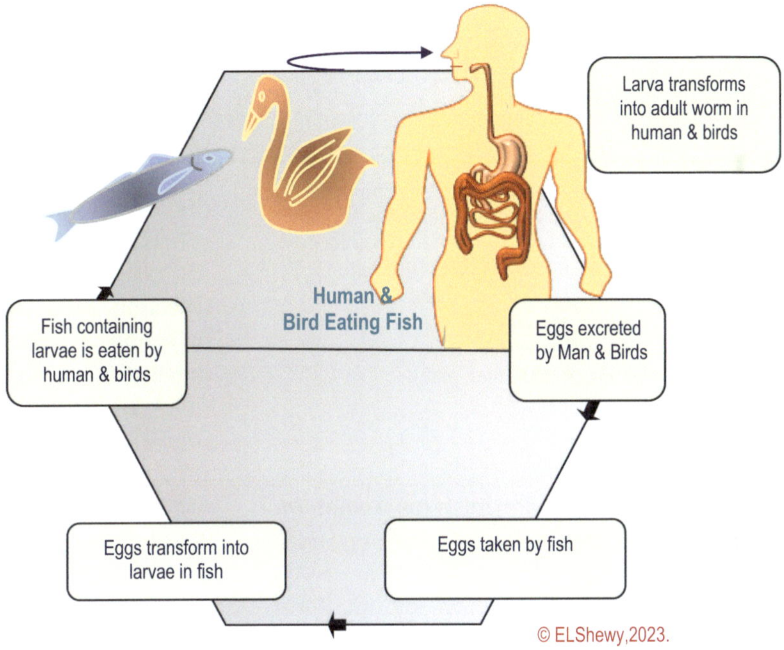

Fig. 2.12 *Capillaria philippinensis*: life cycle

Clinical Picture

Infection with *C. philippinensis* is presented mainly with:

1. Abdominal colic, gurgling of the abdomen.
2. Diarrhea.
3. Marked anorexia, nausea, vomiting, and eventually hypotension.
4. Later, the patient may suffer from cachexia, generalized anasarca, with a visible peristaltic wave over a distended abdomen.

Diagnostic Procedures

Diagnosis of *C. philippinensis* infection depends on:

1. Clinical picture.
2. History of eating undercooked or raw fresh and brackish water fish.
3. Detection of the infected fish in the surrounding water environment.
4. Detection of the characteristic eggs in the patient's stool using routine microscopic stool examination.
5. Microscopic examination of an intestinal biopsy to detect the characteristic eggs/ larvae.

Treatment

In acute capillariasis, fluid and electrolyte replacement with high-protein diet should be given. In chronic cases, anti-helminthic therapy should be given for a long period and should be repeated. The following are commonly used drugs in this respect:

1. **Mebendazole**:
   - 200 mg orally / 2 times daily /20 days.
   - It is the drug of choice and sometimes should be repeated.
   - **SE**: diarrhea and abdominal pain.
2. **Albendazole:**
   - 400 mg daily for 10 days.
   - It is an alternative to Mebendazole.

Prevention and Control

1. Treatment of the infected cases.
2. Abstinence from eating undercooked or raw fresh and brackish water fish.

- **Parasitology:** C. *Philippinensis* is a small nematode worm that prevails mainly in Philippines. It inhabits the intestinal mucosa and lays larvae or eggs. Eggs are barrel in shape with 2 unprotruded mucoid plugs. Infection takes place after eating infected fresh and brackish water fish.
- **Clinical:** Light infection is asymptomatic, but heavy one causes, diarrhea, and malabsorption with cachexia and protein enteropathy.

Review Questions

1. **Which one of the following is not a common mode of infection of *Strongyloides stercoralis?***
   (a) Penetration of the skin by filariform larvae.
   (b) Mucosal penetration by filariform larvae.
   (c) Ingestion of the adult worm.
   (d) Autoinfection.
2. **Culture of human stool is commonly used for diagnosis of.**
   (a) Hookworms.
   (b) *Capillaria philippinensis.*
   (c) *Ascaris lumbricoides.*
   (d) *Hymenolepis nana.*
3. **Which of the following parasites is transmitted through internal autoinfection?**
   (a) *Capillaria philippinensis.*
   (b) Hookworm.
   (c) *Ascaris lumbricoides.*
   (d) *Strongyloides stercoralis.*
4. **In mixed *Ascaris* and *Ancylostoma* infection.**
   (a) Both infections should be treated simultaneously.
   (b) *Ancylostoma* should be treated first before *Ascaris.*
   (c) *Ascaris* should be treated before *Ancylostoma.*
   (d) Only *Ascaris* should be treated.
5. **Intestinal obstruction, and/or appendicitis could be presented during infection with.**
   (a) Hookworms.
   (b) *Strongyloides stercoralis.*
   (c) *Ascaris lumbricoides.*
   (d) *Trichostrongylus colubriformis.*
6. **Infection with *Ascaris lumbricoides* could be acquired through eating food contaminated with.**
   (a) Adult *Ascaris* worm.
   (b) Fully embryonated *Ascaris* eggs.
   (c) *Ascaris* larvae.
   (d) Freshly passed *Ascaris* eggs.
7. ***Ancylostoma duodenale* egg has.**
   (a) A thick mamillated shell.
   (b) A thin shell with segmented ovum.
   (c) Two mucoid plugs.
   (d) An operculum.
8. **Baermann technique is used for.**
   (a) Routine stool examination.
   (b) Detection of larvae in the stool.
   (c) Urine examination.
   (d) Egg counting.
9. **Parasitic pneumonia is noticed during infection with the following except.**
   (a) *Ascaris lumbricoides.*
   (b) *Capillaria philippinensis.*
   (c) *Ancylostoma duodenale.*
   (d) *Strongyloides stercoralis.*

3 Cestode Helminths of the Small Intestine

Test your learning and check your understanding of this book's contents: use the "Springer Nature Flashcards" app to access questions using ▶ https://sn.pub/oyFLIx. To use the app, please follow the instructions in Chap. 1.

Diphyllobothrium latum (*D. latum*)

(Broad Fish Worm)

D. latum is a broad, long tapeworm. Human infection (diphyllobothriasis) is characterized by pernicious anemia with gastrointestinal disorders. **Sparganosis** is another disease condition caused by eating the first intermediate host of *D. latum.*

Geographical Distribution
D. latum prevails in areas where undercooked or pickled salmon fish is commonly used. It is peculiarly recorded in Russia, Canada, Europe, and Japan.

Parasite Morphology

- Adult worm (Fig. 3.1):
  - Segmented body with a spatulate scolex, neck, immature and mature proglottids.
  - 4 to 10 m.
  - Scolex is elongated with two grooves like suckers (bothria) and with no hooks.
  - Mature proglottid has a central rosette-shaped uterus.
- Egg:
  - It is an ovoid in shape and operculated.
  - 50 μm × 70 μm.
  - Yellowish brown. It contains an unembryonated ovum.

Transmission
D. latum infection is transmitted through eating undercooked or pickled salmon and Trout fish. Sparganosis is transmitted through drinking water or eating food containing the first intermediate host (copepods).

Habitat
Adult worms inhabit the small intestines of man, dogs, cats, and other fish-eating animals.

Life Cycle

Final hosts:	Humans
Infective stage:	Plerocercoid
Diagnostic stages:	Unembryonated egg, sometimes the mature segments in stool
Intermediate hosts:	1st, copepod, the second is a freshwater fish (Salmon and Trout fish)

The life cycle of *D. latum* is postulated in Fig. (3.2).

K. ElShewy, *Medical Parasitology*, https://doi.org/10.1007/978-3-031-64736-9_3

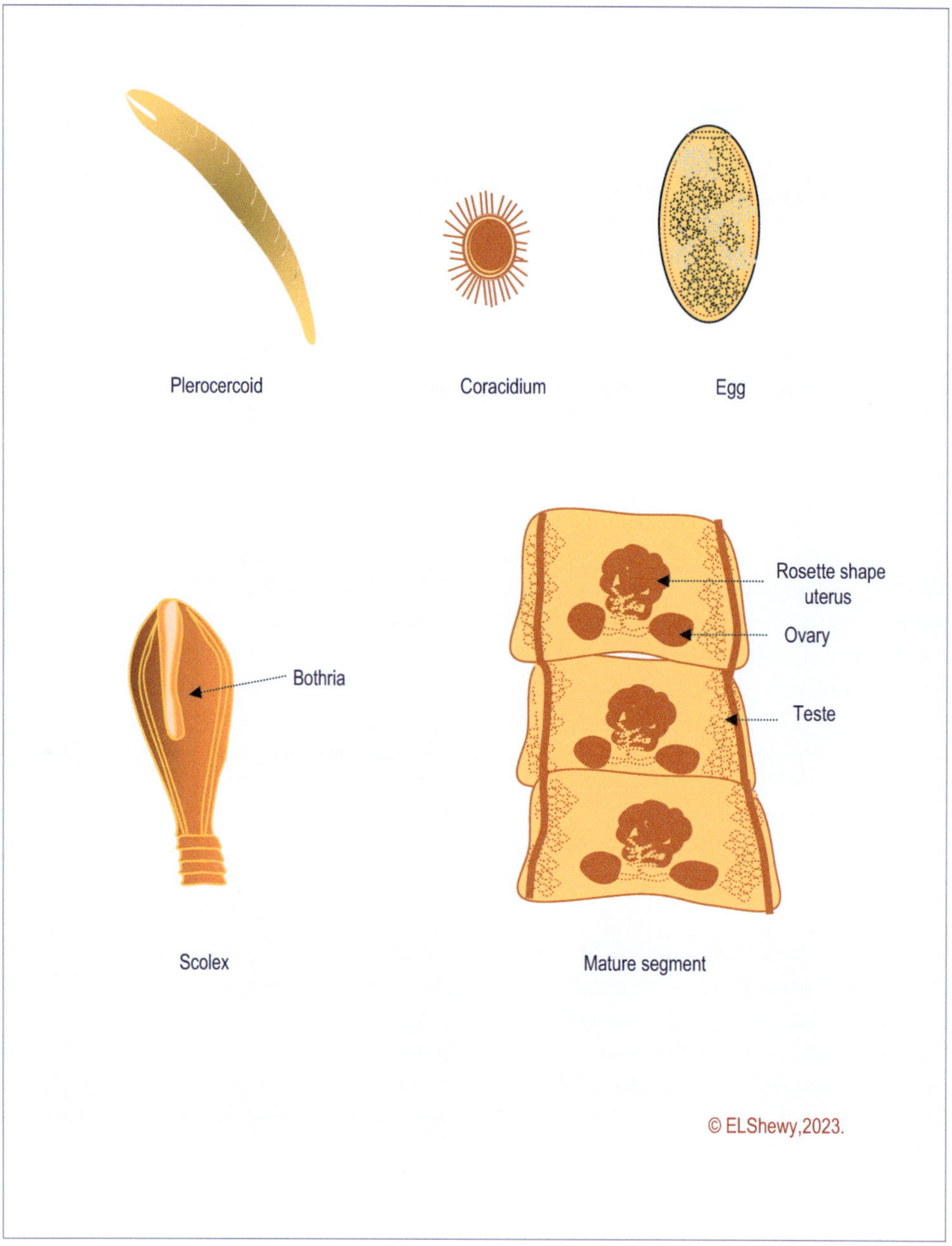

Fig. 3.1 *Diphyllobothrium latum*: morphology

N.B. When a human acts as an intermediate host and accidentally ingests the infected copepod, the procercoid larva develops in his body to Spargnaum, which migrates into the body's subcutaneous tissues. This condition is called **sparganosis**.

Pathogenesis

- The attachment of *D. latum* to the intestinal mucosa causes nonspecific inflammatory reactions.
- If the worm is localized at the proximal part of the jejunum, it causes vitamin B12 deficiency.

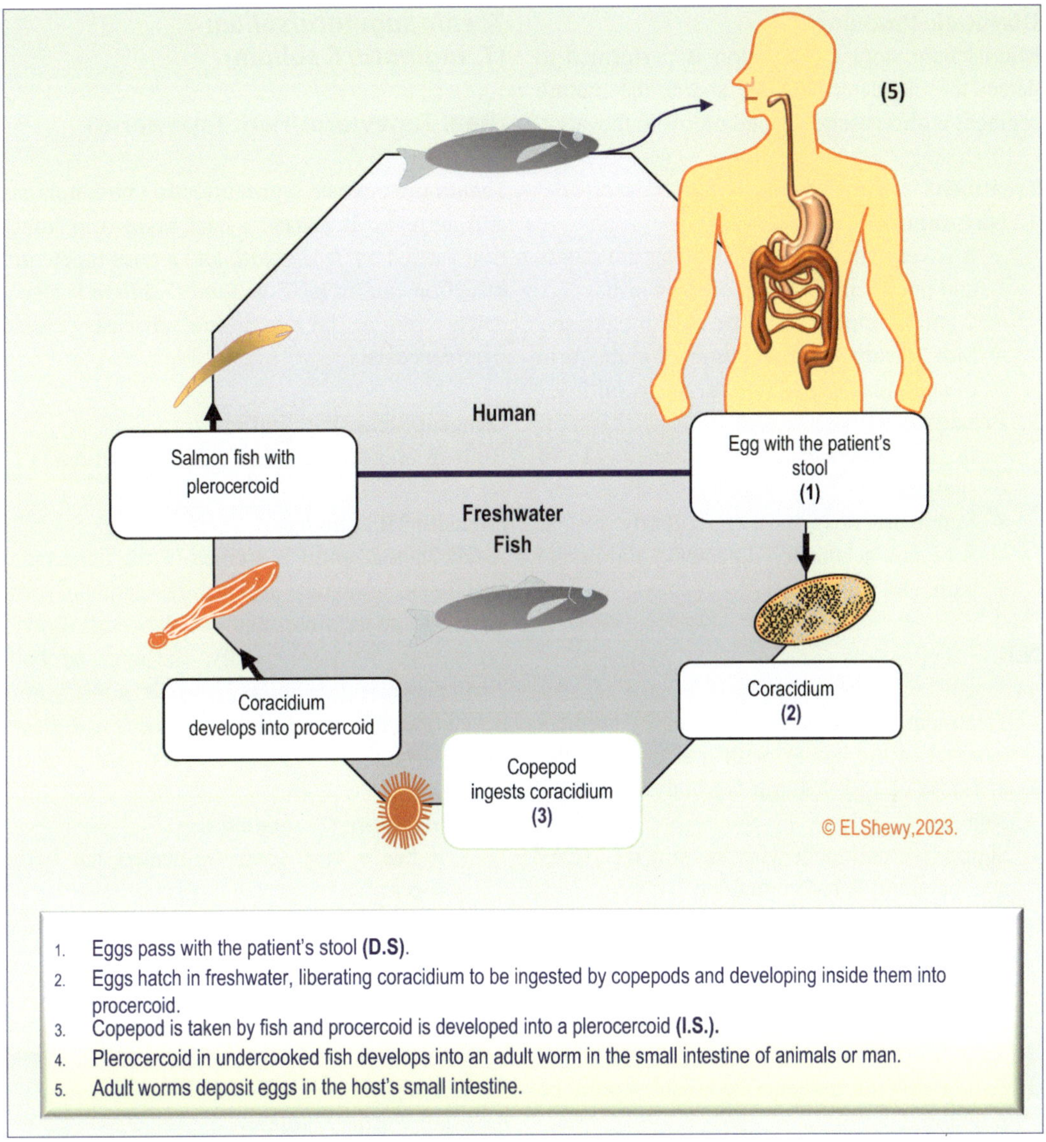

Fig. 3.2 *Diphyllobothrium latum*: life cycle

The worm secretes a factor that disturbs the attachment of vitamin B12 with gastromucoprotein, leading to vitamin deficiency.

- A large number of worms result in intestinal obstruction.

Clinical Picture

- Most of the infection with *D. latum* is asymptomatic.
- In heavy infection, there is diarrhea, malabsorption, abdominal colic, and rarely an intestinal obstruction.
- Vitamin B12 deficiency develops, leading to macrocytic hyperchromic anemia (megaloblastic or pernicious).
- Sparganosis affects many tissues, especially the orbital cavity, causing proptosis.

Diagnostic Procedures
Microscopic stool examination is performed to detect the characteristic egg and/or the mature segment with a rosette-shaped uterus in the stool.

Treatment
1. **Niclosamide**:
   - It is very effective in destroying the scolex and proximal segments of the worm.
   - 2 gm in a single oral dose after breakfast.
   - **Side effects (S.E.)** are minimal in the form of abdominal pain, nausea, and vomiting.
2. **Praziquantel:**
   - It is effective in the treatment of sparganosis.
   - Single oral dose of 10 mg/kg/body weight.
   - **S.E.** are headache, dizziness, abdominal pain, and nausea.

N.B.

- Treatment of sparganosis is done by surgical removal of the larvae, with oral praziquantel in a dose of 120–150 mg /kg/body weight for 2 days.
- Supportive treatment with vitamin B12 (cyanocobalamin) is recommended.

Prevention and Control
1. Proper disposal of human excreta.
2. Proper cooking, freezing, and salting of the fish.
3. Eating raw or undercooked fish should be prohibited.
4. Avoidance of drinking water infected with copepods, e.g., ponds and ditches.
5. Avoidance of the use of animal tissues as local medicinal applicants, e.g., foments.

- **Parasitology:** *D. Latum* is a broad, very long intestinal cestode that needs copepods and fish as intermediate hosts. It is transmitted through eating undercooked or raw salmon fish.
- **Clinical:** No symptoms or enteritis with vitamin B12 deficiency, macrocytic hyperchromic anemia (megaloblastic anemia). Sparganosis when eating larval stages of the worm.

Taenia saginata/solium (*T. saginata/T. solium*)

(Beef Tapeworm/Pork Tapeworm)

Taenia is a cestode worm infecting both humans and animals. It causes a beef tapeworm infection caused by *T. saginata* and a pork tapeworm infection caused by *T. solium*. *T. solium* is also a cause of a larval tissue disease called **cysticercosis**.

Geographical Distribution
Both species have a worldwide distribution.

Parasite Morphology
Adult *Taenia* worm is formed of the head (scolex), neck, immature and mature segments (proglottids); the terminal ones are full of eggs (gravid segments). Morphologically, the eggs of both *Taenia* species are similar. However, adult worms have some morphological differences as follows (Table 3.1):

- Adult worm: (*T. saginata*).
  - It has a very long segmented flat body, 4–8 m.
  - Its scolex has four suckers but with no hooks.
  - Mature proglottid has two lobed ovaries.
  - Gravid proglottid has 15–30 uterine branches, with a genital pore on the lateral margin.
- Adult worm:(*T. solium*).
  - It has a segmented flat body, 3–5 m.
  - Its scolex has four suckers, with a rostellum having two rows of hooks.
  - Mature proglottid has two lobed ovaries, with an accessory lobe.
  - Gravid proglottid has 7–15 uterine branches, with a genital pore on the lateral margin.

Table 3.1 Morphological differences between *T. saginata* and *T. solium*

	T. saginata	*T. solium*
Length	4–8 m	3–5 m
Scolex (head)	 © ELShewy, 2023 No rostellum, no hooks	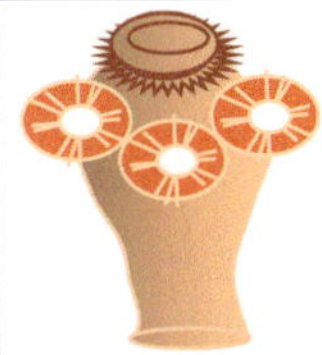 © ELShewy, 2023 There is rostellum with hooks
Mature segment (proglottid)	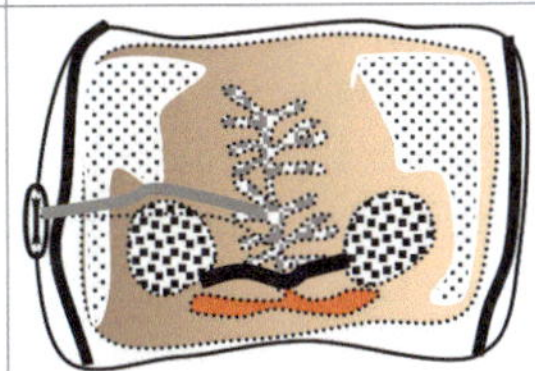 © ELShewy, 2023 Ovary with two large lobes	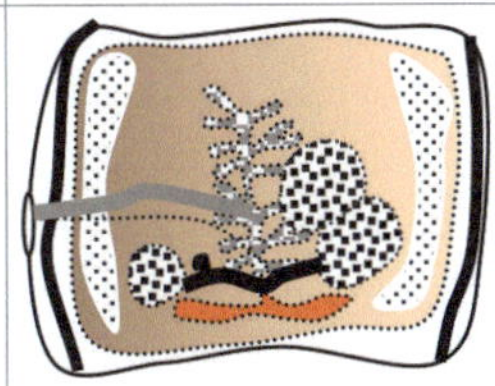 © ELShewy, 2023 Two lobed ovaries, with an accessory lobe
Gravid segment (proglottid)	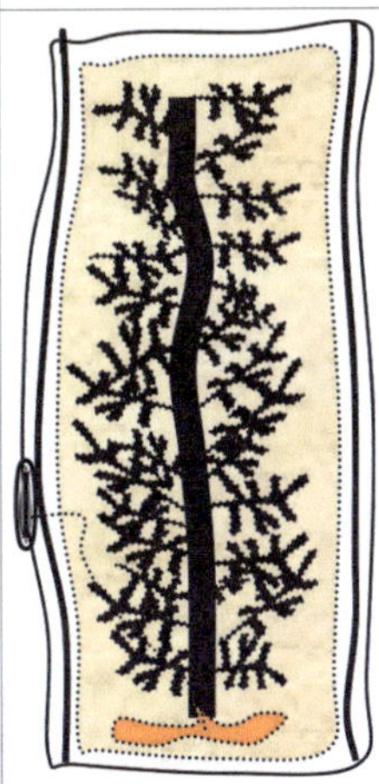 © ELShewy, 2023 15–30 uterine branches	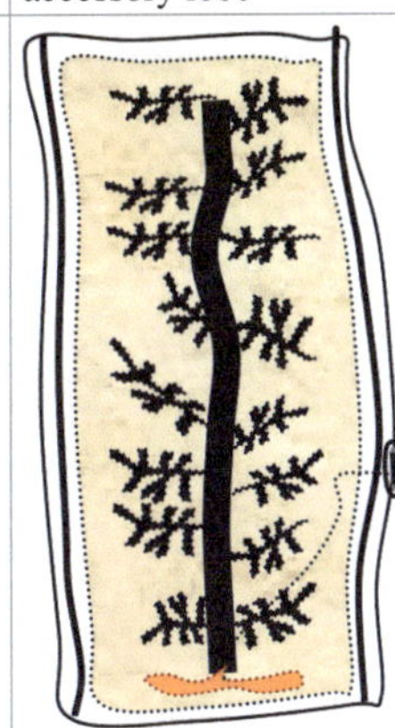 © ELShewy, 2023 7–15 uterine branches
Egg	 © ELShewy, 2023 Spherical in shape, with thick, striated shell, 30–45 μm in diameter It contains 6-hooked embryo (hexacanth oncosphere)	 © ELShewy, 2023 The same as *T. Saginata*

- Egg:
  - Spherical in shape, with a thick, striated shell.
  - 30–45 μm in diameter.
  - Yellowish.
  - It contains a six-hooked embryo (hexacanth oncosphere).

Transmission

Infection with *T. saginata* is transmitted through eating raw or undercooked beef containing cysticercus bovis, while infection with *T. solium* is caused by eating ham infected with cysticercus cellulose. Cysticercosis is caused by ingestion of *T. solium* eggs or through retrograde autoinfection (eggs return to hatch and penetrate the intestine of the same host).

Habitat

Adult worm of *Taenia* inhabits the human small intestine. Cysticercosis of *T. solium* could affect any tissue of the body, especially the brain, muscles, and the eye.

Life Cycle

- Human infection with *T. saginata or solium* is acquired by ingesting undercooked beef or pork containing cysticercus.
- Scolex emerges from the cysticercus and attaches to the wall of the small intestine. It develops into an adult worm and lays eggs.
- Eggs are passed with stool. Eggs are ingested by cows or pigs (intermediate hosts), then hatch releasing oncospheres. Oncospheres penetrate the intestine and develop into cysticercus larvae in different animal tissues (Fig. 3.3).
- If *T. solium* eggs are accidentally ingested by humans, humans act as an intermediate host. Oncospheres released from ingested eggs invade human body tissues, especially the brain, eye, and muscles, causing **cysticercosis** (Fig. 3.4).
- Cysticercosis could also result from retrograde autoinfection, i.e., eggs of *T. solium* return to the human intestine instead of passing out with stool.

Final hosts:	Humans
Infective stage:	Cysticercus (cysticercus bovis for *T. Saginata* and cysticercus cellulosae for *T. Solium)*
Diagnostic stage:	Characteristic eggs and/or gravid segments in the patient's stool
Intermediate hosts:	Cows for *T. Saginata* and pigs for *T. Solium;* in cysticercosis, humans accidentally act as intermediate hosts
Reservoir hosts:	Cows for *T. Saginata* and pigs for *T. Solium*

Pathogenesis

The following are pathologic changes associated with *Taenia* worm infection:

1. The attachment of adult *Taenia* worms to the intestinal mucosa results in little damage to the mucosal lining.
2. In heavy infection, adult worms interfere with the digestion and absorption process and could result in an intestinal obstruction.
3. During *T. solium* infection, cysticercus cellulose could be spread to many tissues. They are commonly found in the central nervous system (CNS), skeletal muscle, and the eye. Invasion of CNS is called **neurocysticercosis** could be associated with the following pathological changes:
   (a) Arachnoiditis, increased proteins in CSF, pleocytosis, and positive serology.
   (b) An intensive inflammation in the meninges with obstructive hydrocephalus.
   (c) A space-occupying lesion with pressure atrophy in the brain tissues and with an increased intracranial pressure.
   (d) Calcification of parenchymal cysticerci.
   - **Cysticercosis in the eye** affects the anterior and/or posterior chambers or the retinal tissues and could lead to edema, hemorrhage, detachment, and vasculitis.
   - **Cysticercosis in the muscles** causes myositis later painless calcified cysts in muscles.

Clinical Picture

1. Most infections are asymptomatic.
2. Heavy infection is characterized by abdominal colic, hunger pain, irregular bowel, and chronic indigestion.

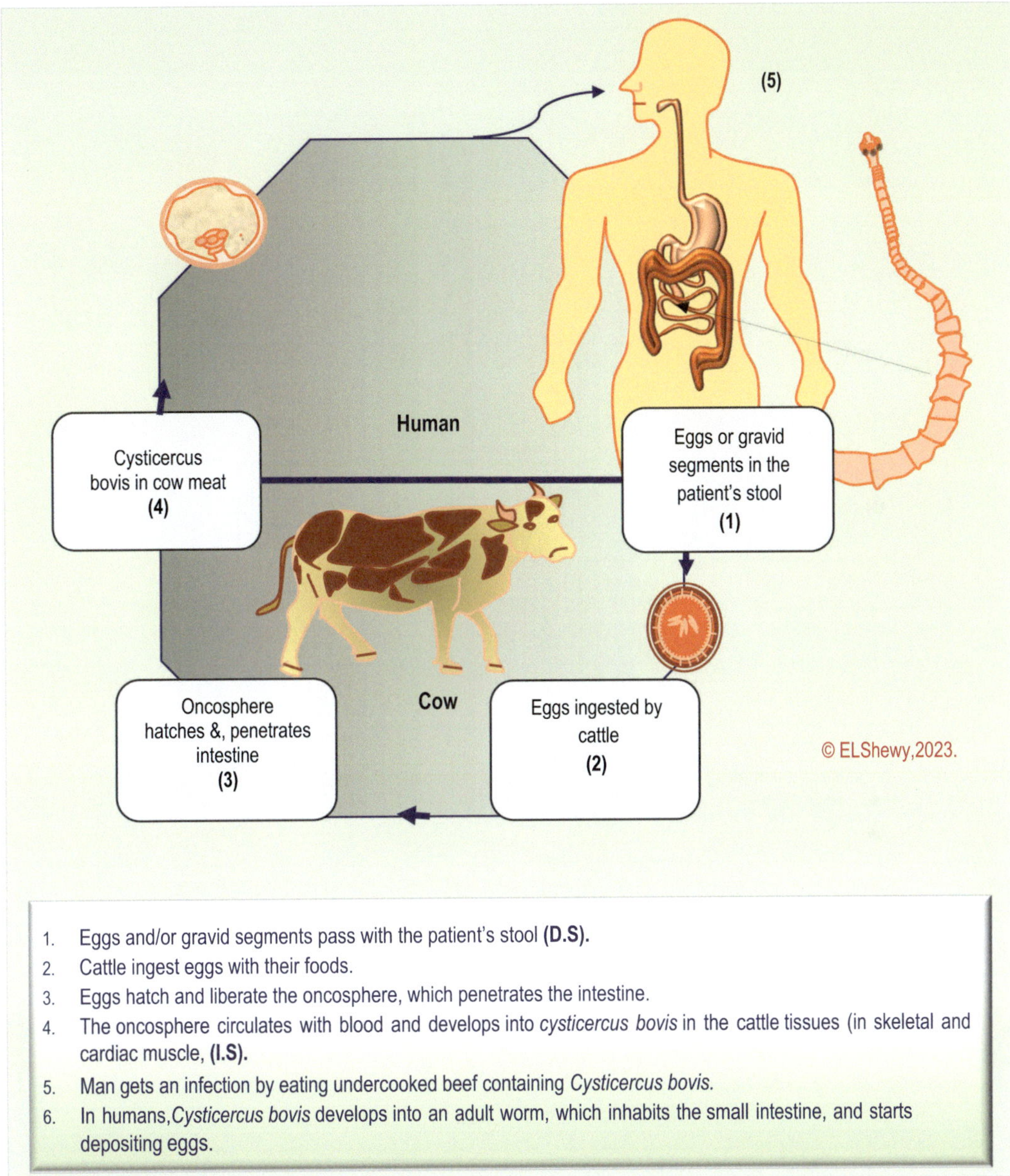

Fig. 3.3 *Taenia saginata*: life cycle

3. Cysticercosis in the CNS is associated with signs and symptoms of the space-occupying lesion (headache, vomiting, psychotic episodes, diplopia, vertigo, and sometimes cranial nerve paralysis).
4. Cysticercosis in the muscle is represented by myositis and painless swellings.
5. Cysticercosis in the eye causes diplopia, a decrease in visual acuity, vasculitis, retinal hemorrhage, and detachment.

Diagnostic Procedures

- Microscopic stool examination is performed to detect the characteristics of eggs.

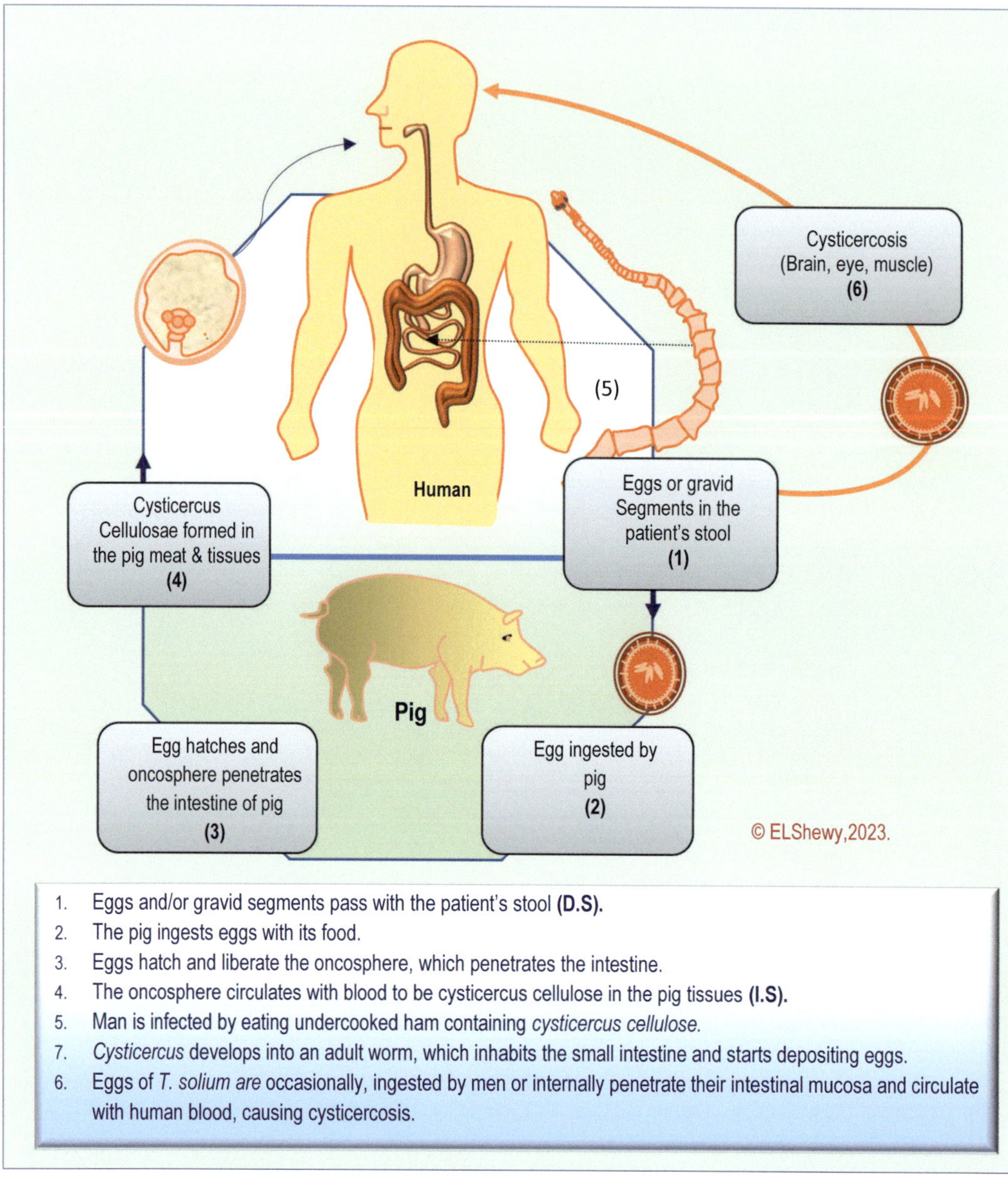

Fig. 3.4 *Taenia solium*: life cycle

- Sometimes mature and/or gravid proglottids are found in the stool and differentiation between the two species of *Taenia* should be done (Table 3.1).
- Diagnosis of **cysticercosis** includes:
  (a) Diet history and food habits.
  (b) Clinically, signs and symptoms of the space-occupying lesion.
  (c) Routine stool examination is ineffective in the diagnosis of cysticercosis.
  (d) A plain X-ray is needed to demonstrate the calcified cysticerci.
  (e) Computerized axial tomography (CAT) scan and magnetic resonance imaging (MRI) demonstrate both the calcified and noncalcified cysticerci in the brain.

(f) Demonstration of the cysticerci in the surgically removed cysts.

(g) Indirect serological methods, such as ELISA, IFA, IHA, and immunoblot assay using *T. solium* antigens, proved to be the **test of choice** for the detection of cysticercosis.

Treatment

Treatment of taeniasis includes the following two drugs:

1. **Niclosamide:**
   - 2 gm in a single oral dose after breakfast.
   - **S.E.** minimal in the form of abdominal pain, nausea, and vomiting.
2. **Praziquantel:**
   - Single oral dose of 10 mg/kg/body weight.
   - **S.E.** headache, dizziness, abdominal pain, and nausea.

Treatment of cysticercosis:

- Surgical removal of the accessible cysts.
- Inaccessible cysts especially in the brain; need symptomatic treatment in the form of corticosteroids and anticonvulsants.
- Praziquantel and Niclosamide should be used before and after surgery.

Prevention and Control

1. Proper cooking of meat.
2. Good sanitary practice and good veterinary care of meat animals.
3. Proper inspection of meat in the slaughters.
4. The use of the porcine vaccine is promising (synthetic 3-protein vaccine (S3PVAC) is an anti-cysticercosis vaccine given to pigs).

- **Parasitology:** *Taenia* is a very long intestinal cestode; it is transmitted by eating undercooked beef or ham; it needs cattle or pig as intermediate hosts, it is diagnosed by its characteristic eggs or gravid segments in the patient's stool.
- **Clinical:** No symptoms or enteritis with hunger pain, chronic indigestion, or occasionally intestinal obstruction. During *T. Solium* infection, cysticercosis is demonstrated in the brain, eye, or skeletal muscle.

Hymenolepis nana (H. nana)

(Dwarf Tapeworm)

H. nana is a cestode tapeworm of the small intestine. It is usually harmless, except in children with heavy infection or in immunocompromised individuals (**opportunistic parasites**).

Geographical Distribution

H. nana has a worldwide distribution with **a markedly high** prevalence rate among young children and immunocompromised individuals.

Parasite Morphology

- Adult worm:
  - It has a segmented body with a length of about 2.5–4 cm.
  - Its scolex has four suckers and an armed rostellum with one row of hooks.
  - Mature segment is wider than long. The gravid segment has a saclike fill with eggs.
- Egg*:*
  - Spherical to oval in shape, with a thin shell.
  - 30–45 μm in diameter. Yellowish in color.
  - It contains a six-hooked embryo (oncosphere) with two polar thickening and 4–8 filaments (fimbriae) in the space between the oncosphere and the outer shell (Fig. 3.5).

Transmission

H. nana is transmitted by many routes (so its high prevalence):

1. Ingestion of the eggs with foods and drinks.
2. Accidental ingestion of arthropods containing cysticercoid.
3. Internal autoinfection in the patient causing a long-lasting persistent infection.
4. External autoinfection (hand to mouth).

Habitat

H. nana inhabits the small intestine of humans.

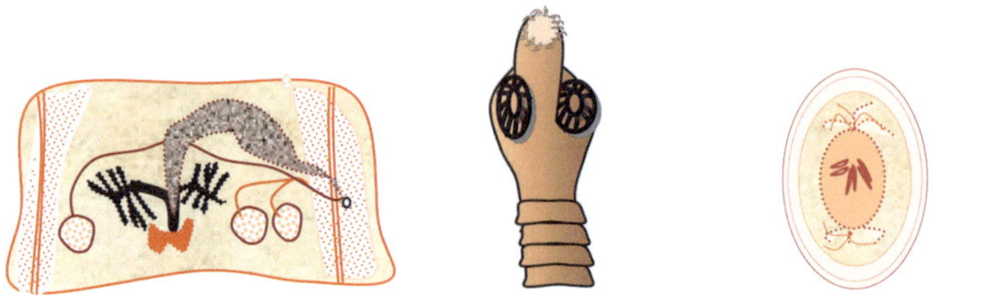

Fig. 3.5 *Hymenolepis nana*: morphology

Life Cycle

The life cycle of *H. nana* is postulated in Fig. (3.6).

Final hosts:	Humans
Infective stage:	Eggs
Diagnostic stage:	Eggs
Intermediate hosts:	No intermediate host (accidentally, beetles and fleas may act as intermediate hosts)

Pathogenesis

H. nana has two phases:

1. **A tissue phase** results from the attachment of the cysticercoid larvae to the intestinal villi. This is usually associated with minimal inflammatory reactions among immunocompetent individuals and with severe fulminate reactions among immunocompromised patients.
2. **An intestinal phase** is due to the presence of adult worms in the intestine, which causes little changes in the intestinal mucosa.

Clinical Picture

1. Hymenolepiasis is mostly asymptomatic, especially in light infection.
2. Heavy infection is associated with abdominal pain, diarrhea, colic, and anorexia.
3. *H. nana* among immunocompromized individuals is associated with severe, intractable diarrhea, fever, abdominal pain, and fulminate extraintestinal spread.

Diagnostic Procedures

Diagnosis of *H. nana* is done by microscopic detection of the characteristic eggs in the stool of the infected patient. The presence of the polar filaments is diagnostic.

Treatment

The following two drugs are commonly used:

1. **Niclosamide**:
   - It is given by mouth after breakfast in a daily dose for 5 days as:
     Four tablets for adults, three tablets for children more than 35 kg, and two tablets for children between 10 and 35 kg.
   - **S.E.** are minimal in the form of abdominal pain, nausea, and vomiting.
2. **Praziquantel**:
   - It is more effective than niclosamide in the treatment of both tissue and luminal phases.
   - Single oral dose of 25 mg/kg/body weight.
   - **S.E.** are headache, dizziness, abdominal pain, and nausea.

Prevention and Control

- **Parasitology:** *H. Nana* is the smallest and commonest cestode; it is transmitted by ingestion of eggs in contaminated foods or through internal or external autoinfection; it does not need an intermediate host. Accidental ingestion of infected insects could cause infection.
- **Clinical:** No symptoms or enteritis with fulminate persistent infection among immunocompromised.

Good sanitary practices are essential in the prevention of infection with *H. nana*.

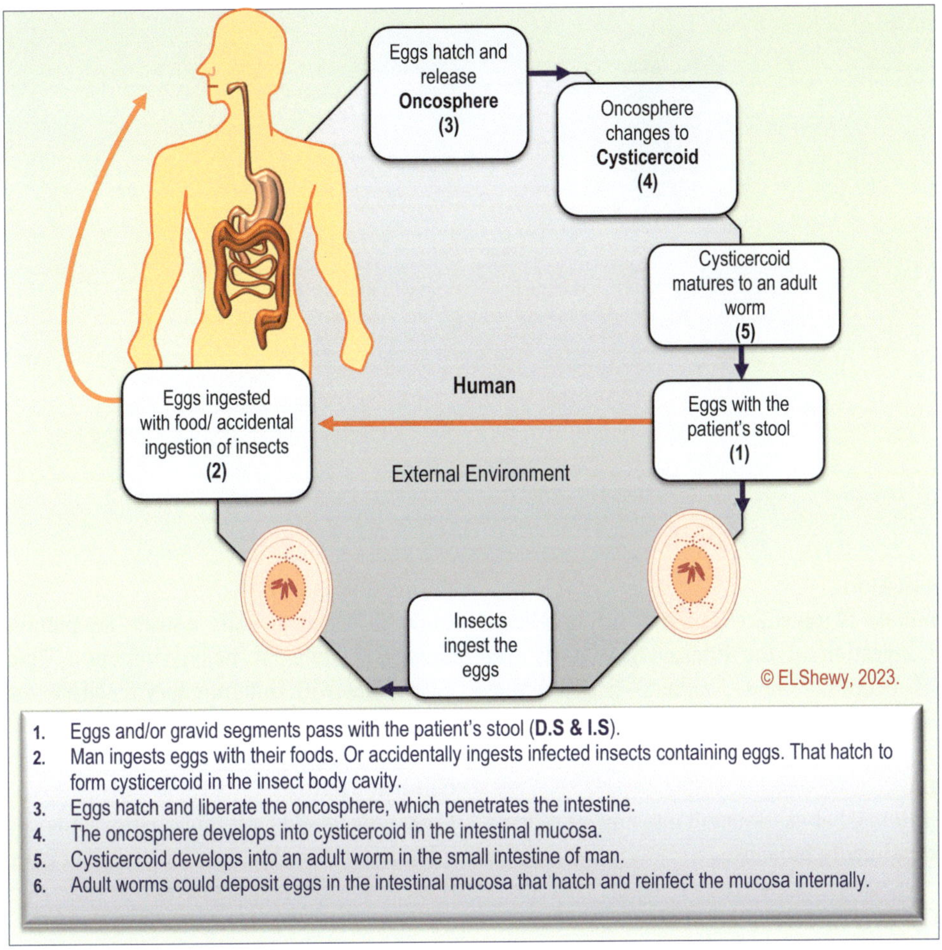

Fig. 3.6 *Hymenolepis nana*: life cycle

Hymenolepis diminuta (*H. diminuta*)

(Rat Tapeworm)

H. diminuta is mainly a tapeworm of rodents, occasionally, it infects humans (**zoonotic parasite**). Human infection is usually asymptomatic except in young individuals with heavy infection or immunocompromised individuals.

Geographical Distribution

H. diminuta has a worldwide distribution.

Parasite Morphology

- Adult worm (Fig. 3.7):
  - It has a segmented body with a length of about 30–60 cm.
  - Its scolex has four suckers and rostellum without hooks.
  - The body segment is wider than long.
- Egg:
  - Spherical to oval in shape, with a thin shell.
  - 50–60 µm in diameter. Yellowish.
  - It contains an oncosphere without polar filaments.

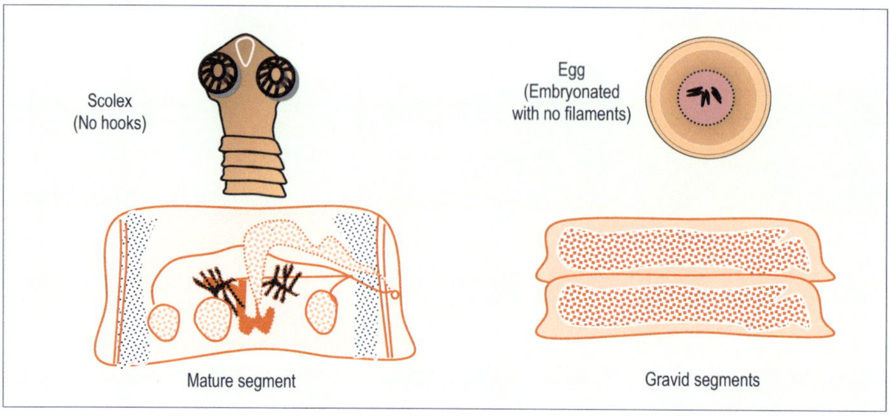

Fig. 3.7 *Hymenolepis diminuta*: morphology

Transmission

H. diminuta is transmitted mainly through accidental ingestion of the intermediate hosts (rat fleas, beetles, and others) containing cysticercoids.

Habitat

H. diminuta inhabits the small intestine of rodents and occasionally humans.

Life Cycle (Fig. 3.8)

- The life cycle of *H. diminuta* starts with a passage of eggs with the patient's stool.
- Insects ingest eggs to hatch inside them, liberating the oncosphere in their body cavity. Oncosphere transforms to crypto cystic cysticercoid.
- Rats and mice (occasionally humans) could swallow the infected insect.
- Cysticercoid develops into an adult worm in the small intestine of animals or humans.

Final hosts:	Rodents, rats, and occasionally humans
Infective stage:	Cysticercoid
Diagnostic stage:	Eggs or gravid segments in the patient's stool
Intermediate hosts:	Beetles and fleas
Reservoir hosts:	Rodents, rats

Pathogenesis

Light infection usually causes no pathological changes, while in heavy infection, intestinal ulcerations with inflammatory reactions could be demonstrated.

Clinical Picture

Light infection is asymptomatic while a heavy infection is associated with abdominal pain, diarrhea, headache, and anorexia.

Diagnostic Procedures

Diagnosis of *H. diminuta* is done by microscopic detection of the characteristic eggs in the stool of the infected patient.

Prevention and Control

Good sanitary precautions with combating rodents.

Dipylidium caninum (*D. caninum*)

(Dog Tapeworm)

D. caninum is mainly a dog tapeworm that could infect humans (**zoonotic parasite**), especially young children causing gastrointestinal disorders (Dipylidiasis).

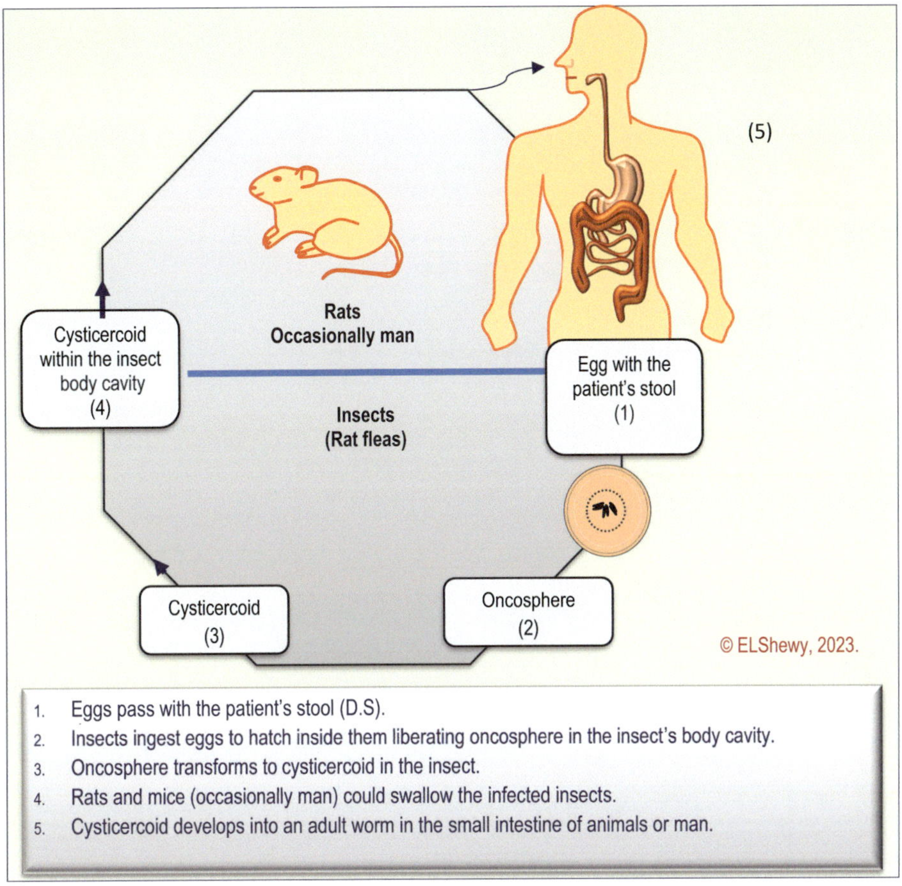

Fig. 3.8 *Hymenolepis diminuta*: life cycle

Geographical Distribution

D. caninum has a worldwide distribution.

Parasite Morphology

- Adult worm:
  - It has a segmented body with a length of about 20–80 cm × 3–4 mm.
  - Scolex has four suckers and an armed rostellum with 1–7 rows of hooks.
  - Mature proglottids (strobila) are long barrel-shaped, with two sets of genitalia.
  - Gravid segments are long barrel-shaped and cucumber seed-like.
- Egg:
  - Spherical in shape, with two shells, inner and outer.
  - 30–45 μm in diameter. Yellowish in color
  - It contains an oncosphere (Fig. 3.9).

Transmission

D. caninum could be transmitted to man during:

1. Accidental ingestion of dog or cat fleas containing crypto cystic cysticercoid. This could happen when there is close contact with dogs or cats.
2. Ingestion of food or drinks contaminated with dog and cat fleas.

Habitat

D. caninum inhabits the small intestine of dogs, cats, and occasionally humans.

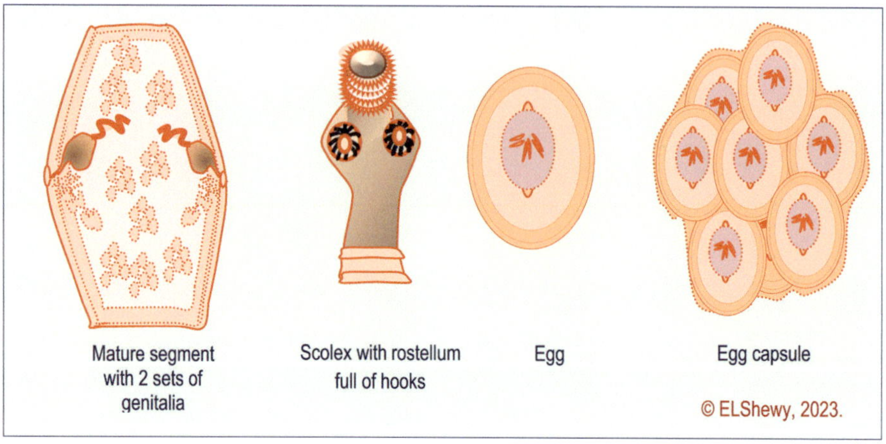

Fig. 3.9 *Dipylidium caninum:* morphology

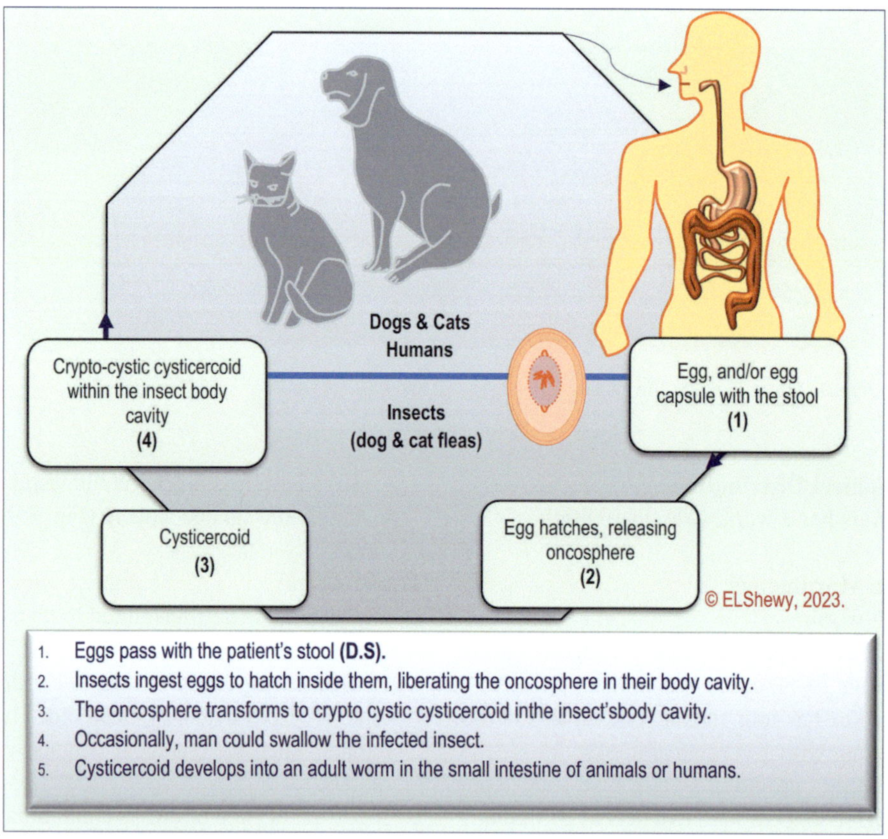

Fig. 3.10 *Dipylidium caninum:* life cycle

Life Cycle

Infection with *D. caninum* is achieved through accidental ingestion of dog or cat fleas containing the infective stage crypto cystic cysticercoid inside them. In the small intestine, cysticercoids are released, attached to the mucosa, and matured into adult worms (Fig. 3.10).

Final hosts:	Dogs, cats, and occasionally humans
Infective stage:	Crypto cystic cysticercoids (cysticercoid without tail)
Diagnostic stages:	Capsule of eggs and/or gravid segments, rarely separate eggs
Intermediate host:	Dog flea (*Ctenocephalus canis*); cat flea (*Ctenocephalus felis*)
Reservoir hosts:	Dogs and cats

Pathogenesis

1. *D. caninum* in light infection does not cause marked pathological changes.
2. In a heavy infection, mucosal inflammation and ulcerations could be detected.
3. Intestinal infection is usually associated with marked eosinophilia and allergic manifestations.

Clinical Picture

Dipylidiasis is represented by gastrointestinal disturbance, malabsorption, diarrhea, and anorexia.

Diagnostic Procedures

Microscopic examination of the stool for detection of egg capsules and or gravid segments, rarely eggs could be detected singly.

Prevention and Control

Good veterinary supervision of pet animals especially dogs and cats and avoidance of close contact with them.

- **Parasitology:** *D. Caninum* is a cestode worm with a rostellum having many rows of hooks. It deposits eggs in a capsule. It mainly infects dogs and cats, but occasionally, it infects humans. It is transmitted through accidental ingestion of eggs or fleas containing the eggs.
- **Clinical:** No symptoms or enteritis with diarrhea, anorexia, and eosinophilia.

Review Questions

1. **A thin-walled egg with a size of 30–45 μm and a fimbriated oncosphere is belonging to**
   (a) *Taenia solium*
   (b) *Diphyllobothrium latum*
   (c) *Hymenolepis nana*
   (d) *Dipylidium caninum*
2. **Ingestion of undercooked beef may cause infection with**
   (a) *Taenia saginata*
   (b) *Taenia solium*
   (c) *Echinococcus granulosus*
   (d) *Heterophyes heterophyes*
3. **Gravid segment of a cestode containing more than 15 uterine branches could belong to**
   (a) *Diphyllobothrium latum*
   (b) *Taenia saginata*
   (c) *Taenia solium*
   (d) *Echinococcus granulosus*
4. **Eggs with hexacanth embryo are produced by**
   (a) Blood trematodes
   (b) Intestinal trematodes
   (c) Nematode worms
   (d) Cestode worms
5. **A cestode that can be identified by the rosette uterus in the gravid segment is**
   (a) *Taenia saginata*
   (b) *Taenia solium*
   (c) *Diphyllobothrium latum*
   (d) *Dipylidium caninum*
6. **Neurocysticercosis could be detected during infection with**
   (a) *Ascaris lumbricoides*
   (b) *Taenia solium*
   (c) *Strongyloides stercoralis*
   (d) *Taenia saginata*
7. **The longest cestode worm infecting human is**
   (a) *Dipylidium caninum*
   (b) *Taenia solium*
   (c) *Diphyllobothrium latum*
   (d) *Hymenolepis nana*
8. **Sparganosis is caused by**
   (a) *Dipylidium caninum*
   (b) *Taenia solium*
   (c) *Diphyllobothrium latum*
   (d) *Hymenolepis nana*
9. **The most common cestode worm infecting children is**
   (a) *Diphyllobothrium latum*
   (b) *Taenia saginata*
   (c) *Taenia solium*
   (d) *Hymenolepis nana*

4 Trematode Helminths of the Small Intestine

Test your learning and check your understanding of this book's contents: use the "Springer Nature Flashcards" app to access questions using ▶ https://sn.pub/oyFLIx. To use the app, please follow the instructions in Chap. 1.

Heterophyes (*H. heterophyes*)

(Small Fish Worm)

H. heterophyes is an intestinal flatworm responsible for heterophyiasis, a fish-transmitted gastrointestinal disease with enteritis, indigestion, and biliary disorders.

Geographical Distribution
H. heterophyes prevails in the Middle East, Egypt, and Turkey.

Parasite Morphology
The morphology of *H. heterophyes* includes (Fig. 4.1):

- Adult worm:
  - Small fluke with a pear-shaped body covered with spines.
  - 0.7 mm to 3 mm.
  - It has three suckers; anterior, ventral, and a third one surrounding the genital pore (genital sucker).
- Egg:
  - Conical in shape with an operculum, double-shelled, with a small nipple-like projection posteriorly.
  - 15 μm × 30 μm
  - Yellowish brown. It contains a miracidium (fully embryonated).

N.B. *Heterophyes* cercaria is single-tailed cercaria with a membrane (lophocercous).

Habitat
Adult worms inhabit the small intestines of dogs, cats, wolves, foxes, and humans.

Transmission
H. heterophyes is transmitted through eating undercooked, poorly salted, or smoked freshwater fish containing encysted metacercaria. Encysted metacercariae are located under the fish scales or in their flesh.

Life Cycle
The life cycle of *H. heterophyes* is postulated in Fig. (4.2).

K. ElShewy, *Medical Parasitology*, https://doi.org/10.1007/978-3-031-64736-9_4

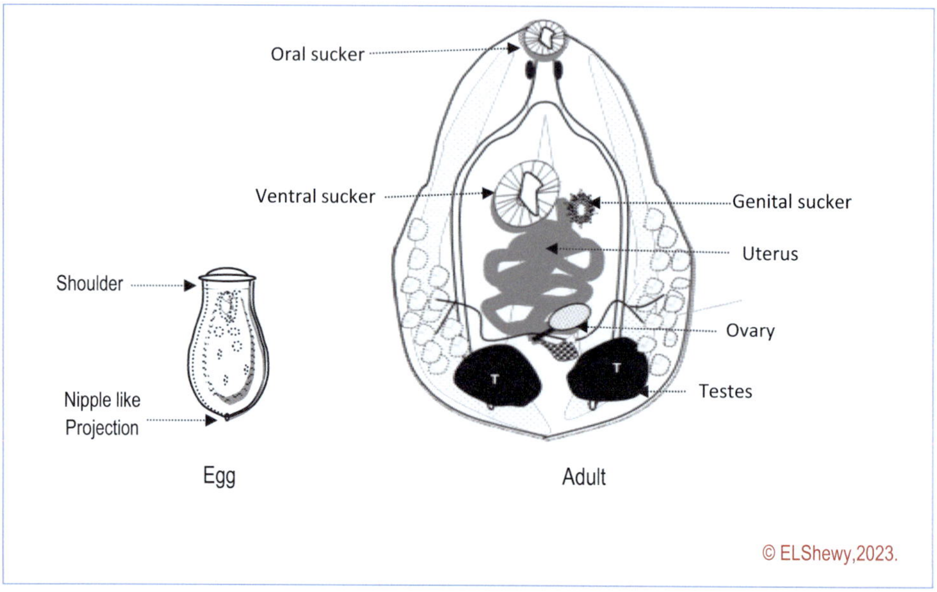

Fig. 4.1 Morphology of *heterophyes*

- Egg deposited with the patient's stool.
- Freshwater snails take the egg to hatch and release miracidium, which develops into cercaria.
- Cercaria is released and swims in water to encyst on freshwater fish (encysted metacercaria) **under** the fish's scales and between flesh fibers.
- Human infection occurs by ingestion of undercooked or incompletely salted (less than 10 days) freshwater fish. Encysted metacercaria is released and matured into adult *Heterophyes* worm.
- Adult worm inhabits duodenum and biliary passages and fixes itself through its suckers.

Final hosts:	Humans and fish-eating animals.
Infective stage:	Encysted metacercaria in undercooked freshwater fish.
Diagnostic stage:	Eggs containing fully mature miracidium.
Intermediate hosts:	1st is freshwater snail (*Pirenella conica*); the second is freshwater fish (*Mugil*, *tilapia*) (paratenic hosts).
Reservoir hosts:	Fish-eating animals.

Pathogenesis

The attachment of adult worms to the intestinal mucosa results in little changes. However, a heavy infection may produce marked inflammation with mild mucosal ulceration.

Clinical Picture

1. Light infection is asymptomatic.
2. Heavy infection is associated with vague abdominal pain, chronic intermittent diarrhea, indigestion, nausea, and vomiting.
3. Rarely, worms can infect the biliary passages, causing cholangitis and jaundice.
4. Occasionally, the eggs could pass in the circulation, causing embolic manifestations. (This occurs in rare conditions when the genital sucker is attached to a blood vessel, and released eggs pass through this vessel into circulation).

Diagnostic Procedures

Diagnosis of *H. heterophyes* is done by detecting the characteristic eggs in the stool of the patients.

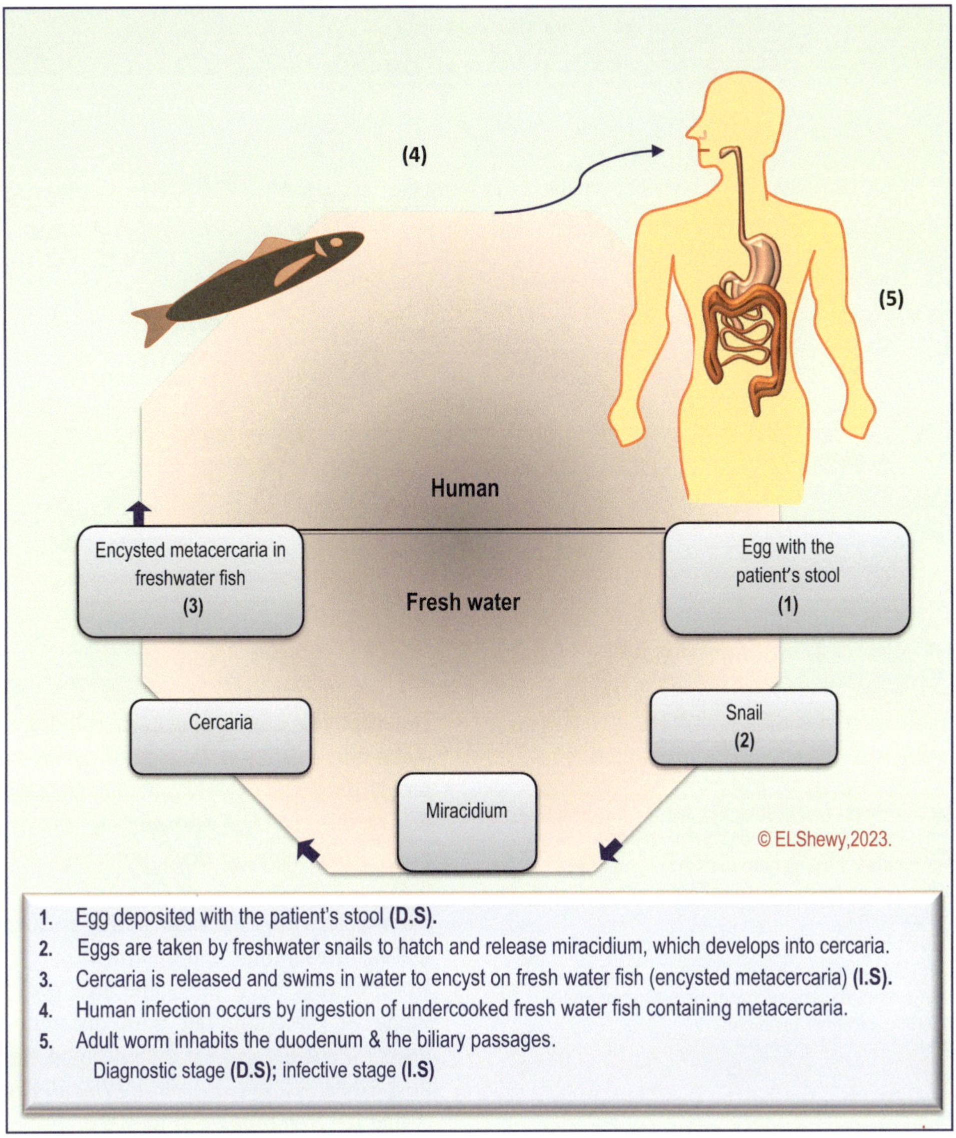

Fig. 4.2 *H. heterophyes and Metagonimus yokogawai* (*Heterophyids*)*:* life cycle

- A routine microscopic stool examination is usually used.
- Adult worms are rarely seen in stool.

Treatment

1. **Praziquantel:**
   - It is a drug of choice, given as 30 mg/kg twice daily for a single day.
2. **Tetrachloroethylene:**
   - It is an alternative to praziquantel.
3. **Niclosamide:**
   - Four tablets once as a single dose is effective.

Prevention and Control

1. Proper cooking, smoking, and salting of the fish.
2. Proper disposal of human excreta.
3. Periodic examination of fishermen and mass treatment.

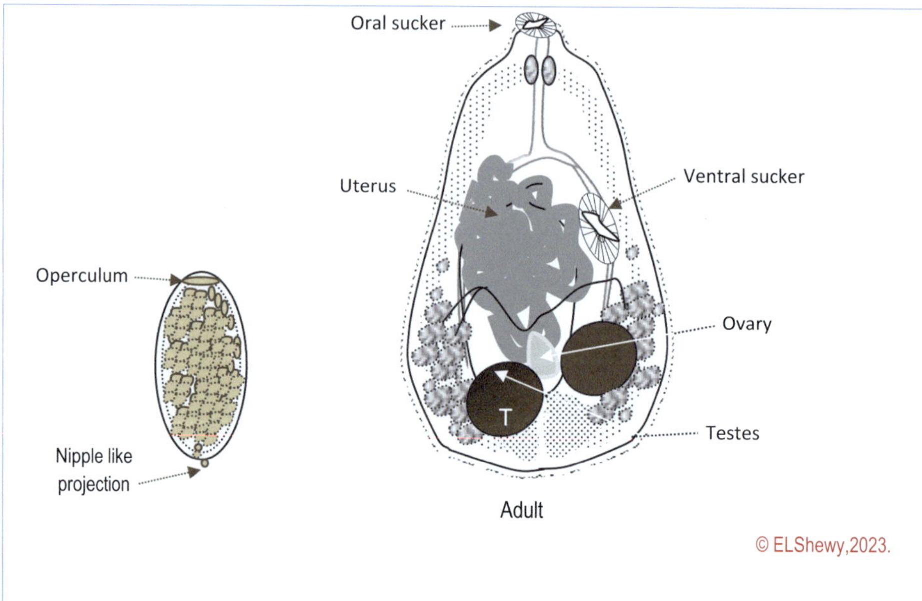

Fig. 4.3 Morphology of *Metagonimus yokogawai*

4. Eradication of snails.
5. Eating raw fish should be prohibited.

- **Parasitology:** *Heterophyes* is a small intestinal trematode fluke that needs a snail as an intermediate host. It is transmitted by eating undercooked freshwater fish containing encysted metacercaria.
- **Clinical:** *Heterophyes* cause gastrointestinal disturbance only in heavy infection: Diarrhea, indigestion, nausea and vomiting. Occasionally, jaundice and liver problems.

Metagonimus yokogawai (*M. yokogawai*)

(Japanese Fluke)

– *M. yokogawai* looks like *Heterophyes* worm, both called *Heterophyids*. Both worms are almost similar and responsible for gastrointestinal disease in the form of enteritis and indigestion with biliary problems.
– *M. yokogawai* prevails mostly in the Far East, China, Korea, and Japan, as well as in Manchuria, the Balkan states, and Spain.
– Heterophyids are similar in morphology with some minor differences, e.g., *M. yokogawai* is larger than *Heterophyes*, while *Heterophyes* has a third sucker surrounding the genital pore, *Metagonimus* does not (Fig. 4.1 and 4.3).
– Both worms have the same pathogenesis, clinical manifestations, and treatment.
– Both worms have the same life cycle and mode of infection, but with different snail hosts (*Semisulcospira* is the intermediate snail host for *M. yokogawai, while Pirenella conica* is that for *Heterophyes* (Fig. 4.2).

Fasciolopsis buski (*F. buski*)

(Giant Intestinal Fluke)

F. buski is a large fluke inhabiting the small intestine; it causes gastrointestinal disturbance in the form of abdominal colic, diarrhea, and occasionally intestinal obstruction.

Geographical Distribution
F. buski is detected mainly in China, Vietnam, Thailand, and Indonesia.

Fig. 4.4 *Fasciolopsis buski*: adult and egg morphology

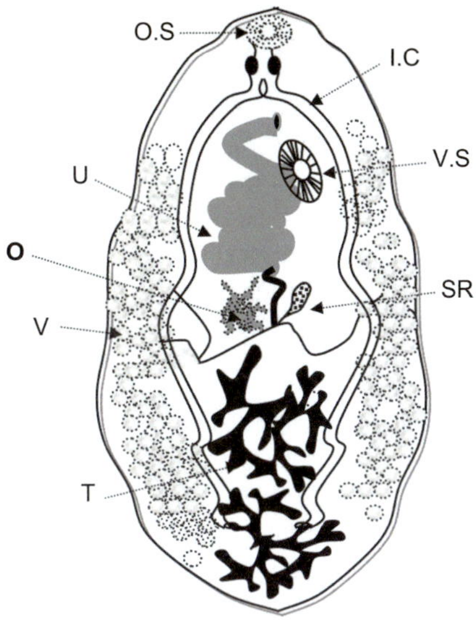

O. S = oral sucker, V. S = ventral sucker; G.S = genital sucker, T= test, O= ovary, U= uterus, V= vitellaria, S.R = seminal receptacle, I.C=intestinal ceca.

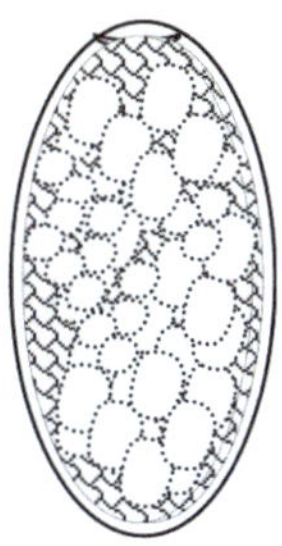

Parasite Morphology

F. buski is considered the largest trematode fluke. The worm is hermaphroditic, and its eggs pass with the patient's stool (Fig. 4.4).

- Adult worm:
  - It has a large oval body covered with spines.
  - It has two suckers (anterior and ventral).
  - It is 7 cm in length and 1.5 cm in width (largest fluke).
- Egg:
  - Oval in shape with an operculum and double shell.
  - 140 μm × 80 μm, transparent.
  - Not embryonated when passed (immature).

Habitat

F. buski inhabits the small intestine of humans and herbivorous animals.

Transmission

F. buski is transmitted through ingestion of the metacercaria encysted on freshwater vegetation that prevails in areas of the worm's distribution, e.g., bamboo shoots or water chestnuts.

Life Cycle

The unembryonated egg of *F. buski* passes with stool in freshwater canals, and after 1 month, it becomes mature (containing miracidium). Mature egg hatches and miracidium are released in the water and ingested by the first intermediate host (a freshwater snail, *Segmentina*) to develop inside into a cercaria. Cercaria is released in water to be encysted (encysted metacercaria) on the freshwater vegetation (second intermediate host). After ingestion by the final host, the encysted metacercaria is released and developed into adult worms in the small intestine (adult worm lives up to 6 months) (Fig. 4.5).

Final hosts:	Humans and herbivorous animals.
Infective stage:	Encysted metacercaria in freshwater vegetations.
Diagnostic stage:	Eggs in the patient's stool.
Intermediate hosts:	First is a freshwater snail (*Segmentina*); the second freshwater vegetation.
Reservoir hosts:	Pigs.

Pathogenesis

1. Light infection with *F. buski* causes minor changes in the intestinal mucosa. While in heavy infection, a marked local inflammation, with ulcerations and hemorrhage, is usually demonstrated.
2. Intestinal obstruction could happen if there is a large number of worms.
3. Long-standing infection results in malabsorption and hypo-albuminuria, which could be responsible for marked edema and ascites.
4. Eosinophilia is very common.

Clinical Picture

1. Light infection usually passes unnoticed.
2. Heavy infection with *F. buski is* associated with abdominal pain, suggestive of duodenal ulcer.
3. Diarrhea, with a light yellowish-colored stool containing many undigested foods due to malabsorption.
4. Vitamin B12 deficiency.
5. Intestinal obstruction in very heavy infection.
6. Edema and ascites are occasionally seen.

Diagnostic Procedures

Diagnosis of *F. buski* is done by detection of the characteristic eggs in the stool of the patients. A routine stool examination is usually used. Care must be taken to distinguish between the liver fluke (*Fasciola hepatica*) and *F. buski.*

Treatment

1. **Praziquantel:**
   - It is a drug of choice.
   - It is given as 25 mg/kg/body weight three times for a single day.
2. **Niclosamide:**
   - It is an alternative to praziquantel.
   - It is not absorbed from the gastrointestinal tract, so it has no side effects.
   - It is given as four chewable tablets in the morning and should be chewed thoroughly and then swallowed (0.5 gm/tablet).

Prevention and Control

1. Proper human fecal disposal.
2. Snail eradication.
3. Prevention of eating raw vegetation without proper cleaning.
4. The use of human excreta as fertilizers should be prohibited.

- **Parasitology:** *F. Buski* is the largest intestinal trematode fluke; it needs a snail as an intermediate host. It is transmitted through eating contaminated freshwater vegetation containing encysted metacercaria. The worm is confined to China, Indonesia, and Thailand.
- **Clinical:** *F. Buski* causes gastrointestinal disturbance only in heavy infection: Yellowish-colored diarrhea, severe abdominal pain simulating duodenal ulcer, intestinal obstruction, and vit. B12 deficiency.

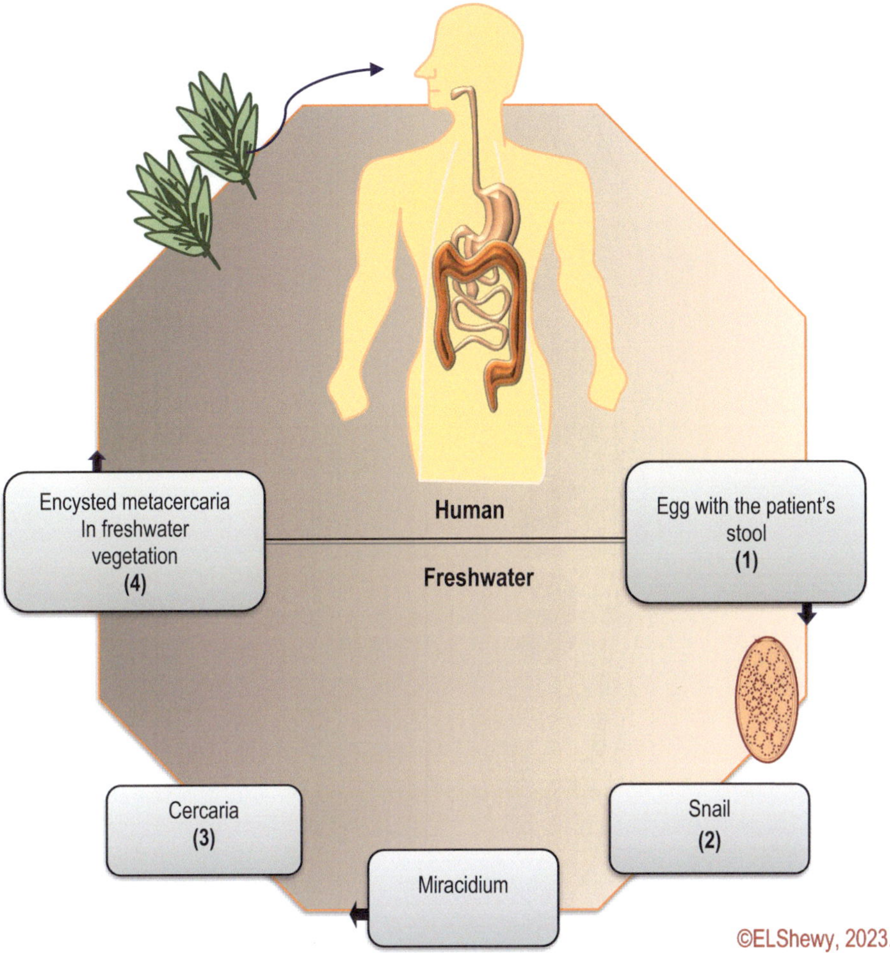

1. Egg deposited with the patient's stool. **(D.S)**
2. Egg is taken by freshwater snails to hatch and release miracidium, which develops into cercaria.
3. Cercaria is released and swims in water to be encysted on freshwater vegetation (encysted metacercaria) **(I.S)**
4. Human infection occurs by ingestion of freshwater vegetation containing metacercaria.
5. Adult worm inhabits the duodenum & the biliary passages.

Fig. 4.5 *Fasciolopsis buski*: life cycle

Echinostoma ilocanum (E. ilocanum)

(Garrison's Fluke)

E. ilocanum is a parasite of wild and domestic animals. It occasionally infects humans, causing echinostomiasis, an abdominal disease characterized by diarrhea and abdominal pain with minor intestinal ulcerations.

Geographical Distribution

E. ilocanum prevails mainly in Japan, the Philippines, Indonesia, and in some areas of Asia,

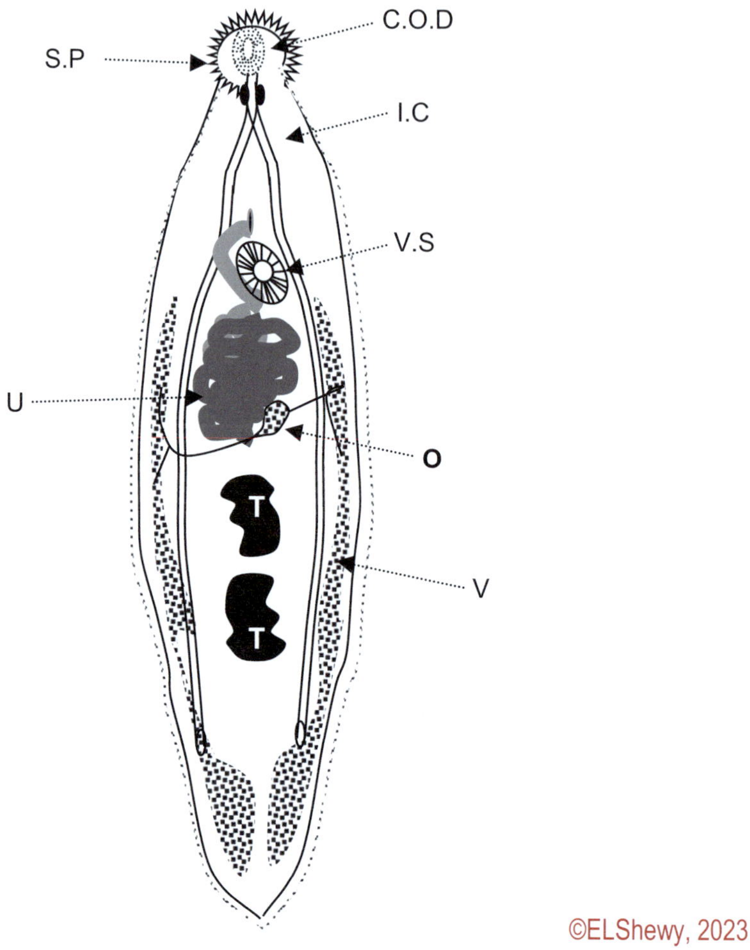

C.O. D= circumoral disc, S. P= spinous processes, V. S= ventral sucker; T= testes, O= ovary, U= uterus, V= vitellaria, I.C= intestinal ceca.

Fig. 4.6 *Echinostoma ilocanum*: adult morphology

especially among those eating undercooked or raw snails, fish, and amphibians.

Parasite Morphology

Adult worm of *E. ilocanum* ranges between 0.8 and 1 cm in length and 0.2 cm in width. It has a collaret of spines on a disk surrounding the oral sucker. Eggs are excreted unembryonated. They are similar to *F. buski* eggs, so diagnosis should not depend on the detection of the eggs but rather on the clinical history (Fig. 4.6).

Transmission

E. ilocanum is transmitted through eating undercooked or raw freshwater fish containing encysted metacercaria.

Habitat

Adult worms attach themselves to the mucosa of the small intestine of rodents, carnivores, rodents, and humans.

Life Cycle

Little is known about the actual life cycle; however, it includes a freshwater snail as an intermediate host and a fish or a freshwater shellfish as another intermediate host. However, the worm goes through different hosts.

- The egg passes with stool and stays infective for 5 months. It hatches and releases miracidium, which is taken by first host (snail).
- Cercaria is released from the snail and taken by the second host.
- Cercaria changes into metacercaria in the third host (fish).
- Humans and other final hosts get the infection through eating encysted metacercaria in undercooked or raw fish.

Pathogenesis

In light infection with *E. ilocanum* worms, minor changes occurred to the intestinal mucosa. However, in heavy infection, the worms may produce mucosal inflammation with relevant ulceration because of inflammatory and traumatic effects caused by the spined collar disc of the worms.

Clinical Picture

Mild infection passed unnoticed. Heavy infection is represented by diarrhea, abdominal pain, and intestinal ulceration.

Diagnostic Procedures

There is great variation in the *Echinostoma* eggs, so diagnosis does not depend on the detection of the eggs but rather on the clinical history and travel history.

Treatment

1. **Praziquantel:**
   - 40 mg/kg/body weight (b.w), a single dose is a curative in most of the cases.
2. **Tetrachloroethylene:**
   - 0.1 ml/kg, a single dose is given on an empty stomach.

- **Parasitology:** *E. Ilocanum* is a trematode fluke sized 0.8 to 1 cm. It inhabits the small intestine. It has a collaret of spine on a disk around the oral sucker. It is transmitted through eating freshwater fish containing encysted metacercaria. It is confined to Japan, the Philippines, and Indonesia.
- **Clinical:** *E. Ilocanum* causes gastrointestinal disturbance with diarrhea and intestinal ulceration.

Review Questions

1. **The smallest-sized trematode egg belongs to**
   (a) *Schistosoma hematobium*
   (b) *Heterophyes heterophyes*
   (c) *Schistosoma mansoni*
   (d) *Fasciola hepatica*
2. **Freshwater fish could transmit the following worm**
   (a) *Taenia solium*
   (b) *Heterophyes heterophyes*
   (c) *Schistosoma mansoni*
   (d) *Fasciola hepatica*
3. **Encysted metacercaria could transmit infection with**
   (a) *Taenia solium*
   (b) *Heterophyes heterophyes*
   (c) *Schistosoma mansoni*
   (d) *Hymenolepis nana*
4. **Most infection with heterophyes is manifested by**
   (a) Severe abdominal colic
   (b) Indigestion and diarrhea
   (c) No symptoms
   (d) Vomiting and abdominal distension

5 Protozoan Parasites of the Small Intestine

Test your learning and check your understanding of this book's contents: use the "Springer Nature Flashcards" app to access questions using ▶ https://sn.pub/oyFLIx. To use the app, please follow the instructions in Chap. 1.

Giardia lamblia

(*G. lamblia, G. intestinalis*)

Giardia lamblia is a flagellated waterborne protozoan parasite. It is responsible for giardiasis, a gastroenteritis disorder with malabsorption and fat-soluble vitamin deficiency, especially in young and immunocompromised individuals.

Geographical Distribution

- *G. lamblia* is a common intestinal protozoan parasite with a worldwide distribution. It has eight genotype assemblages from A to H.
- Its prevalence rate could reach up to 25%, mainly of A and B assemblages.
- It is an important cause of traveler's diarrhea and waterborne infection that could cause epidemic infections.
- There is little or no host specificity for the *Giardia* parasite, so that it could infect many hosts, most of them are considered reservoir hosts (rats, rodents, and beavers).

Parasite Morphology

G. lamblia has two morphological forms: a trophozoite active form and a cyst inactive form (Fig. 5.1).

- Trophozoite:
  - Leaf-like bilaterally symmetrical body, 9–20 μm in length.
  - It is convex dorsally with a ventral shallow sucking disc and four pairs of flagella.
  - It has two nuclei with central axostyle and two curved paramedian bodies (parabasal bodies).
  - It has the motility of a falling leaf.
- Cyst form:
  - Ovoid in shape with a thick cyst wall.
  - 8–12 μm length × 6–10 μm width.
  - Its cytoplasm is shrunk from the cell wall at one side.
  - Mature cyst has four nuclei.

Transmission

G. lamblia could be transmitted through the following:

1. Ingestion of contaminated food and drinking water contaminated with *Giardia* cysts.
2. Hand-to-mouth contamination of the same patient (external autoinfection).
3. Mechanical transmission of cysts through flies.

K. ElShewy, *Medical Parasitology*, https://doi.org/10.1007/978-3-031-64736-9_5

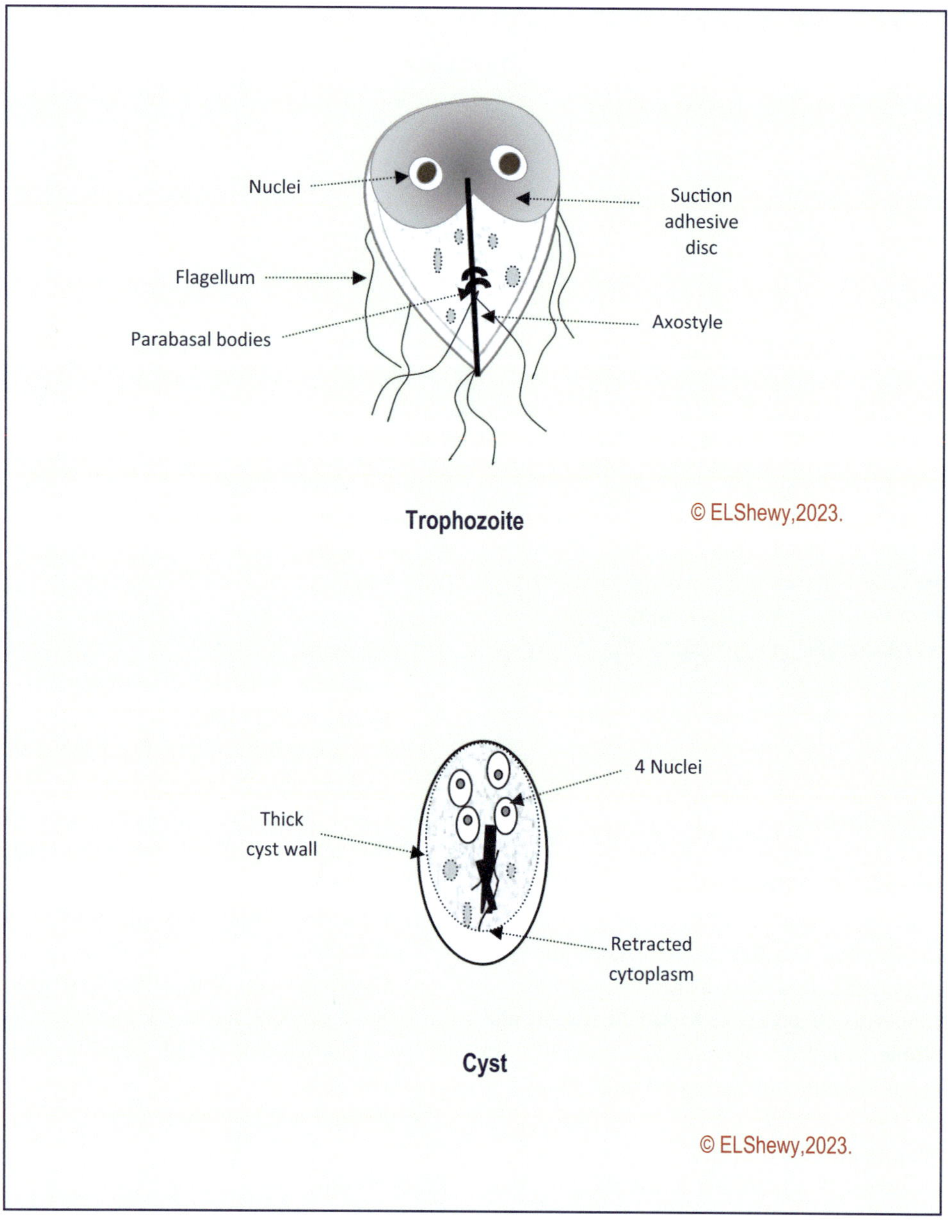

Fig. 5.1 *Giardia lamblia:* morphology

Habitat

Trophozoites of *G. lamblia* inhabit the mucosa of the upper small intestine (duodenum and jejunum). Sometimes, they could be seen in gallbladder and biliary passages.

Life Cycle

The life cycle of *G. lamblia* is postulated in Fig. (5.2):

- Ingested *G. lamblia* cysts pass through the stomach to reach the duodenum, where excystation and release of trophozoites occur.
- Released trophozoites start multiplication by binary fission.
- Encystation occurs when trophozoites pass down to the large intestine.
- Both trophozoites and cysts could be found in the stool; however, cysts could withstand

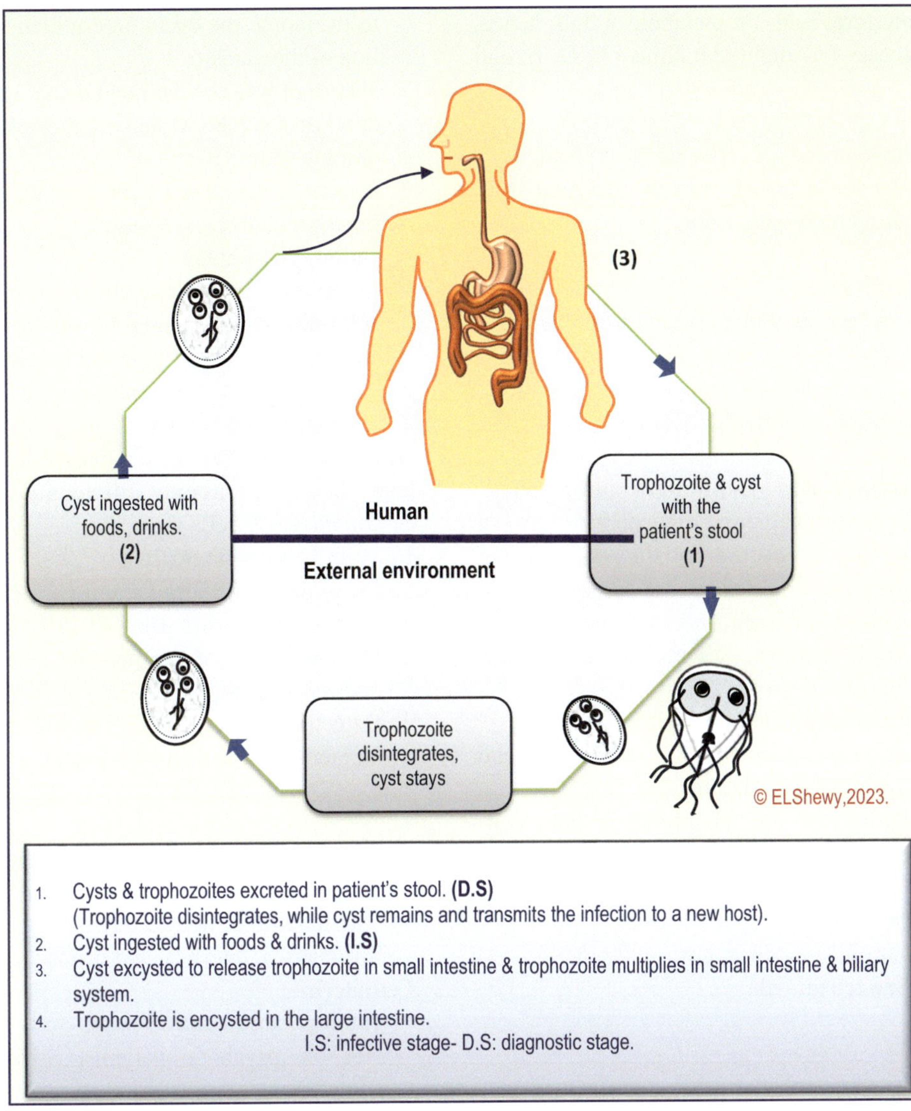

1. Cysts & trophozoites excreted in patient's stool. **(D.S)** (Trophozoite disintegrates, while cyst remains and transmits the infection to a new host).
2. Cyst ingested with foods & drinks. **(I.S)**
3. Cyst excysted to release trophozoite in small intestine & trophozoite multiplies in small intestine & biliary system.
4. Trophozoite is encysted in the large intestine.

I.S: infective stage- D.S: diagnostic stage.

Fig. 5.2 *Giardia lamblia*: life cycle

unfavorable external conditions more than the fragile trophozoites, which disintegrate within 1/2 an hour outside the body.

Final hosts:	Humans, mice, rats, and beavers.
Infective stage:	Cyst.
Diagnostic stages:	Cyst and trophozoite stages in the stool.
Intermediate host:	None.
Reservoir hosts:	Beavers, rats, and mice.

Pathogenesis

Attachment of *G. lamblia* to the intestinal mucosa is associated with various pathological changes sequenced as follows:

1. Minimal villous atrophy with or without shortening of the intestinal villi.
2. Marked villous atrophy with mucosal necrotic changes and hypercellular inflammatory reaction at lamina propria.
3. Deep tissue affection, especially with the animal strains (*G. muris*).

4. It interferes with fat metabolism and, hence, decreases fat-soluble vitamins (A, D, E, and K).

N.B. Attachment of *Giardia* trophozoites is helped by the adhesive sucking disc and by a duodenal secretion of a lectin.

Clinical Picture

Giardiasis affects young children more than adults, and most cases are asymptomatic. However, according to the severity of the infection, there are two clinical forms:

A. **Uncomplicated symptomatic giardiasis:**
   1. A prepatent period (the period between getting an infection and the appearance of the parasites) of about 10–40 days.
   2. Anorexia, nausea, and vomiting.
   3. Mild diarrhea, abdominal colic, and flatulence.
B. **Complicated chronic giardiasis:**
   1. Hypoproteinemia and hypogammaglobulinemia.
   2. Folic acid and fat-soluble vitamin deficiencies (vitamins A, D, E, and K).
   3. Fatty dyspepsia with greasy stool (steatorrhea) and lactose intolerance.
   4. Chronic malabsorption syndrome.
   5. Stunted growth.

N.B.

1. Infection with *Giardia* is found to be more common among individuals with blood group A, achlorhydria, and malnutrient individuals and is associated with bacterial colonization of the duodenum.
2. Secretory IgA proved to have an important protective role against giardiasis.
3. Human milk showed a protective role against giardiasis in breast-feeding infants.

Diagnostic Procedures

- **Direct:**
  1. Microscopic stool examination of wet mount, stained and concentrated stool samples. Several samples should be taken to overcome the cyclic intermittent excretion of the parasite.
  2. The stool was concentrated using formalin ethyl acetate (FEA) and staining with trichrome stain.
  3. Duodenal aspiration with examination of the aspirate microscopically.
  4. Culture of samples.
  5. Enterotest (entero capsule or string capsule) for the sampling of the duodenal mucus (ingestion of gelatin capsule on an empty stomach and examination of the content on glass slide).

  N.B. Many nonpathogenic intestinal flagellates should be carefully differentiated from *G. lamblia* (Fig. 5.3).
- **Indirect Immunological:**
  1. Detection of the specific anti-*Giardia* antibodies in the serum using ELISA.
  2. Demonstration of the parasite using IFA and specific antibodies.
  3. Detection of specific *Giardia* coproantigens (antigens in stool) using dipstick, e.g., Triage.

Treatment

1. **Metronidazole**:
   - 250 mg/3 times daily/5 days.
   - **S.E.** nausea, diarrhea, and headache.
2. **Quinacrine:**
   - 100 mg/3 times daily/5 days.
   - Not for psychotic patients. **S.E.**: CNS stimulations.
3. **Nitazoxanide:** 500 mg/3 times daily/3 days.
4. **Tinidazole** (single oral dose of 1 gm, cure rate up to 90%).
5. **Furazolidine.**
6. **Paromomycin** is the drug of choice in pregnant women (500 mg/3 times/7 days).

Prevention and Control

1. Strict health monitoring of the common and municipal water sources.
2. Following sanitary precautions, especially for food handlers.
3. Proper disposal of human excreta.
4. Safe and healthy water supply, with filtration and chemical treatment.

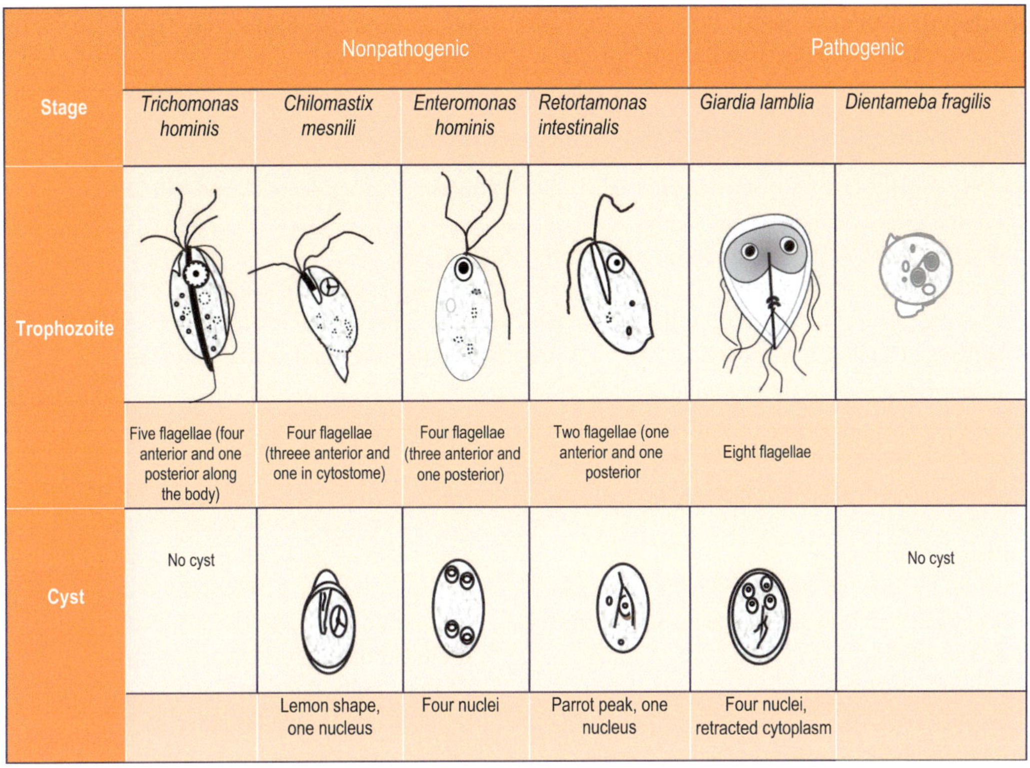

© ELShewy, 2023.

Fig. 5.3 Pathogenic and nonpathogenic intestinal flagellates detected in human stool

5. Treatment of symptomatic as well as asymptomatic cases.

- **Parasitology**: *Giardia* is an intestinal flagellate; it has a leaf-like trophozoite with two nuclei and eight flagellae. Its cyst is infective, oval in shape with four nuclei, and retracted cytoplasm. It is a water-borne parasite.
- **Clinical:** Malabsorption, fatty dyspepsia, vit. A, D, E, K deficiency, and diarrhea. Young children are affected more. Breast milk is protective against giardiasis.

Nonpathogenic Intestinal Flagellates

A group of intestinal flagellates commonly detected in human stool. They have the same life cycle and mode of infection as *G. lamblia*. They are harmless but usually misdiagnosed as pathogenic species. Proper differentiation between them and the pathogenic ones could save time and cost and decrease the side effects of unnecessary treatment. The most common nonpathogenic intestinal flagellates include:

1. *Trichomonas hominis.*
2. *Chilomastix mesnili.*
3. *Enteromonas hominis.*
4. *Retortamonas intestinalis.*

Cryptosporidium Parvum (C. parvum)

C. parvum is an intestinal *Apicomplexan* water-borne parasite infecting both humans and animals **(zoonotic)**. In immunocompromised individuals, *C. parvum* causes a severe, even fatal, gastrointestinal disease **(opportunistic).**

Geographical Distribution

1. There are many species of *Cryptosporidium*; both *C. parvum and C. hominis* are the most common species infecting humans.
2. *C. parvum* has a worldwide distribution, especially among immunocompromised individuals, the elderly, and in daycare centers.
3. It causes waterborne epidemics in many parts of the world.
4. The prevalence of cryptosporidiosis ranges between 3 and 35% worldwide.
5. *Cryptosporidium* is the second most common cause of enteritis after *Giardia* worldwide.
6. *Cryptosporidium* has no host specificity, so it could infect many vertebrate hosts.

Parasite Morphology

Cryptosporidium has an oocyst with four sporozoites inside and it is excreted as mature, ready to infect once outside.

- Oocyst:
  - It is rounded in shape.
  - Its size ranges between 2 and 7 μm.
  - It has a thin oocyst wall.
  - It contains four spindle-shaped sporozoites, each 2–4 μm.

N.B. There are many other species with different sizes isolated from different hosts, e.g., *C. muris*, *C. bovis*, *C. meleagridis*, and *C. fragile*.

Transmission

1. The feco-oral route especially among farm workers handling infected young calves.
2. Contamination of foods with oocysts.
3. Water-borne (swimming in contaminated lakes, pools, and drinking).
4. Internal cyclic autoinfection (in the immunocompromised individuals).
5. External autoinfection (hand to mouth) from the same patient.

Habitat

- *Cryptosporidium* is considered an intracellular extra-cytoplasmic parasite, as it forms a parasitophorous vacuole at the epithelial cells, but it does not invade the cytoplasm of the host cells.
- It inhabits the epithelial lining of the gastrointestinal tract.
- It may be detected in the respiratory epithelium, biliary passages, and gall bladder.

Life Cycle

Following their ingestion, *Cryptosporidium* oocysts excyst and release their sporozoites to invade the intestinal epithelial cells. Sporozoites transform into merozoites that undergo two multiplication cycles as follows (Fig. 5.4):

1. **Asexual cycle (schizogony):** results in the formation of schizont containing eight merozoites (first generation). These merozoites undergo another schizogony to form second generation schizonts (four merozoites) which reinfect other epithelial cells.
2. **Sexual reproduction (sporogony/gametogony):** some of the resulting merozoites form micro and macro gametocytes. Macro and microgametes form a zygote, which transforms into an oocyst with four sporozoites that pass to the external environment with the patient's stool.

Final hosts:	Humans, cattle, sheep, and goats.
Infective stage:	Oocysts.
Diagnostic stage:	Oocysts in the stool.
Intermediate host:	None.
Reservoir hosts:	Cattle, sheep, and goats.

Pathogenesis

1. *C. parvum* infecting the epithelial lining of the gastrointestinal tract causes minimal changes in the epithelial architecture.
2. An enterotoxin antigen with *Cholera*-like action is proven to be excreted during intestinal infection.
3. A marked disturbance in the water and electrolyte balance causes prolonged diarrhea.
4. In immunocompromised patients with lowered immune reaction, *C. parvum* acts as an opportunistic parasite, causing uncontrolled severe water and electrolyte loss with severe dehydration and occasionally biliary and respiratory illness.

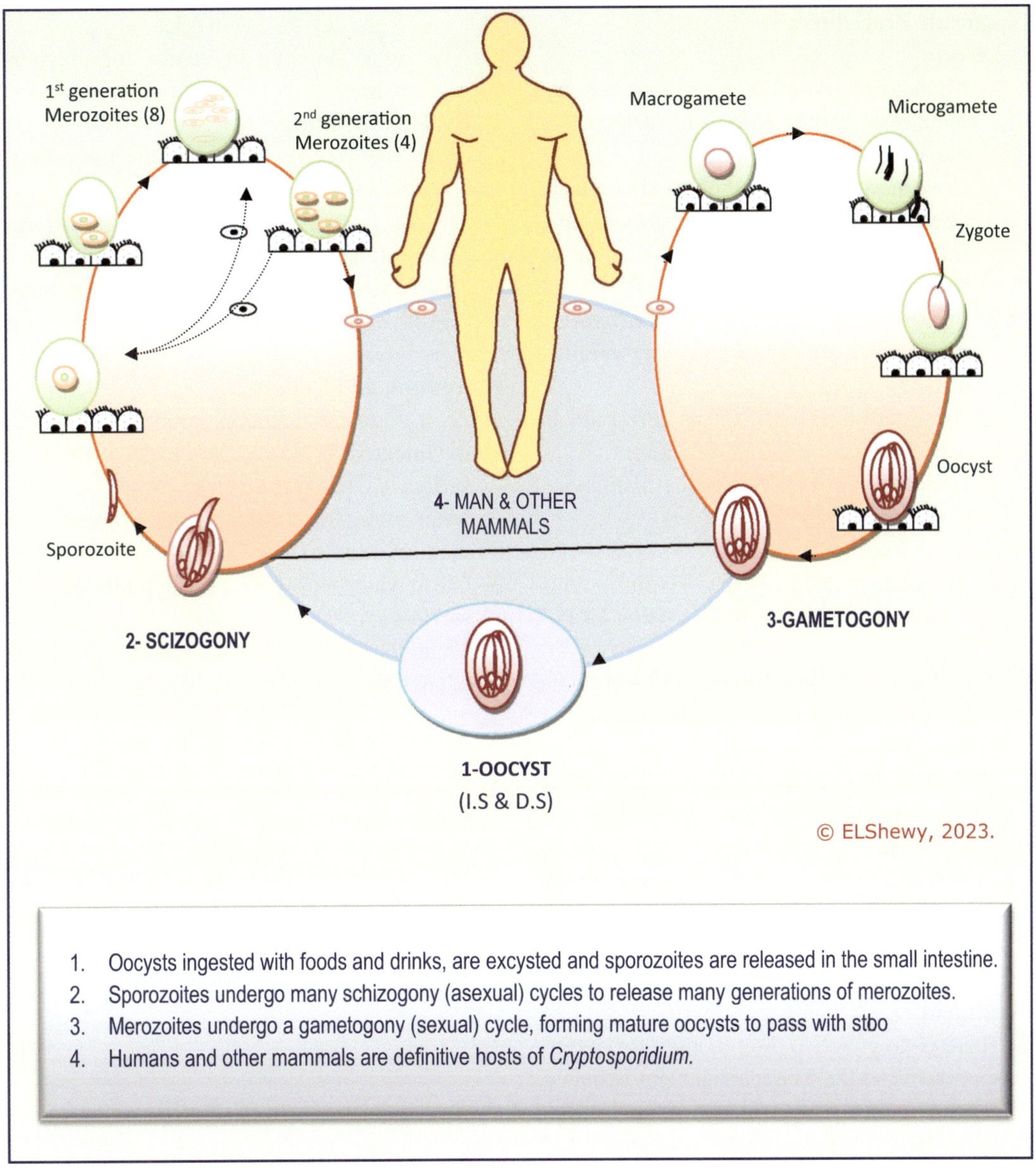

Fig. 5.4 *Cryptosporidium:* life cycle

Clinical Picture

1. Infection with *C. parvum* in immunocompetent healthy individuals is usually asymptomatic, or it may cause mild self-limited diarrhea.
2. Diarrhea is the main symptom of cryptosporidiosis; however, constipation, dyspepsia, nausea, vomiting, weight loss, and dehydration can be seen in cryptosporidiosis.
3. Malabsorption, indigestion, flatulence, and distension.
4. Cryptosporidiosis in immunocompromised or malnourished infants is represented by severe prolonged diarrhea, with severe dehydration that might end fatally.

Diagnostic Procedures

- **Direct:**
  1. Microscopic stool examination of wet mounts or stained smears for detection of the characteristic oocysts.
  2. Concentration of the stool specimen using Sheather's floatation concentration method (high concentration of sucrose solution used for floatation of the oocysts).
  3. Staining of the fecal smear with **Kinyoun's acid-fast staining**, iodine, or modified acid-fast stain.
  4. An entero-test (entero-capsule) could be used to detect obscured infection.
  5. Intestinal biopsy followed by histological examination can be used.
- **Indirect Immunological:**
  1. Detection of the serum anti-Cryptosporidium antibodies using ELISA and IFA.
  2. Detection of the specific antigen in the patient's stool (copro-antigens) using CIE and ELISA.
  3. **Triage dipstick** is a commercially available dipstick for the detection of three protozoan parasites: *Giardia, Cryptosporidium,* and *Entamoeba* in the stool. It has a fair sensitivity, especially in surveys and field studies.

Treatment

- There is no effective drug therapy for cryptosporidiosis so far. The management is mainly depending on the replacement of the fluid and electrolytes lost in severe dehydration.
- Immunotherapy using hyperimmune colostrum has been tried with animals.
- The following drugs are routinely used in the treatment of cryptosporidiosis:
  1. **Nitazoxanide:**
     - 500 mg/2 times daily/3 days for adults, 100 mg (1–3 years), and 200 mg (4–11 years).
     - It is the most recent and effective drug for cryptosporidiosis.
  2. **Spiramycin** (Rovamycin):
     - 3 gm/day for 2–3 weeks.
     - It is the drug of choice for pregnant women.
  3. **Somatostatin:**
     - 300 μg/subcutaneous injection, three times a day.
  4. A combination of **Nitazoxanide, Paromomycin, and Azithromycin** proved effective in immunocompromised patients with cryptosporidiosis.

Prevention and Control

1. As a zoonotic infection, proper handling of the infected calves and their excreta should be seriously considered.
2. Veterinary care of the farm animals with treatment of infected cases.
3. Clean water supply with enough filtration and sieving.
4. Treatment of the infected cases with health education.

- **Parasitology**: *Cryptosporidium* is a water-borne *apicomplexan* zoonotic parasite. Its oocysts in contaminated foods or drinks transmit infection to humans. Diagnosis is done mainly by the detection of oocysts in stool stained with Kenyon's acid-fast and concentrated with Sheather's floatation method.
- **Clinical:** Cryptosporidiosis is mostly asymptomatic in healthy hosts, but a severe persistent infection with severe dehydration and water loss is common in immunocompromised.

Cystoisospora belli (C. belli)

Cystoisospora belli (previously known as *Isospora belli*) is an intestinal *Apicomplexan* parasite that infects both humans and animals. Cystoisosporiasis is usually asymptomatic or it causes mild gastroenteritis. In immunocompromised patients, infection is severe and may be disseminated **(opportunistic).**

Geographical Distribution

There are many *Cystoisospora* species with worldwide spread. *C. belli* is the most common species infecting humans, especially among immunocompromised individuals and homosexuals. *C. suis* is the most common species in pigs.

Parasite Morphology

Oocyst of *C. belli* is immature when first passed. It takes 2–5 days in the external environment to be mature and infective (Fig. 5.5):

- Immature Oocyst:
  - It is oval in shape with tapered ends and a thick oocyst wall.
  - 20–30 μm.
  - It contains two protoplasmic masses.
- Mature Oocyst:
  - It is an oval is shape with tapered ends and a thin wall.
  - 20–30 μm long.
  - It contains two sporocysts, each with four sporozoites.

Habitat

Cystoisospora species inhabit the intestinal epithelium of humans and animals, where both the sexual and asexual stages can be detected.

Transmission

Infection with *C. belli* is transmitted through the feco-oral routes, contamination of food or water and among homosexuals.

Life Cycle

The life cycle of *C. belli* is postulated in Fig. (5.5).

Final hosts:	Humans.
Infective stage:	Mature oocyst with two sporocysts and four sporozoites in each.
Diagnostic stage:	Oocysts in the stool.

Clinical Picture

1. Cystoisosporiasis is usually asymptomatic especially in immunocompetent individuals.
2. In immunocompromised patients, cystoisosporiasis is manifested with chronic persistent enteritis, nausea, vomiting, abdominal colic, indigestion, and diarrhea.

Diagnostic Procedures

1. Microscopic examination of stool specimens to detect both mature and immature oocysts using:
   - Sheather's floatation concentration method.
   - Kinyoun's acid-fast staining.
   - Modified acid-fast stain.
2. Entero-test could be used for getting samples.

Treatment

Asymptomatic patients usually need no treatment. In case of severe infection or infection of immunocompromised individuals, the following drugs could be used in the treatment of cystoisosporiasis:

1. **Trimethoprim and sulfamethoxazole** (Bactrim, Septra, or Cotrim).
2. **Sulfadiazine-pyrimethamine.**
3. **Metronidazole.**
4. **Tetracycline.**

Prevention and Control

Good personal hygiene and standards of sanitary conditions should be adopted.

- **Parasitology**: Oocysts of *cystoisospora* are immature when first passed. They need 4–5 days to mature. Mature oocysts in contaminated foods or drinks transmit infection. Diagnosis is made mainly by detecting oocysts in stool stained with Kinyoun's acid-fast and concentrated with Sheather's floatation method.
- **Clinical:** Cystoisosporiasis is mostly asymptomatic. In immunocompromised, severe gastroenteritis is manifested.

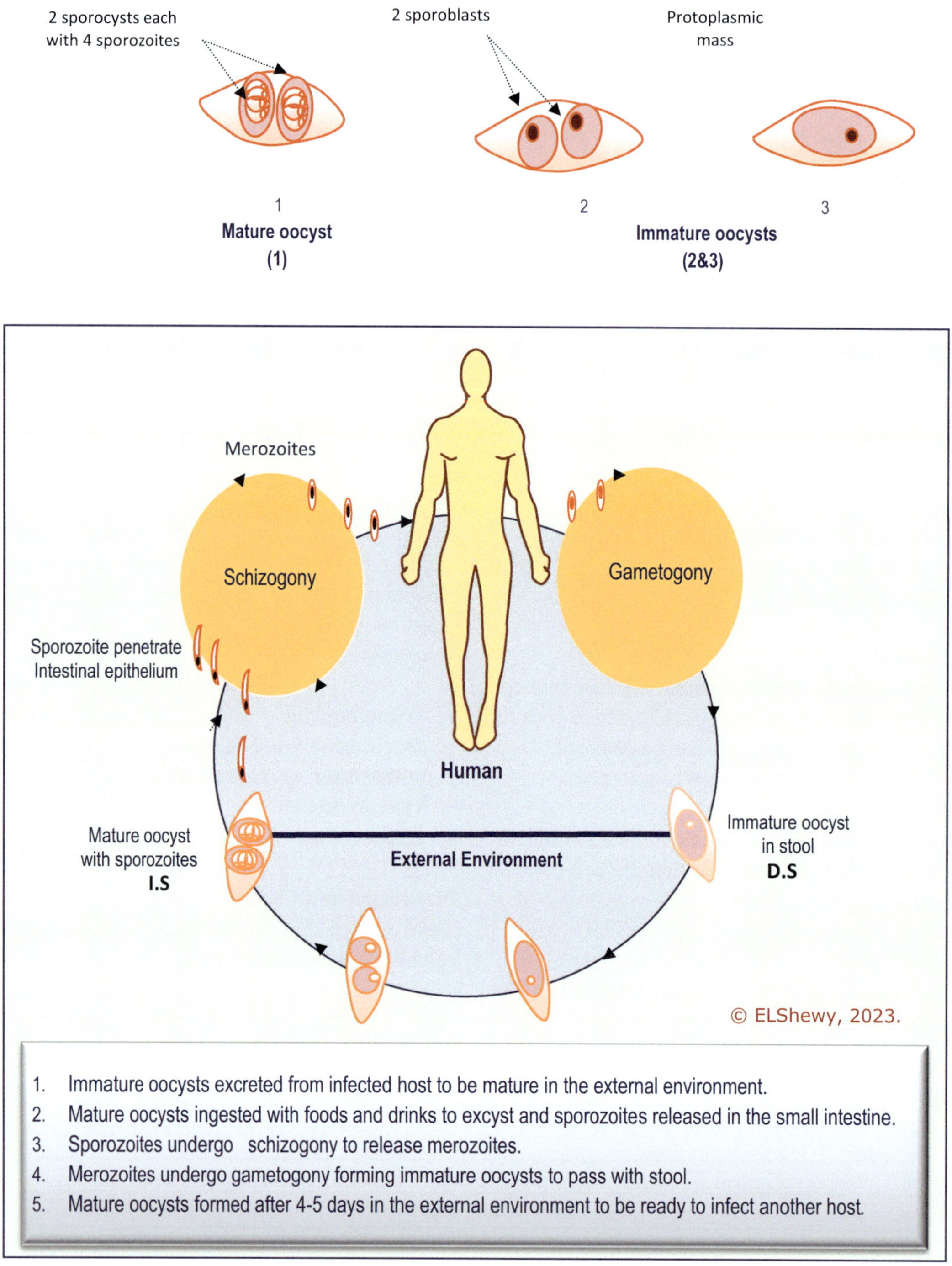

Fig. 5.5 *Cystoisospora belli*: morphology and life cycle

Cyclospora cayetanensis (*C. cayetanensis*)

C. cayetanensis is an intestinal *Apicomplexan* parasite. It causes a disease similar to that of cryptosporidiosis. It causes gastroenteritis, which might be severe and fatal among immunocompromised individuals **(opportunistic).**

Geographical Distribution

C. cayetanensis has a worldwide distribution, especially among immunocompromised individuals and homosexuals.

Parasite Morphology

C. cayetanensis has two morphological forms: an immature oocyst unsporulated when freshly passed in the stool and a mature oocyst, which is infective and contains two sporocysts; each contains two sporozoites (Fig. 5.6):

- Immature Unsporulated Oocyst:
  - Spherical.
  - 8–10 μm. Thin wrinkled walled.
  - It contains a central protoplasmic mass (morula) with six to nine granules.
  - It takes 2 weeks to mature in the external environment.
- Mature Sporulated Oocyst:
  - Spherical.
  - 8–10 μm. Thin wrinkled walled.
  - It contains two sporocysts, each with two sporozoites.

Habitat

C. cayetanensis inhabits the small intestine of the hosts (humans and other mammals).

Transmission

Transmission of *C. cayetanensis* is similar to that of *Cystoisospora*. The feco-oral route is the main route of transmission.

Life Cycle

The life cycle of *C. cayetanensis* is not well understood. However, it appears to be similar to that of *Cystoisospora* species (Fig. 5.7).

Final hosts:	Humans and other mammals.
Infective stage:	Mature oocyst containing two sporocysts, each with two sporozoites.
Diagnostic stage:	Oocysts in the stool.

Clinical Picture

1. Among immunocompetent individuals, infection with *C. cayetanensis* is mostly mild or asymptomatic.
2. Clinically, cyclosporiasis is similar to that of cryptosporidiosis.
3. After an incubation period of 2–7 days, manifestations of gastroenteritis with diarrhea, myalgia, anorexia, weight loss, fatigue, and vomiting with chronic dyspepsia.
4. The clinical picture tends to be severe in immunocompromised patients.

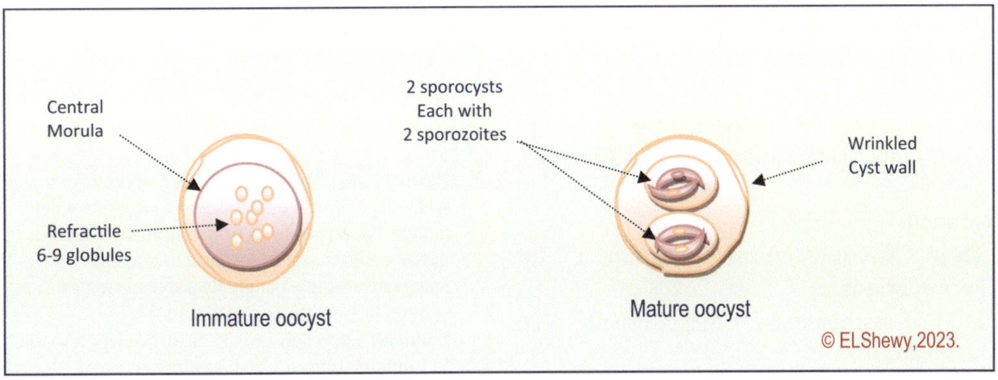

Fig. 5.6 *Cyclospora cayetanensis*: morphology

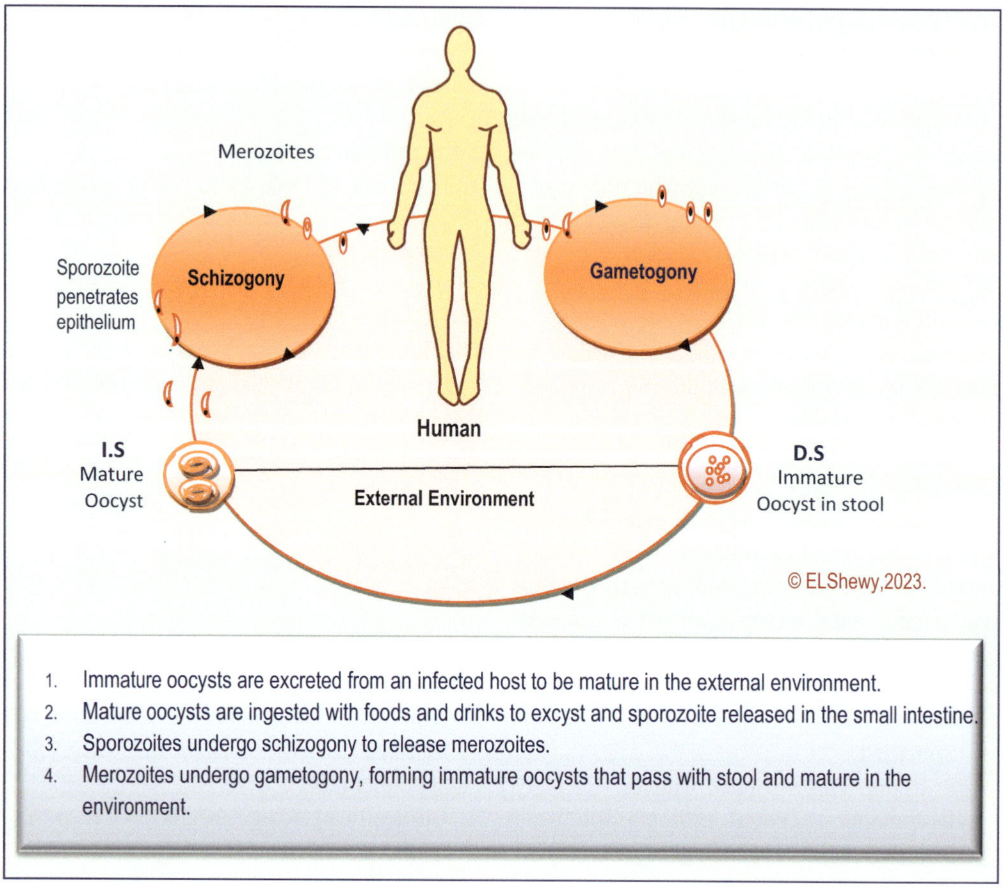

Fig. 5.7 *Cyclospora cayetanensis*: life cycle

Diagnostic Procedures

1. Microscopic demonstration of the characteristic oocysts in the stool either using direct wet mount or after concentration and staining techniques. Acid-fast stain is commonly used for staining *Cyclospora*.
2. IFA and Epi-fluorescence.

N.B. The common *Apicomplexan* parasites infecting humans are illustrated in Fig. (5.8).

Treatment

1. There is no effective treatment for cyclosporiasis.
2. Mild or asymptomatic patients usually need no treatment.
3. In case of severe infection or in immunocompromised individuals, the following drugs could be used in the treatment of cyclosporiasis:
   - **Trimethoprim and sulfamethoxazole** (Bactrim, Septra, or Cotrim).
   - **Sulfadiazine-pyrimethamine.**

Prevention and Control

Good personal hygiene and standards of sanitary conditions should be adopted.

- **Parasitology**: The oocyst of *Cyclospora* is immature when deposited; it contains two sporocysts when mature. Mature oocysts in contaminated foods or drinks could transmit infection to humans. Diagnosis is made mainly by the detection of oocysts in stool stained with acid-fast.
- **Clinical:** Cyclosporiasis is mostly asymptomatic. In immunocompromised, severe persistent infections with gastroenteritis are common presentations.

	Tissue	Intestinal			
	Toxoplasma	*Cystoisospora*	*Cyclospora*	*Cryptosporidium*	*Sarcocystis*
In The Stool	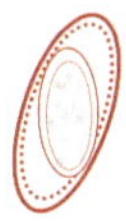 10-12 μ Thick wall Oval Protoplasmic mass Immature	 20-30 μ Thin wall Protoplasmic mass Immature	 8-10 μ Spherical 6-9 granules Central morula Immature	 4-6μ Spherical 1 sporocyst with 4 sporozoites Mature	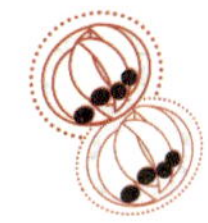 25-33 μ 2 sporocysts with 4 sporozoites in each. Very thin wall Mature
In External Environment	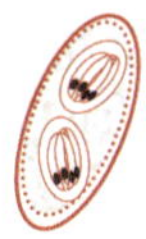 10-12μ Thick wall 2 sporocysts, with 4 sporozoites in each.	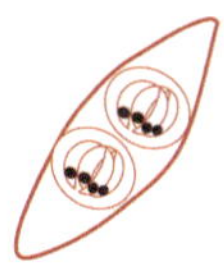 20-30 μ Thin wall 2 sporocysts, with 4 sporozoites in each.	 8-10 μ Spherical 2 sporocysts Each with 2sporozoites	The same	The same

Fig. 5.8 Common human *Apicomplexan* parasites

Review Questions

1. ***Cryptosporidium parvum*** **is located in the epithelium as an.**
   (a) Extracellular parasite.
   (b) Intracellular intracytoplasmic parasite.
   (c) Intracellular extra cytoplasmic parasite.
   (d) Extracellular intracytoplasmic parasite.
2. **Which one of the following parasites could be transmitted through hand to mouth?**
   (a) Hookworm.
   (b) *Cyclospora cayetanensis.*
   (c) *Strongyloides stercoralis.*
   (d) *Diphyllobothrium.*
3. **Which of the following parasites could cause an epidemic of water-borne gastroenteritis?**
   (a) *Strongyloides stercoralis.*
   (b) Hookworm.
   (c) *Giardia lamblia.*
   (d) *Trichostrongylus colubriformis.*
4. **The following intestinal flagellate has only two flagellae.**
   (a) *Chilomastix mesnili.*
   (b) *Retortamonas intestinalis.*
   (c) *Trichomonas hominis.*
   (d) *Giardia lamblia.*
5. ***Giardia lamblia*** **infection could be presented with the following except.**
   (a) Fatty dyspepsia.
   (b) Water-soluble vitamins deficiency.
   (c) Flatulence.
   (d) Diarrhea.
6. ***Giardia lamblia*** **infection could be diagnosed by the following except.**
   (a) Microscopic stool examination.
   (b) Enterotest (entero-capsule).
   (c) Blood film.
   (d) IFA.

7. **Oocyst is the infective stage of the following parasites except.**
   (a) *Cyclospora cayetanensis.*
   (b) *Cryptosporidium.*
   (c) *Giardia lamblia.*
   (d) *Cystoisospora.*

8. **Which of the following parasites is ready to infect when it is freshly passed with stool?**
   (a) *Cyclospora cayetanensis.*
   (b) *Cryptosporidium and Cyclospora cayetanensis.*
   (c) *Cryptosporidium and Giardia lamblia.*
   (d) *Cryptosporidium and Isospora.*

6 Protozoan Parasites of the Large Intestine

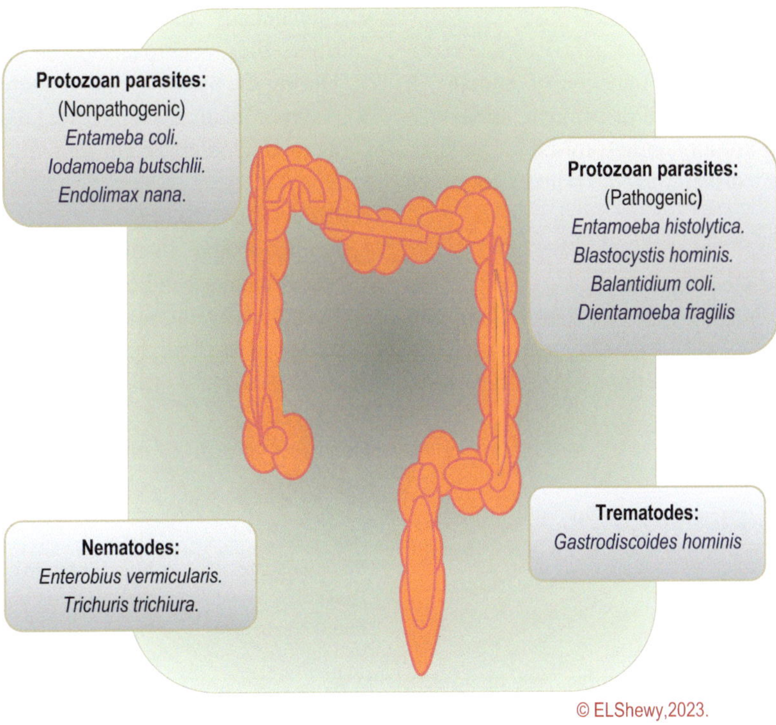

Parasites of the Large Intestine

K. ElShewy, *Medical Parasitology*, https://doi.org/10.1007/978-3-031-64736-9_6

Test your learning and check your understanding of this book's contents: use the "Springer Nature Flashcards" app to access questions using ▶ https://sn.pub/oyFLIx. To use the app, please follow the instructions in Chap. 1.

Entamoeba histolytica

(*E. histolytica*)

E. histolytica is a protozoan parasite that principally inhabits the large intestine of humans. It might cause intestinal and/or extraintestinal amebiasis. Among immunocompromised individuals, it causes a disseminated brutal form of amebiasis.

Geographical Distribution
E. histolytica has a worldwide distribution, particularly in areas with poor hygienic conditions and a lack of clean water supply, e.g., slums, jails, and overcrowded areas.

Parasite Morphology
E. histolytica presents in two morphological forms: an active trophozoite and an inactive static cyst form (Fig. 6.1):

- Trophozoite:
  - Shape: ameboid (irregular and unstable), 10–60 μm.
  - Cytoplasm: finely granular, containing bacteria, yeast, and RBCs.
  - A single nucleus with peripheral chromatin dots is distributed equally as fine beads, and a small central, compact karyosome is also present.
  - Movement is progressive with clear pseudopods.
- Cyst:
  - Rounded in shape with a cyst wall, 10–15 μm.
  - Cytoplasm: It has elongated chromatoid bodies with blunted ends. Immature cysts may contain glycogen vacuoles.
  - Nucleus: the same as in trophozoite form.
  - A mature cyst has four nuclei, while an immature cyst contains less than four nuclei (1–4).

Transmission
1. Waterborne transmission.
2. Ingestion of foods contaminated with cysts.
3. Transmission through the same patient (hand-to-mouth external autoinfection).
4. Mechanical transmission through insects (houseflies, cockroaches).
5. During the use of human manure as fertilizer.
6. Asymptomatic carriers (cyst passers) working as food handlers (main source).

Habitat
- *E. histolytica* inhabits the large intestine (cecum and ascending colon).
- Occasionally, it could be detected extra intestinally in the liver, lungs, and brain.

Life Cycle
The life cycle of *E. histolytica* is postulated in Fig. (6.2).

Final hosts:	Humans.
Infective stage:	Mainly mature cyst.
Diagnostic stage:	Cyst and trophozoite stages.
Intermediate hosts:	None.

N.B.

- Occasionally, *E. histolytica* trophozoites invade the wall of the colon and pass into the circulation, i.e., extraintestinal amebiasis.
- Many nonpathogenic amebae inhabit the large intestine, and their life cycle and mode of infection are similar to *E. histolytica*.

Pathogenesis
Pathological changes during intestinal amebiasis are only caused by trophozoites. The following are the mechanisms and sequences of these changes (Fig. 6.3):

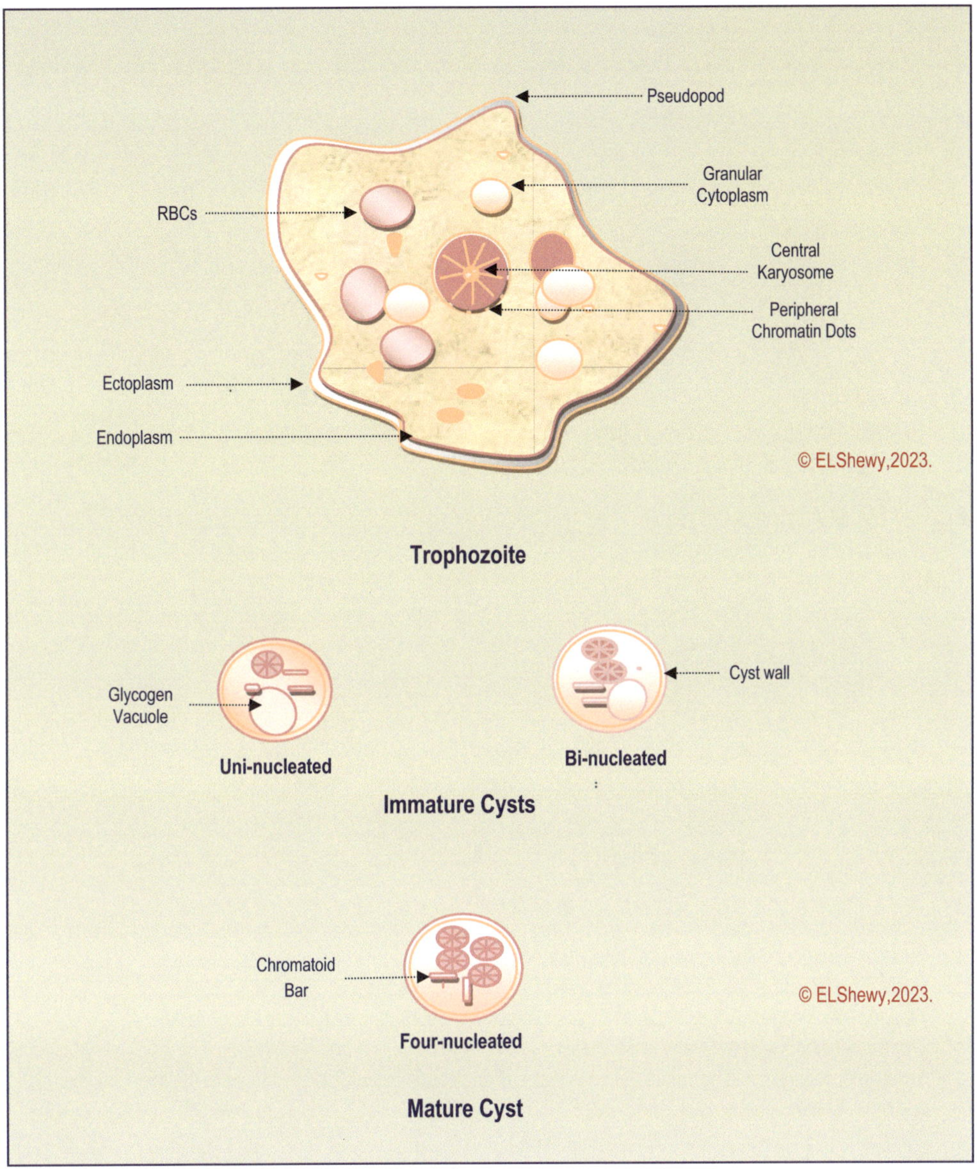

Fig. 6.1 *Entamoeba histolytica:* morphology

1. **Attachment** of trophozoites to the intestinal epithelial cells through receptors.
2. Trophozoites start to **invade** the target intestinal cells at the primary site of invasion (the cecal area and the recto-sigmoid region). Later, any part of the colon could be affected.
3. After invasion, trophozoites cause cell **necrosis**, followed by cell death, which leads to ulceration of the affected intestinal mucosa through the following factors:
   (a) Secretion of proteolytic enzymes.
   (b) Enterotoxins production.
   (c) Contact-dependent cell killing.
   (d) Phagocytosis of the killed mucosal cells by trophozoites.
4. The resulting amebic ulcers are multiple flask-shapes with yellowish centers,

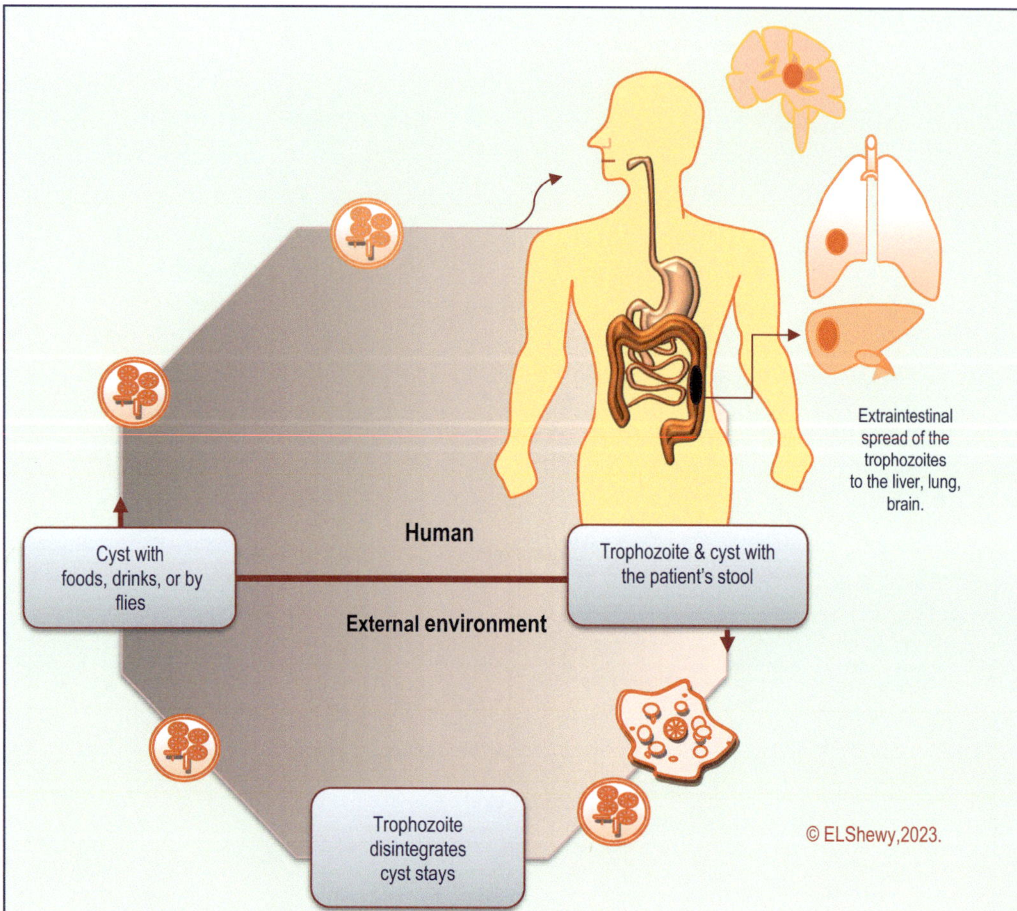

1. Cysts of *Entamoeba* are ingested with foods & drinks. **(I.S.).**
2. Cysts are excysted in the small intestine, releasing many trophozoites.
3. Trophozoites colonize in the cecum and ascending colon, where they multiply by binary fission.
4. Trophozoites encyst in the lower large intestine.
5. Some trophozoites could perforate the intestinal wall and cause extraintestinal spread to the liver, lungs, brain and other tissues.
6. Cysts and trophozoites are excreted in stool. **(D.S.)**
7. Trophozoites disintegrate while cysts resist and transmit the infection to another host.

Fig. 6.2 *Entamoeba histolytica*: life cycle

erythematous peripheries, and healthy mucosa in between.

5. In some circumstances, trophozoites **perforate** the intestine, especially at the ulcer sites, and they are disseminated through the blood to the liver, lungs, and brain.

Clinical Picture

Amebiasis could be manifested as:

(a) **Asymptomatic amebiasis** (85–95% of the cases)**:**
The infected persons may be either cyst passers (carriers) or non-cyst passers.

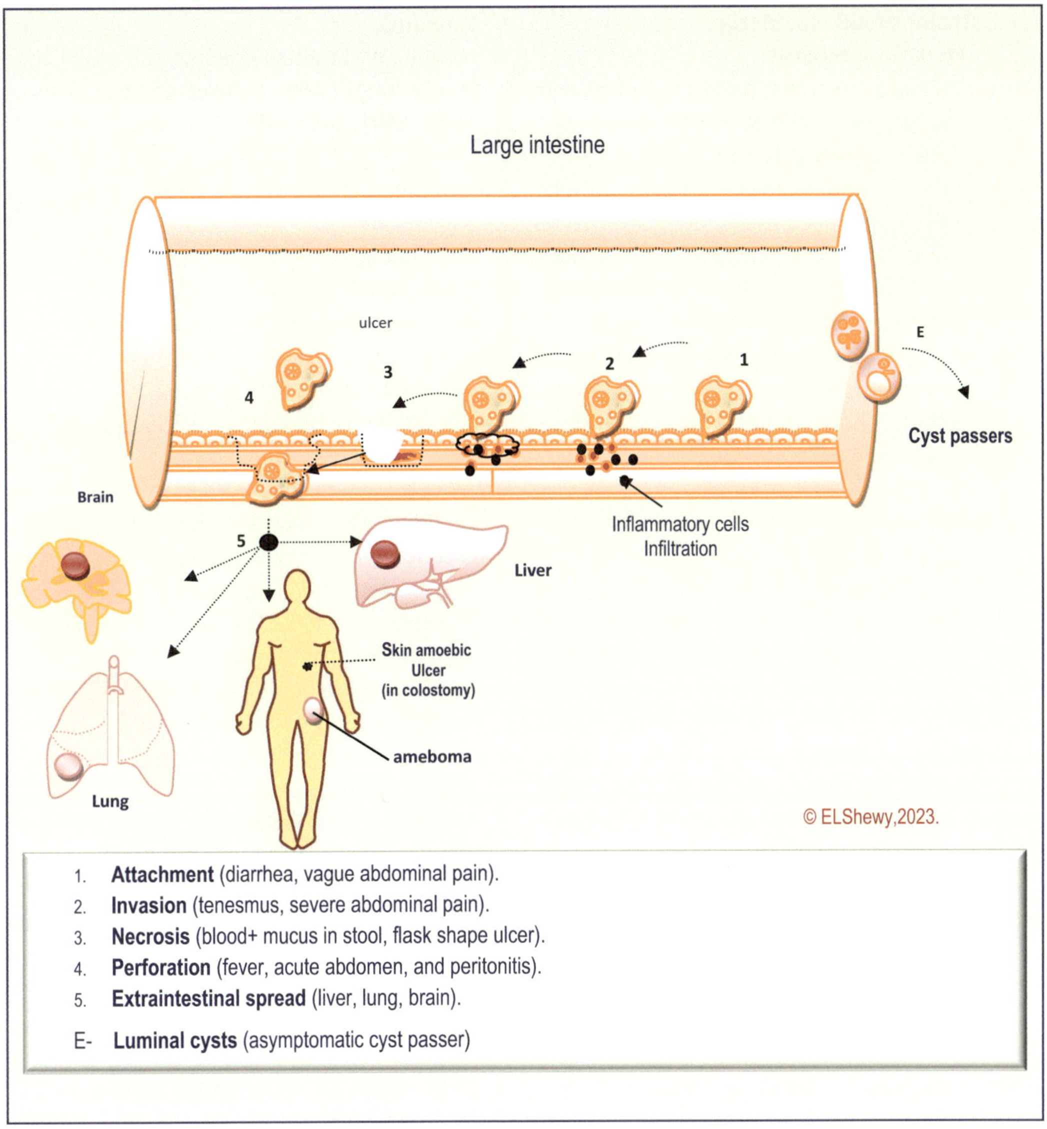

Fig. 6.3 *Entamoeba histolytica*: pathogenesis

(b) **Symptomatic amebiasis** (5–15% of the cases)**:**

1. Uncomplicated intestinal amebiasis*:* Vague abdominal pain, cramps, flatulence, diarrhea or constipation, anorexia, and chronic fatigue.
2. Acute intestinal amebiasis: Characterized by diarrhea, abdominal cramps, tenesmus (painful, difficult defecation), with passage of blood and mucus in stool (amebic dysentery).
3. Complicated amebic colitis (amebic ulcer): Severe abdominal cramps, diarrhea, fever, tenderness, and abdominal rigidity.
4. Intestinal amebic perforation (perforated ulcer): Perforated amebic ulcer causes peritoneal irritation, abdominal rigidity, distension, ileus, and gases.
5. Amebic granuloma (ameboma): It is a chronic granulomatous lesion. It occurs at the cecal or recto-sigmoidal region.

(c) **Extraintestinal amebiasis** (very rare):
1. **Hepatic amebiasis:**
   - Amebic hepatitis: Liver tenderness and enlargement with fever and sweating.
   - Amebic liver abscesses: Single or multiple abscesses, liver tenderness, severe and continuous night fever. Aspiration from the ulcer revealed pus with a chocolate brown color (anchovy sauce pus).
2. **Pulmonary amebiasis***:*
   A right lower lobe pneumonitis could result from erosion of a sub-diaphragmatic amebic abscess. The patient suffers from fever, dyspnea, and cough with sputum that may contain amebae.
3. The brain, pericardium, and spleen could be sites for amebic abscesses.
4. **Amebic skin:**
   Skin could be a site for amebic ulcer, in case of colostomy or during draining of hepatic abscess.

Diagnostic Procedures

In addition to the clinical picture, the following procedures are used for detection of *E. histolytica* infection:

- **Direct:**
  1. Microscopic stool examination of:
     - Wet mount smear.
     - Concentrated stool samples, e.g., using formalin ethyl acetate (FEA).
     - Permanent smear stained with trichrome stain.
  2. Other samples suitable for microscopic examinations include sigmoidoscopic biopsy and aspirate from a liver abscess.
  3. Culture of samples in **Diamond's** medium.
- **Indirect Immunological:**
  1. Detection of the specific anti-amebic antibodies by IHA, ELISA, and IFA.
  2. Detection of amebic antigens in the stool samples (copro-antigens) by Counter immuno-electrophoresis (CIE) or dipstick strips.
- **Molecular diagnosis:**
  Polymerase chain reaction (PCR) is used for diagnosis and differentiation between the pathogenic and nonpathogenic strains.
- **Imaging:**
  Imaging of the amebic lesions (intestinal ameboma, lung, brain, or liver abscesses) using sonar, MRI, and CAT.

Treatment
1. **Tinidazole:**
   - 2 gm/day/3 days.
2. **Iodoquinol** (Enterovioform):
   - Contraindicated in pregnancy and liver diseases.
3. **Diloxanide furate** (Furamide):
   - 500 mg/3 times daily for 3 days course.
   - Good for cyst passers.
4. **Metronidazole** (Flagyl):
   - 500–750/3 times daily/for 7 days course.
   - Used for acute and asymptomatic amebiasis. Its side effects are nausea and diarrhea.
5. **Emetine hydrochloride:**
   - 1 mg/kg/body weight for 5 days through S.C. or I.M. administration
   - No longer than 5 days course.
6. **Nitazoxanide:**
   - 500–750 mg/3 times/day for 5 days.

Prevention and Control
1. Following sanitary precautions especially for food handlers, medical certificate (cyst passers should not work as food handlers), personal cleanness, washing hands, and cutting nails.
2. Proper disposal of the human excreta.
3. Safe and healthy water supply, with filtration and chemical treatment.
4. Treatment of all cases of amebiasis, including (symptomatic and asymptomatic).

- **Parasitology**: *E. Histolytica* trophozoite is ameboid in shape; it has a single nucleus and granular cytoplasm with RBCs. The cyst is spherical with 1–4 nuclei. The mature cyst is an infective stage. Infection by feco-oral, flies, and contaminated foods.
- **Clinical:** Most infections are asymptomatic. Symptomatic amebic dysentery with extraintestinal spread to the liver, lungs, and brain. A flask-shaped ulcer is pathognomic in intestinal amebiasis.

Amebic Confounders

1. This group of nonpathogenic amebae and blood cells could be found in the stool and misdiagnosed as the pathogenic *E. histolytica*.
2. Differentiation should be done between them to save time, cost, and side effects of unnecessary drugs. The following are some of these confounders:
   - *Entamoeba coli.*
   - *Entamoeba dispar.*
   - *Endolimax nana.*
   - *Iodamoeba butschlii.*
   - *Blastocystis hominis* (? protozoan parasite, one of the *Stramenopiles*).
   - Macrophages and RBCs.

N.B. All nonpathogenic amebae mentioned above have the same life cycle and mode of infection as *E. histolytica*.

Tables 6.1, 6.2 and 6.3 illustrate the morphological differences between *E. histolytica* and all of these confounders.

Table 6.1 Morphological differences between *E. histolytica* and macrophages

	E. histolytica	*Macrophage*
Nucleus	Single nucleus with tiny and central karyosome Equally distributed beadlike peripheral chromatin bodies	Large kidney shape nucleus with no peripheral chromatin bodies
Ratio of nuclear material to cytoplasm N/C	1:10	1:4
Contents	RBCs, some debris, no polymorphic leukocytes (PML)	Usually contain ingested debris, RBCs and PML
Size	10–60 μm	35–55 μm

Table 6.2 Morphological differences between *E. histolytica* and *E. coli*

	E. coli	*E. histolytica*
TROPHOZOITE	Sluggish multidirectional movement.	Rapid unidirectional movement.
	No RBCs in the cytoplasm.	There are many RBCs.
	Cytoplasm contains bacteria, yeast, and debris.	Smooth finely granular cytoplasm.
	Large and eccentric karyosome, with uneven and blotchy peripheral chromatin bodies.	Karyosome is tiny and central with equally distributed beadlike peripheral chromatin bodies.
	Size 15–50 μm.	Size 10–60 μm.
CYST	Mature cyst contains 8 nuclei.	Mature cyst contains 4 nuclei.
	Karyosome is large ad eccentrically placed.	Karyosome is small and central.
	Peripheral chromatin is coarse and uneven.	Peripheral chromatin is delicate and even.
	Chromatoid bodies have pointed, splintered ends.	Chromatoid bodies have smooth, rounded ends.
	Size 10-25 μm.	Size 10–15 μm.

Table 6.3 Common intestinal and oral amebae (trophozoites, cysts, and their nuclear morphology)

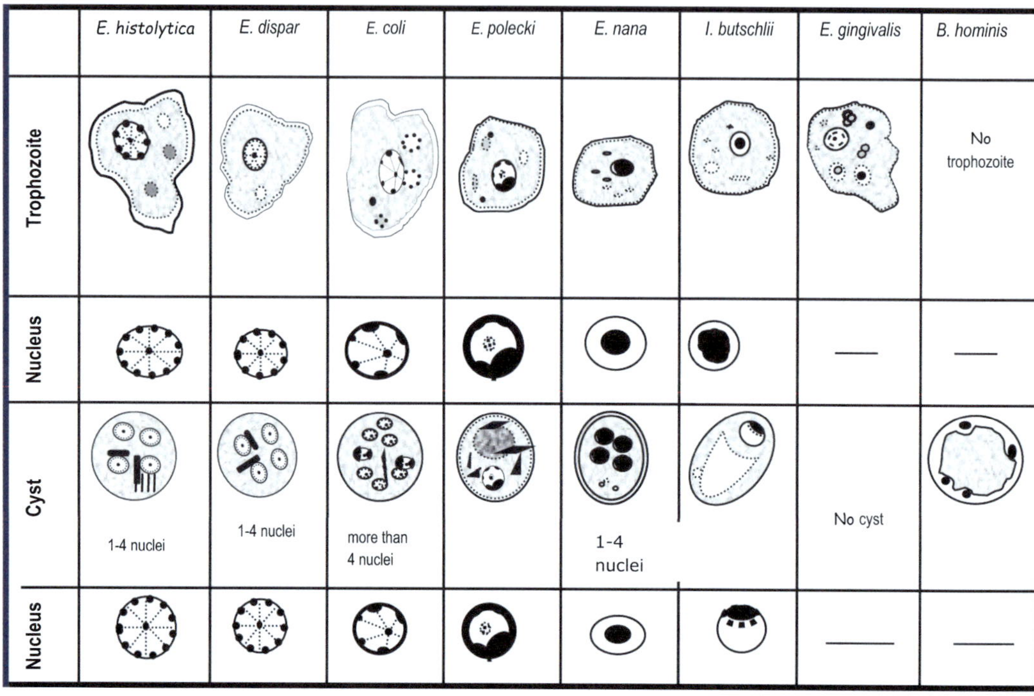

	E. histolytica	*E. dispar*	*E. coli*	*E. polecki*	*E. nana*	*I. butschlii*	*E. gingivalis*	*B. hominis*
Trophozoite								No trophozoite
Nucleus							—	—
Cyst	1-4 nuclei	1-4 nuclei	more than 4 nuclei		1-4 nuclei		No cyst	
Nucleus							—	—

Blastocystis hominis

(*B. hominis*)

B. hominis is frequently detected in human stool. It was unclassified; however, lately, it has been considered an ameba. It is potentially pathogenic, as it is associated with many different diseases.

Geographical Distribution
B. hominis has a worldwide distribution, especially in overcrowded areas with poor hygiene.

Parasite Morphology
B. hominis is detected in four basic forms: central body, granular, ameba, and cyst forms (Fig. 6.4). These basic forms are in common with the following characteristics:

1. Size ranges from 6 to 50 μm.
2. No cell wall.
3. They have outer membranes with vesicles and pores.
4. The nucleus has a crescent-shaped concentration of chromatin.
5. The shape is not always fixed.

The four basic forms of *B. hominis* are:

- **Central body form (vesicular form):**
  - It is a predominating form in fecal samples.
  - It has a central body that occupies more than 2/3 of the cell volume.
  - It has many nuclei at the rim of the cytoplasm.
  - Its main function is the reproduction.
- **Granular form:**
  - It is detected in fresh stool.
  - Its cytoplasm is filled with many granules.
  - Granules are mainly mitochondria.
- **Ameba form (trophozoite):**
  - It has a multilobed or convoluted body.
  - It looks like WBC.
  - It has no central body.
- **Cyst form:**

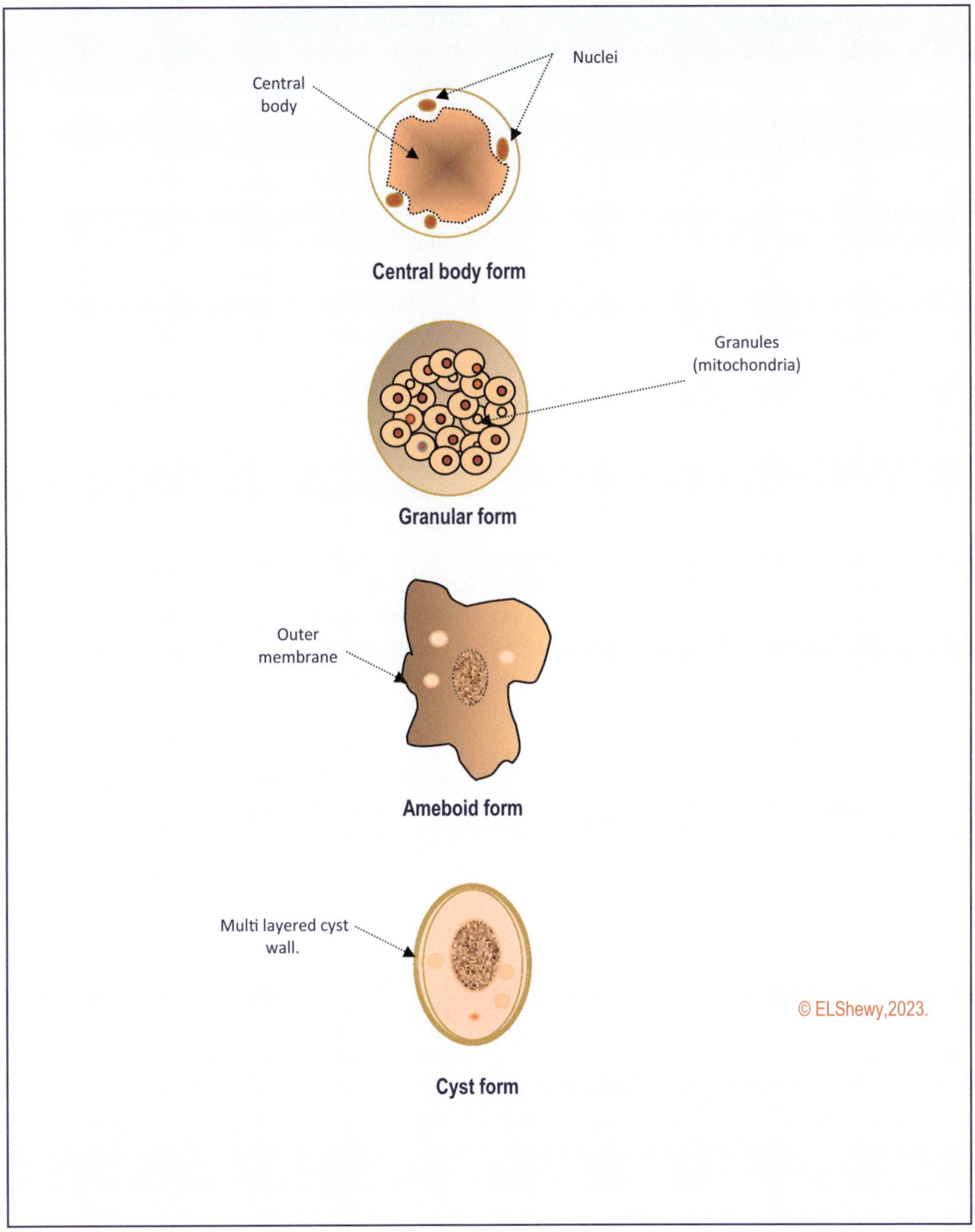

Fig. 6.4 *Blastocystis hominis*: morphology

- It has many layered cyst walls, which are seen in stored stool samples and not in culture.
- 3–5 µm. It can withstand external unfavorable conditions and gastric acidity.

Transmission
B. hominis is transmitted through:

1. Feco-oral route.
2. Flies (mechanically).
3. Contaminated foods and drinks.

Habitat
B. hominis inhabits the large intestine of humans.

Life Cycle
The suggested life cycle of *B. hominis* has been postulated as follows: The cyst form is the infective stage, which is changeable to other forms in the host. The central body form is essential for the start of reproduction, which may occur through:

1. Schizogony, through binary fission (one cell divided into two by mitotic division of the nucleus, then each forms a separate cell body).
2. Sporulation (endodyogony) (one cell nucleus undergoes multiple divisions to give many nuclei inside the same cells).
3. Both schizogony and sporulation give rise to many central body forms.

Pathogenesis
An argument is still running on the pathogenicity of *B. hominis*. However, the intensity of infection is found to be a determining factor for the association of infection with *B. hominis* and the presence of symptoms. High parasite load (number of parasites/high power field) in the stool is usually associated with significant clinical manifestations.

Associated clinical manifestations with *B. hominis* are mainly gastrointestinal (diarrhea, nausea, vomiting, distension, and malabsorption).

Diagnostic Procedures
Microscopic stool examination is used for the detection of *B. hominis* using the following:

- Direct wet mount specimens.
- Concentrated stool specimens.
- Permanent smear stained with iron hematoxylin or trichrome stain.
- Immunofluorescence.
- Culture of fecal specimen on a biphasic egg/serum medium under anaerobic conditions is useful in the detection of *B. hominis*.

N.B. It is crucial to quantify the intensity of *B. hominis* infection in the fecal samples. Higher intensity is usually associated with disease and hence the need for treatment.

Treatment
1. **Trimethoprim + Sulfamethoxazole.**
2. **Cotrimazole.**
3. **Metronidazole.**

Prevention and Control
1. Following sanitary precautions, especially for food handlers.
2. Proper disposal of the human excreta, with frequent handwashing.
3. Safe and healthy water supply, with filtration and chemical treatment.

Balantidium coli (*B. coli*)

B. coli is the only ciliated pathogenic human protozoan parasite. It inhabits the large intestine of humans and some animals **(zoonotic parasites).** It is the causative agent of intestinal balantidiasis (gastroenteritis with dysenteric manifestations). *B. coli* could have an extraintestinal spread.

Geographical Distribution
B. coli is prevalent in areas with poor sanitary conditions, overcrowding and at the breeding farm of pigs.

Parasite Morphology

B. coli has two morphological forms: trophozoite and cyst (Fig. 6.5).

- Trophozoite:
  - It is an oval with an anterior tapering and a funnel-shaped cytostome.
  - Size: 30–100 μm.
  - Cytoplasm: granular with a contractile vacuole.
  - Nucleus: two; a large kidney-shaped macronucleus and a small round micronucleus.
  - The outer body surface is covered with fine cilia.

- Cyst:
  - It is an oval to spherical.
  - Size: 50–70 μm.
  - Cytoplasm: granular with a contractile vacuole.
  - Nucleus: as in trophozoite.
  - Thick wall without cilia.

Transmission

Balantidiasis is transmitted through feco-oral route, contaminated foods, drinks, and flies.

Habitat

B. coli inhabits the cecum, colon, and terminal ileum of humans and pigs.

Life Cycle

The life cycle of *B. coli* is postulated in Fig. (6.6).

- The ingestion of *B. coli* cysts is followed by their excystation, which releases trophozoites.
- Trophozoites attach to the mucosal lining and may penetrate the intestinal wall.
- Trophozoites are encysted at the lower parts of the large intestine, and cysts pass with stool.

Pathogenesis

The following are the pathological sequences during balantidiasis (Fig. 6.7):

1. *B. coli,* after **attachment** to the mucosal lining, secret a hyaluronidase enzyme, which facilitates tissue **invasion.**
2. Tissue invasion is followed by **secondary bacterial infection** (needs antibiotics for treatment).
3. Later, an inflammatory cellular reaction appears around the trophozoites, which may lead to necrotic changes or even mucosal ulcerations.
4. Lately, and in rare conditions, **perforation** and an extraintestinal spread to the peritoneum, liver, and pleura could occur.

Clinical Picture

Most infection with *B. coli* is asymptomatic; however, balantidiasis could be manifested with:

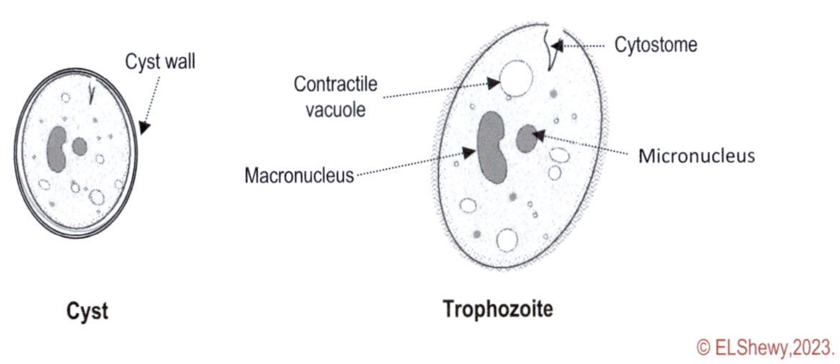

Fig. 6.5 *Balantidium coli*: morphology

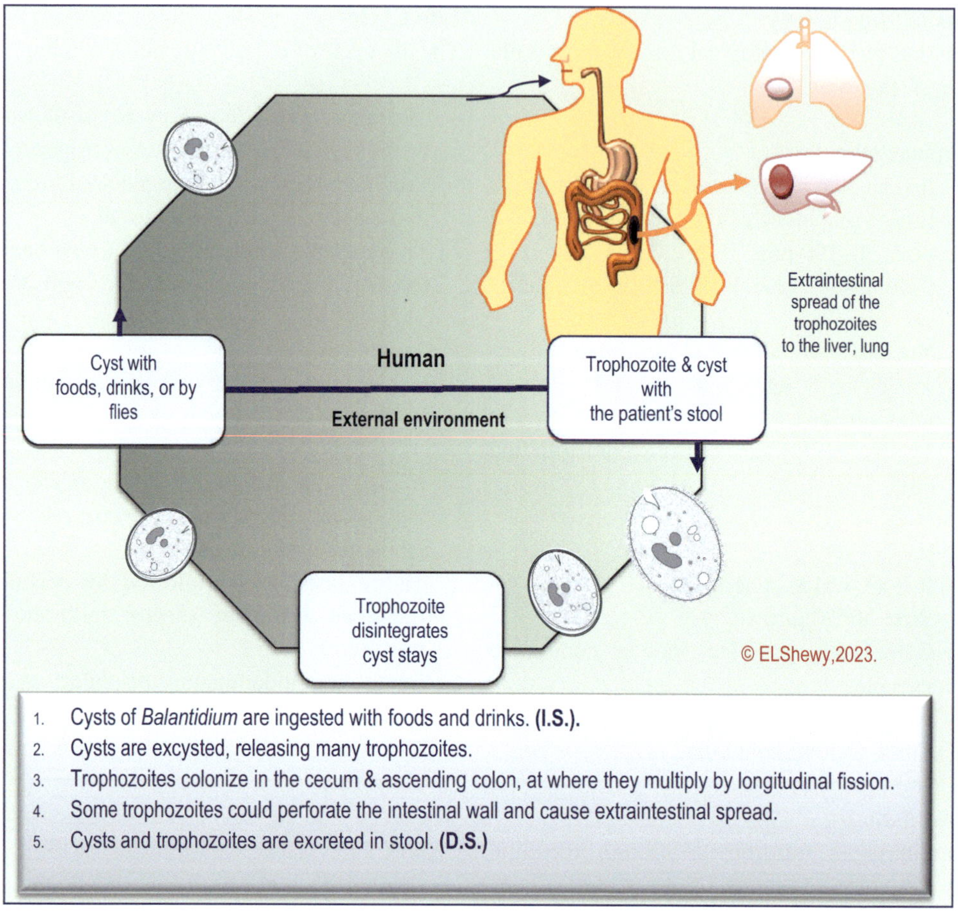

Fig. 6.6 *Balantidium coli:* life cycle

1. Mild diarrhea.
2. Dysentery.
3. Appendicitis and peritonitis in case of ulceration and perforation.
4. The extraintestinal system spreads to the adjacent lymph nodes, liver, and pleura.

Diagnostic Procedures

The main diagnostic procedure for the detection of *B. coli* is microscopic examination of the stool specimens and identification of the characteristic trophozoite and/or cyst.

Treatment

During balantidiasis, an antibiotic should be added because of the associated secondary bacterial infection:

1. **Oxytetracycline** (Terramycin):
   - 500 mg three times daily for 10 days for the adults.
   - 40 mg/kg/day for children older than 8 years.
2. **Iodoquinol:**
   - 650 mg three times daily for 20 days.
3. **Nitazoxanide:**
   - 500–750 mg/3 times/day for 5 days.

Prevention and Control

1. Following sanitary precautions, especially for food handlers.
2. Proper disposal of the human excreta.
3. Sanitary standards and veterinary care in pig farms.

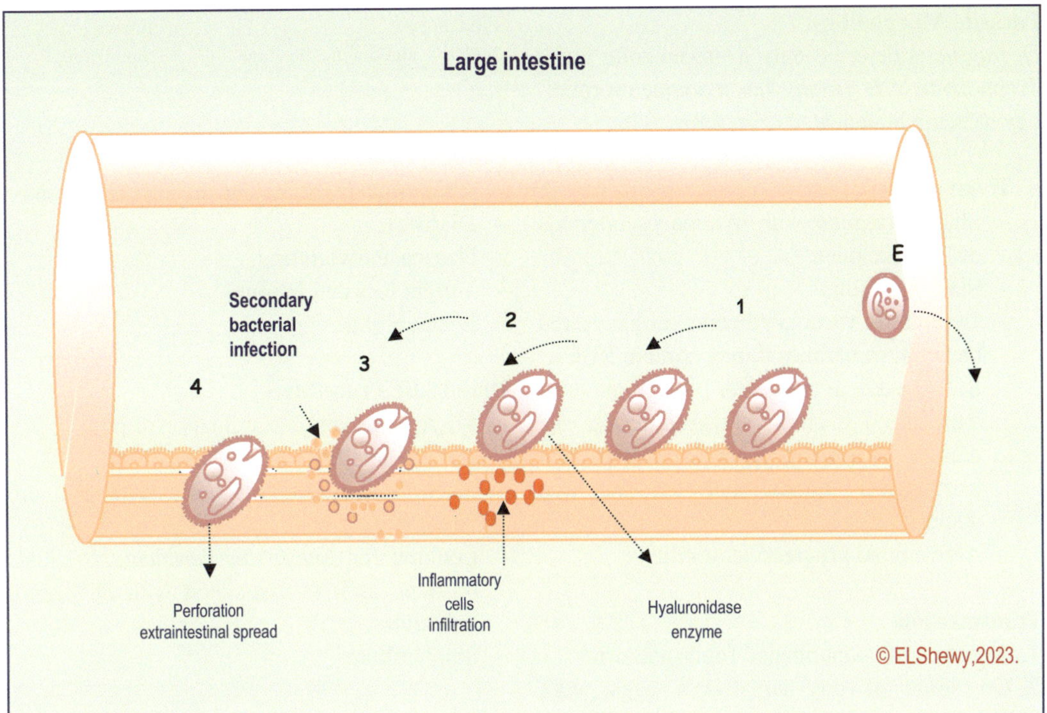

1. **Attachment** through receptors.
2. **Invasion** through hyaluronidase enzyme, with inflammatory cell infiltration (abdominal colic, tenderness).
3. **Secondary bacterial infection** with epithelial necrosis (dysenteric manifestations).
4. **Perforation** and extraintestinal spread (peritoneum, liver).

E- Cyst.

Fig. 6.7 *Balantidium coli:* pathogenesis

4. Safe and healthy water supply, with filtration and chemical treatment.
5. Treatment of all cases of balantidiasis, including the symptomatic and asymptomatic.

- **Parasitology**: *B. Coli* is a ciliated protozoan parasite; it has trophozoite and cyst forms. Each has two nuclei: a macronucleus, kidney-shaped, and a micronucleus small. Pigs are reservoir hosts. Infection is mainly an oral ingestion of the cyst.
- **Clinical:** It causes enteritis with dysenteric manifestations, and occasionally, an extraintestinal spread to the liver is demonstrated.

Dientamoeba fragilis (*D. fragilis*)

Although its name, *D. fragilis,* is not an ameba, it is considered one of the flagellated parasites. It infects humans and animals. It is responsible for many gastrointestinal symptoms, particularly among immunocompromised individuals.

Geographical Distribution

The prevalence of *D. fragilis* ranges between 1 and 20% all over the world, especially in areas with poor sanitation and among homosexual and immunocompromised individuals.

Parasite Morphology

D. fragilis is detected only in trophozoite form. Trophozoite of *D. fragilis* has a unique morphology of being binucleated (Fig. 6.8):

- Trophozoite:
  - Shape: irregular with hyaline pseudopods or leaf-like ones.
  - Size: 7–12 µm.
  - Cytoplasm: vacuolated containing ingested bacteria, sometimes it may contain RBCs.
  - Binucleated, at the center of each nucleus, there are 4–8 chromatin granules in one mass. No peripheral chromatin. There is an extranuclear spindle between the two nuclei.
  - Active rapid progressive motility.

Transmission

1. Ingestion of contaminated foods and drinks.
2. On some of the nematode's eggs, e.g., *Enterobius vermicularis*.
3. Hand-to-mouth autoinfection.

Habitat

Mucosal crypts of the human large intestine.

Life Cycle

D. fragilis has a simple life cycle; its trophozoite inhabits the mucosal crypts of the large intestine, where it multiplies by binary fission and passes with stool outside the body to repeat the life cycle (Fig. 6.8).

Final hosts:	Humans.
Infective and diagnostic stage:	Trophozoite.

Clinical Picture

- Abdominal pain, flatulence, and distension.
- Diarrhea.
- Nausea and vomiting.
- Weight loss and fatigue.
- Eosinophilia.

Diagnostic Procedures

- Microscopic stool examination of fresh or preserved samples.
- Multiple stool samples should be taken on different days.
- Looking for *Enterobius vermicularis* should be done as it is associated with *D. fragilis* infection.
- Stool culture.

Treatment

1. **Iodoquinol.**
2. **Oxytetracycline.**
3. **Metronidazole.**
4. **Ornidazole.**

Prevention and Control

1. Following sanitary precautions for food and drinks, cutting nails, and washing hands.
2. Treatment of the associated oxyuriasis.
3. Sanitary water supply with proper sewage disposal.

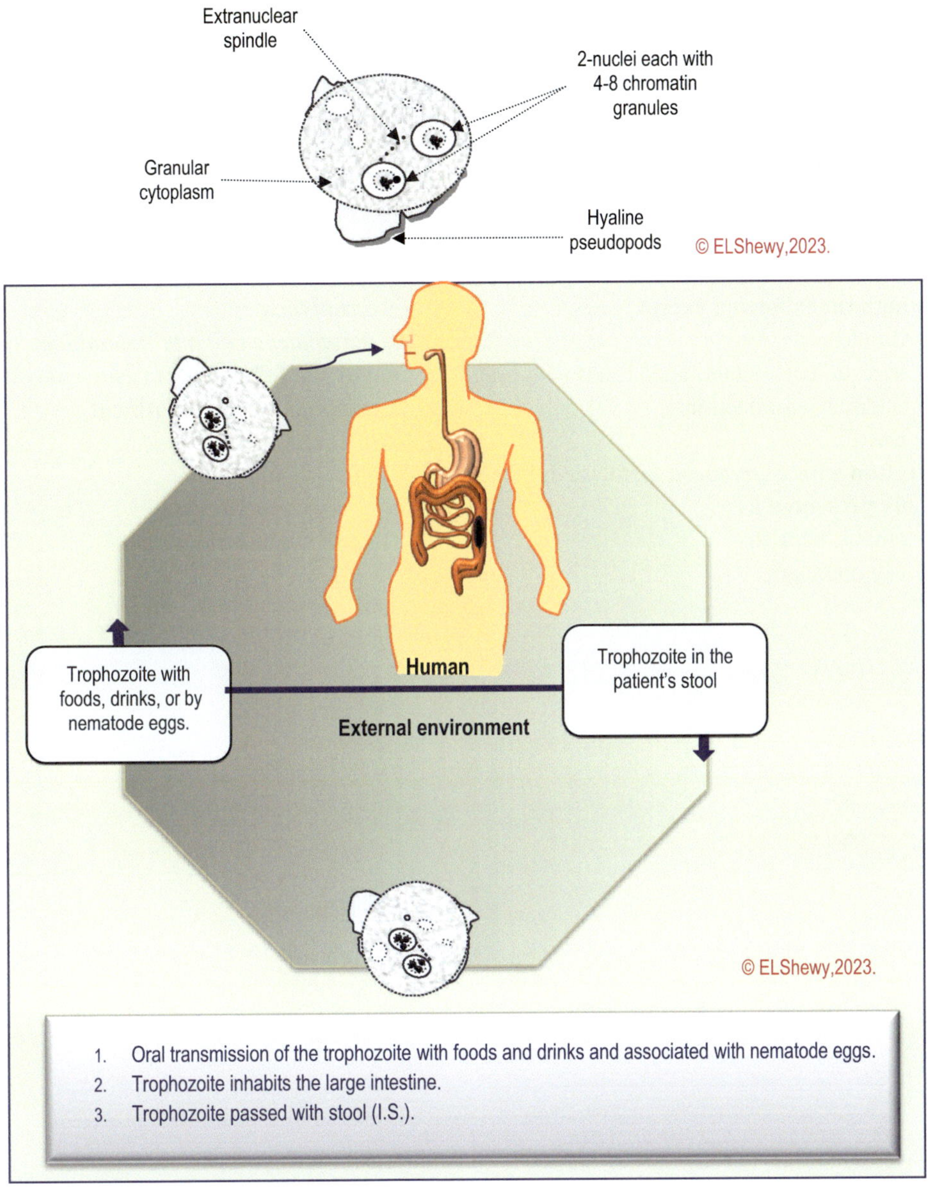

Fig. 6.8 *Dientameba fragilis*: morphology & life cycle

Review Questions

1. ***E. histolytica*** **could affect the intestinal mucosa through the following factors except**
   (a) proteolytic enzymes
   (b) ulceration
   (c) secondary bacterial infection
   (d) phagocytosis of the intestinal cells
2. **The following are four or more than four nucleated cysts, except cysts of**
   (a) *E. histolytica*
   (b) *Endolimax nana*
   (c) *Balantidium coli*
   (d) *E. coli*
3. **Flask shape intestinal ulcers are caused by**
   (a) *Entamoeba coli*
   (b) *Entamoeba histolytica*

(c) *Iodamoeba butschlii*
(d) *Balantidium coli*

4. **Pathogenesis in balantidiasis is due to the following except**
   (a) hyaluronidase enzyme
   (b) cholinesterase enzyme
   (c) secondary bacterial infection
   (d) necrosis of the intestinal epithelium
5. ***Entamoeba histolytica* could be transmitted through the following except**
   (a) droplets
   (b) ingestion of contaminated foods with cyst
   (c) internal autoinfection
   (d) flies
6. **Infection with *Entamoeba histolytic* is commonly presented as**
   (a) amebic hepatitis
   (b) asymptomatic
   (c) perforated ulcer
   (d) liver abscess
7. **The following two parasites could have an extraintestinal spread**
   (a) *Entamoeba coli and Entamoeba histolytica*
   (b) *Entamoeba histolytica and Balantidium coli*
   (c) *Giardia lamblia and Entamoeba histolytica*
   (d) *Entamoeba coli and Balantidium coli*
8. **One of the following parasites necessitates an antibiotic for its treatment**
   (a) *Giardia lamblia*
   (b) *Balantidium coli*
   (c) *Iodamoeba butschlii*
   (d) *Entamoeba histolytica*

7 Helminths of the Large Intestine

Test your learning and check your understanding of this book's contents: use the "Springer Nature Flashcards" app to access questions using ▶ https://sn.pub/oyFLIx. To use the app, please follow the instructions in Chap. 1.

Enterobius vermicularis (*E. vermicularis*)

(Pinworm, Oxyuris Worm)

E. vermicularis is a nematode worm inhabiting the large intestine. The worm looks like the pin, so its name. *E. vermicularis* has a high prevalence rate, especially among young children, and causes **oxyuriasis**.

Geographical Distribution
E. vermicularis is one of the most common helminthic infections all over the world in temperate and tropical regions. It affects mainly preschool children.

Parasite Morphology
Male and female worms of *E. vermicularis* have the appearance of the pin with the following differences (Fig. 7.1):

- Adult worm (♀):
  - It has a rounded body with a finely pointed tail.
  - 10–13 mm in length.
  - It has a long vagina and uterus loaded with eggs.
  - Vulva opening in the middle of the body.
- Adult worm (♂):
  - Its body is rounded and ventrally curved.
  - 2–5 mm in length.
  - It has one spicule at its posterior end.
- Egg:
  - It is flattened from one side and curved on the other (D-shaped), with a thin shell.
  - 25–55 μm, transparent or yellowish brown.
  - It is passed embryonated and ready to infect.

Transmission
1. Ingestion of infective eggs with foods and beverages.
2. External autoinfection (hand to mouth; feco-oral) from the same host.
3. Internal autoinfection (retroinfection): eggs hatch inside the patient and reinfect him again.
4. Mechanical transmission of the worm's eggs through insects and flies.
5. Familial or group spread through sharing toilet seats, linens, and clothes.

K. ElShewy, *Medical Parasitology*, https://doi.org/10.1007/978-3-031-64736-9_7

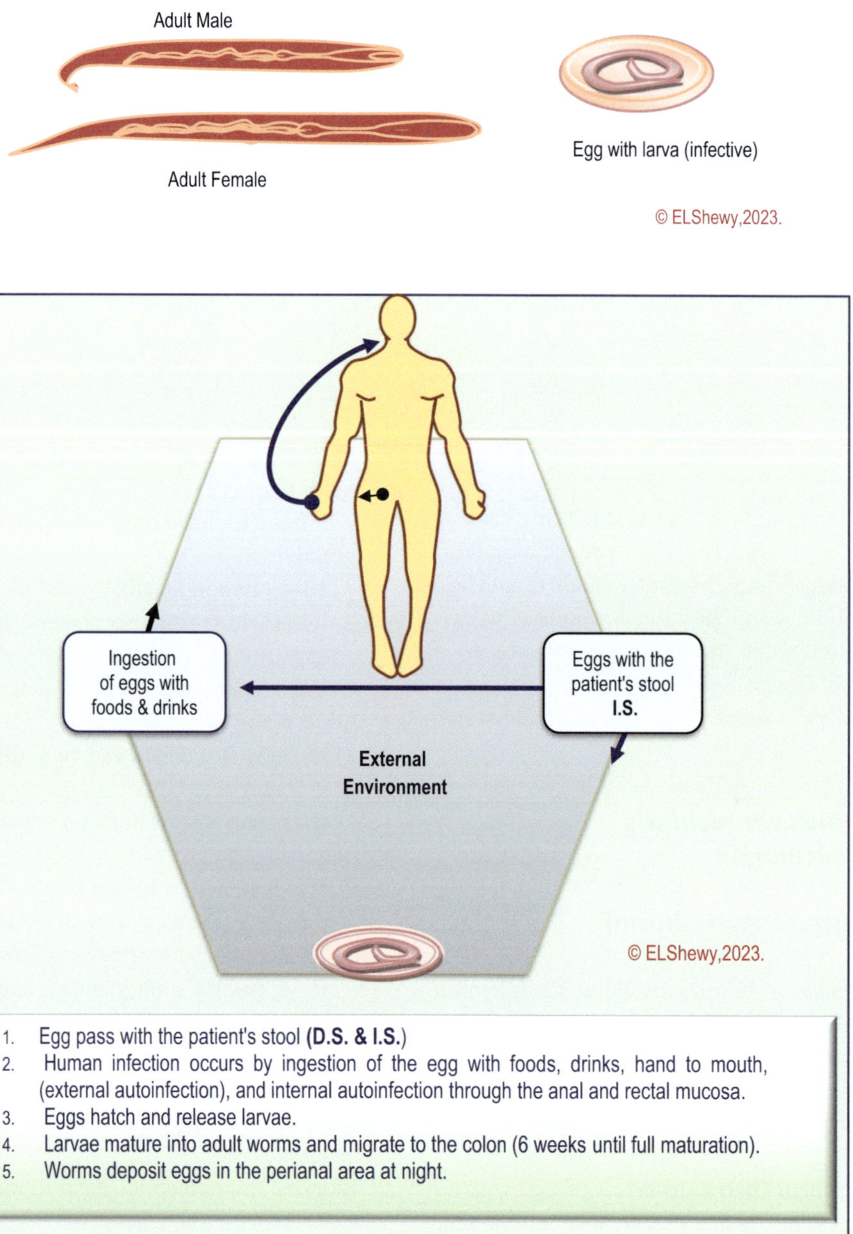

Fig. 7.1 *Enterobius vermicularis*: morphology and life cycle

Habitat

Adult worm of *E. vermicularis* inhabits the cecum and the adjacent portions of the large intestine.

Life Cycle

The life cycle of *E. vermicularis* is postulated in Fig. (7.1).

- At night, an adult gravid female migrates down to the anal and the perianal areas, where it deposits eggs.
- Eggs are fully embryonated and ready to infect when first passed.
- Ingested eggs hatch and release larvae.
- Larvae mature and migrate to the colon (6 weeks until full maturation to adults).

Final hosts:	Humans.
Infective stage:	Fully embryonated egg containing larva.
Diagnostic stage:	Eggs in the stool; sometimes, the adult worms could be detected in the stool of the host.
Intermediate hosts:	None.

Infection with *E. vermicularis* takes place in three main ways:

1. Embryonated eggs in the external environment are transmitted to a new host through the consumption of contaminated food, beverages, and insects.
2. By the same host through fingers to mouth infection (**external autoinfection**).
3. The laid eggs at the anal mucosa hatch, and the larvae go back to the same host (**internal autoinfection, retroinfection**).

Pathogenesis

Although *E. vermicularis* is considered as a harmless worm, however, the following could be encountered:

1. In allergic patients, especially children, the perianal and anal area could be affected by a local inflammatory reaction.
2. The attachment of the worms to the intestinal mucosa causes a mild inflammatory reaction.
3. The appendix could be invaded by *E. vermicularis,* triggering appendicitis.
4. Occasionally, eggs are laid in the peritoneal cavity, causing chronic peritonitis.

Clinical Picture

Oxyuris or **pinworm** infection, in most cases, is asymptomatic. However, the following symptoms could be detected, especially among children:

1. Pruritus ani, especially at night.
2. Invasion of the vagina in females may lead to a local irritation with vaginitis.
3. Insomnia due to perianal itching.
4. Nocturnal enuresis.

Diagnostic Procedures

- Diagnosis of oxyuriasis is mainly made through microscopic recognition of the characteristic eggs in the stool, especially in the **early morning** before washing.
- Sometimes, eggs could be found in urine and adult worms could be detected in the stool.
- Recovery of the eggs from the perianal region using transparent adhesive tape in the morning and before defecation. The tape is examined under the microscope for the presence of deposited eggs (**Scotch adhesive tape**).
- Morning perianal **cotton swab** wetted with saline to be examined microscopically.

Treatment

The following two drugs are commonly used for oxyuriasis:

1. **Pyrantel pamoate**:
   - 11 mg/kg/body weight (maximum 1 gm).
   - Single dose to be repeated after 2 weeks.
2. **Mebendazole**:
   - 100 mg orally/2 times daily/3 days for both adults and children.
   - Single dose to be repeated after 2 weeks.
3. **Albendazole**:
   - 400 mg orally.
   - Single dose to be repeated after 2 weeks.
4. **Local application of antipruritic preparations.**

N.B.

- All family members should receive the same treatment to prevent the recurrence.
- Treatment should be repeated once or twice every 2 weeks.

Prevention and Control

1. Following the same preventive rules for all helminthic infections, e.g., washing hands after the toilet, cutting nails, linens and under wares should be boiled.
2. Toilet seats should be cleaned properly.
3. Children are encouraged to stop the habit of finger-sucking.

4. Treatment of the infected cases and their surroundings (familial treatment).
5. Repetition of the treatment.

- **Parasitology:** *E. Vermicularis* is a pin-like worm inhabiting the large intestine. Infection is mainly through ingestion of the embryonated eggs and internal and external autoinfection in the same host. It is diagnosed by morning adhesive tape or perianal swab. Eggs could be found in the urine.
- **Clinical:** Usually asymptomatic, pruritus ani with nocturnal enuresis.

Trichuris trichiura (*T. trichiura*)

(Whip Worm, Hair Tail Worm)

T. trichiura is a nematode of human large intestine. It looks like a whip with a long anterior end (hairtail). It causes dysenteric-like manifestations with mild anemia among infected individuals.

Geographical Distribution

T. trichiura infection is prevalent worldwide, and it is more common in tropical areas and areas with poor sanitation. The life span of the worm is 4–8 years.

Parasite Morphology

Both male and female worms of *T. trichiura* are similar to a whip with a simple mouth armed with a stylet that helps in penetration (Fig. 7.2).

- Adult worm (♀):
  - It is 4–5 cm in length.
  - Its anterior end is thread-like with a blunt, rounded posterior end.
- Adult worm (♂):
  - It is smaller than the female.
  - Its tail (posterior end) is curved.
  - Its posterior end has one spicule protected by a retractile sheath with spines.
- Egg:
  - Shape: barrel with bipolar, mucoid plugs at the ends covered with three-layer shell.
  - Size: 50–55 × 20–25 μm.
  - Color: brownish yellow.
  - Content: unembryonated when passed (not infective).

Transmission

T. trichiura is transmitted mainly through the ingestion of mature eggs with contaminated foods or drinks. Eggs could also be transmitted by flies and by using human stools as fertilizer.

Habitat

T. trichiura inhabits the colon, especially at the cecum.

Life Cycle

The life cycle of *T. trichiura* is postulated in Fig. (7.3).

Final hosts:	Humans.
Infective stage:	Fully embryonated egg.
Diagnostic stage:	Characteristic barrel-shaped egg with two mucoid plugs.
Intermediate hosts:	None.

Pathogenesis

The following pathological changes are demonstrated during trichuriasis:

1. Interlacing of the anterior end of the *T. trichiura* causes little inflammatory changes in the colonic mucosa.
2. In heavy infection, there is a marked mucosal inflammation in the colon.
3. Rectal edema resulting from extensive inflammation could be responsible for rectal prolapse, especially in young children and very old patients.
4. Bacterial invasion of colonic mucosa is facilitated by worm penetration.
5. *T. trichiura* could cause minimal blood loss at a rate of 0.005 ml/worm/day.

Clinical Picture

1. Abdominal pain and distension.
2. Dysenteric manifestations, i.e., bloody or mucoid diarrhea, tenesmus (painful difficult defecation), weight loss, and weakness.
3. Prolapse of the rectum is occasionally seen, especially in children and old patients.
4. Mild to moderate anemia.

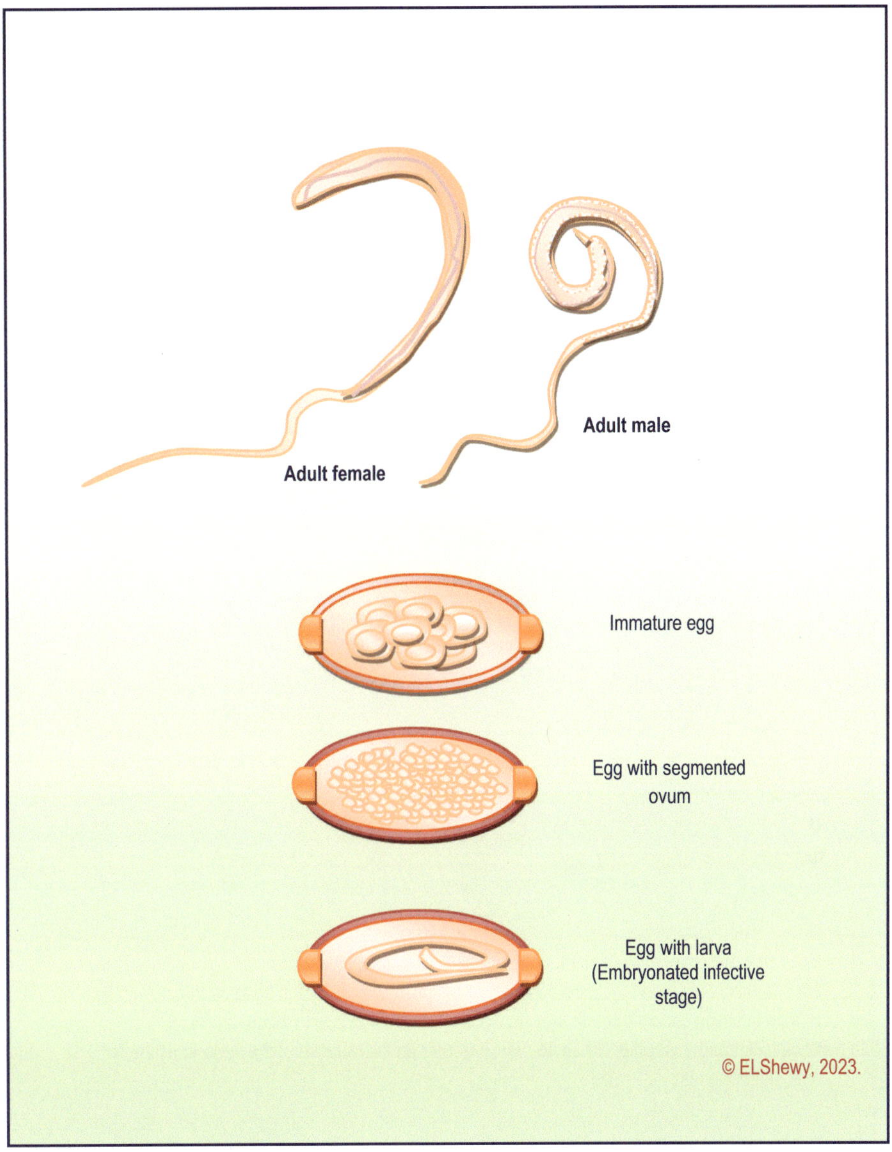

Fig. 7.2 *Trichuris trichiura*: morphology

Diagnostic Procedures

1. Stool examination after a **zinc sulfate** concentration technique is an ideal way to detect the characteristic eggs of *T. trichiura.*
2. **Kato's** thick smear is used for the determination of the intensity of infection and monitoring of the therapy.

Treatment

The following two drugs are commonly used for the treatment of trichuriasis:

1. **Mebendazole:**
   - 100 mg, two times /day for 4 days. It gives a cure rate of 90%.

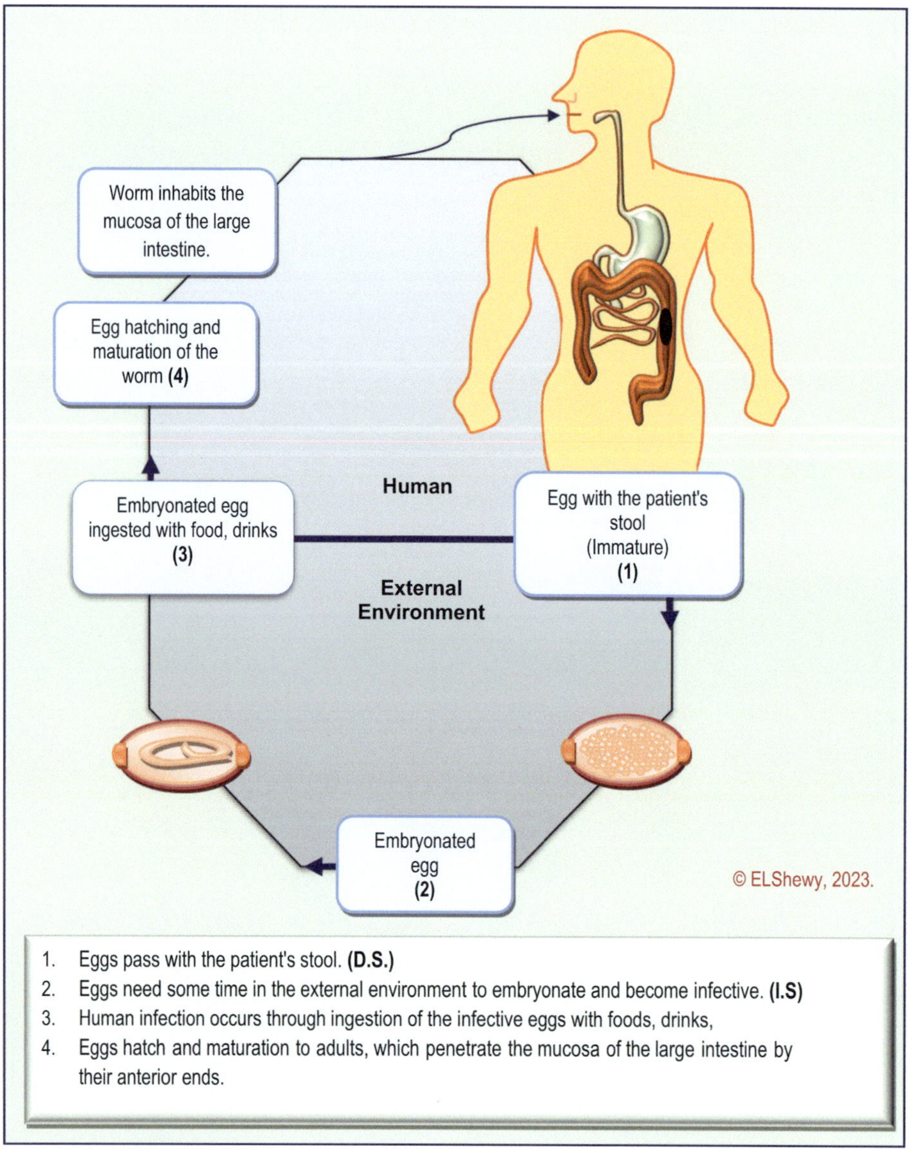

Fig. 7.3 *Trichuris trichiura*: life cycle

- A single dose of 600 mg could be used for mass treatment.

2. **Loperamide**:
   In addition to Mebendazole, loperamide may help in increasing the contact time between a drug and the parasite in the intestine.

Prevention and Control

1. Avoid using a human stool as fertilizer.
2. Healthy and safe water supply.
3. Appropriate sewage disposal.
4. Combat flies and insects.

Gastrodiscoides hominis (*G. hominis*)

(Colonic Fluke)

G. hominis is a zoonotic trematode fluke that infects a wide range of vertebrate animals and occasionally infects humans. It is responsible for gastrodiscoidiasis, a sort of colitis with abdominal disturbance, and mucoid diarrhea.

Geographical Distribution

G. hominis is common in Assam (India), Southeast Asia, Thailand, Vietnam, and Malaysia. It is mainly a parasite of pigs, and occasionally it infects humans.

Parasite Morphology

The adult worm of *G. hominis* looks like *H. heterophyes* but with the following criteria (Fig. 7.4):

- Adult worm:
  - A large oval fluke with anterior and ventral suckers.
  - The ventral sucker is enlarged, forming an acetabulum.
  - Its size is 5 mm × 15 mm.
- Egg:
  - It is oval in shape and has a double shell.
  - 150 μm × 75 μm.
  - Greenish brown.
  - Not embryonated when passed (immature).

Transmission

G. hominis is transmitted through eating vegetation containing encysted metacercaria. Sometimes, freshwater fish could be the second intermediate host of *G. hominis* and eating undercooked or raw fish containing metacercaria could transmit infection.

Habitat

Adult *G. hominis* inhabits the large intestine of humans and animals. It attaches to the mucosa of the cecum and ascending colon.

Fig. 7.4 *Gastrodiscoides hominis*: adult and egg morphology

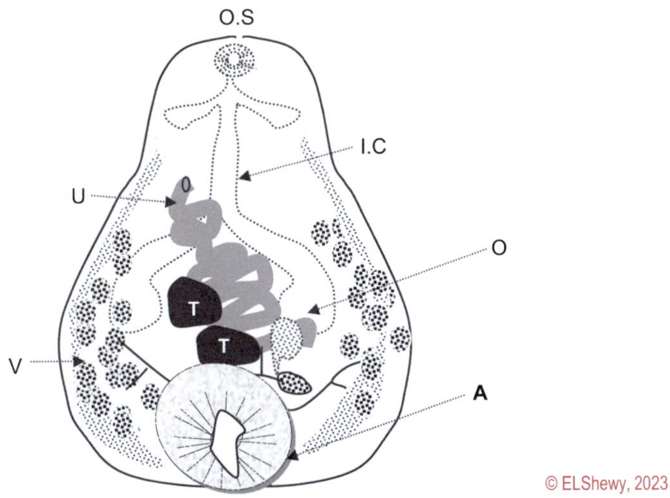

O.S.= oral sucker, V= vitellaria, T= tests, O= ovary, U= uterus, A= acetabulum, I.C.= intestinal ceca.

Life Cycle

The life cycle of *G. hominis* is not well understood. However, it could be similar to that of other related worms, e.g., *Heterophyids* and *Fasciola*. However, freshwater fish or plants could be the second intermediate hosts of the *G. hominis* (Fig. 7.5).

Clinical Picture

Because of the minimal inflammatory effect of the worm, only mucoid diarrhea is a common presentation of gastrodiscoidiasis. In severe infection, bloody mucoid diarrhea is observed.

Diagnostic Procedures

G. hominis is diagnosed through the detection of the characteristic eggs in the patient's stool.

Treatment

Tetrachloroethylene and **Praziquantel** are two drugs commonly used in the treatment of gastrodiscoidiasis.

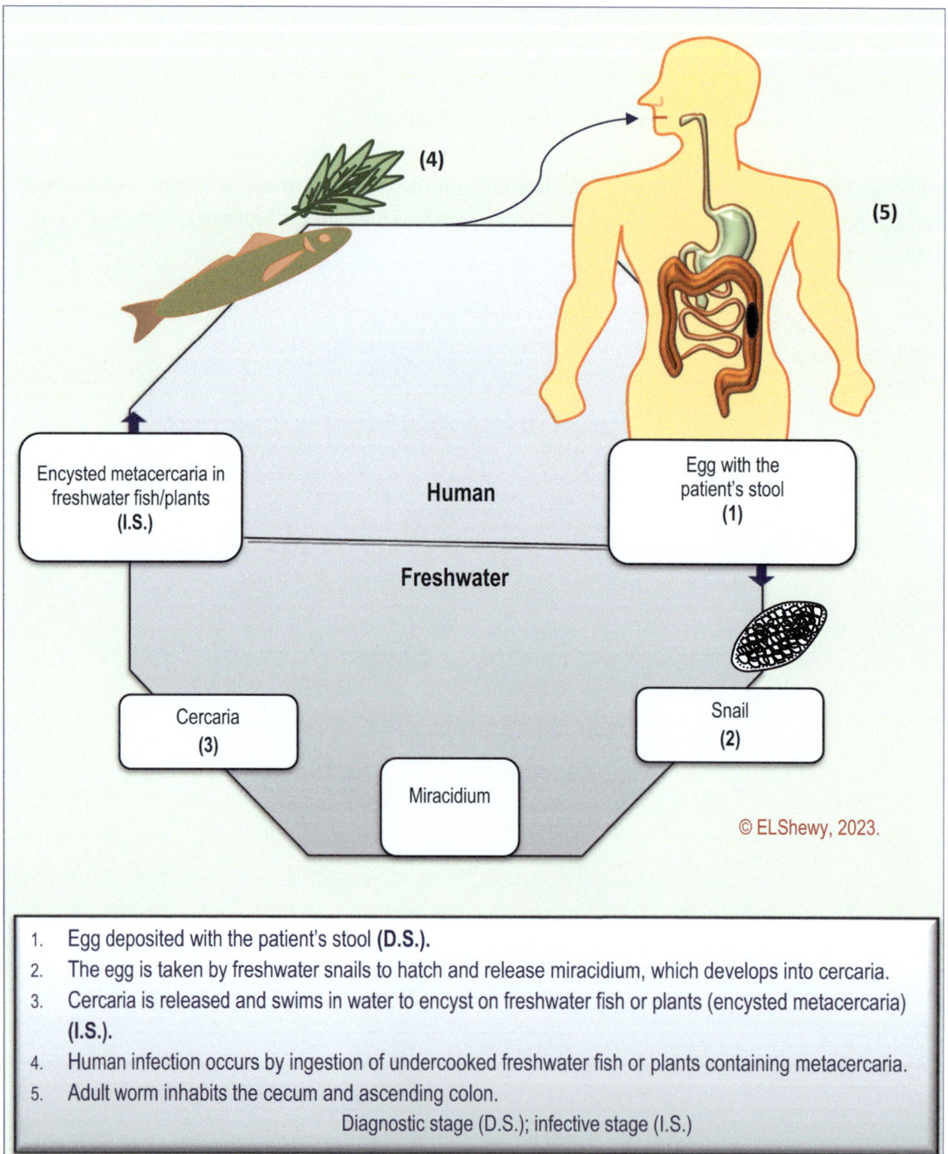

Fig. 7.5 *Gastrodiscoides hominis*: life cycle

- **Parasitology:** *G. Hominis* is an intestinal trematode fluke that inhabits the cecum and ascending colon. It is similar to the *Heterophyes* worm but with a big acetabulum. It is transmitted through eating contaminated freshwater vegetation and fish containing encysted metacercaria. It is confined to Assam (India), Southeast Asia, and Malaysia.
- **Clinical:** *G. Hominis* causes gastrointestinal disturbance with bloody mucoid diarrhea.

Review Questions

1. **Which of the following is not commonly used in the diagnosis of *Enterobius vermicularis?***
   (a) direct stool examination
   (b) adhesive tape
   (c) urine examination
   (d) morning perianal swaps
2. **Rectal prolapse could be a complication of infection with**
   (a) *Ascaris lumbricoides*
   (b) *Hymenolepis nana*
   (c) *Trichuris trichiura*
   (d) *Echinococcus granulosus*
3. **Nocturnal enuresis is a common complaint during infection with**
   (a) *Trichuris trichiura*
   (b) *Enterobius vermicularis*
   (c) *Hymenolepis nana*
   (d) *Ancylostoma duodenale*
4. **Dysenteric manifestation could be noticed during infection with**
   (a) *Enterobius vermicularis*
   (b) *Hymenolepis nana*
   (c) *Trichuris trichiura*
   (d) *Ancylostoma duodenale*
5. **During oxyuriasis, perianal skin swap should be taken in the morning because**
   (a) adult worm causes nocturnal anal itching
   (b) adult worm lays eggs at night
   (c) swab is easier to be taken in the morning
   (d) eggs are laid in the morning

8 Liver Parasites

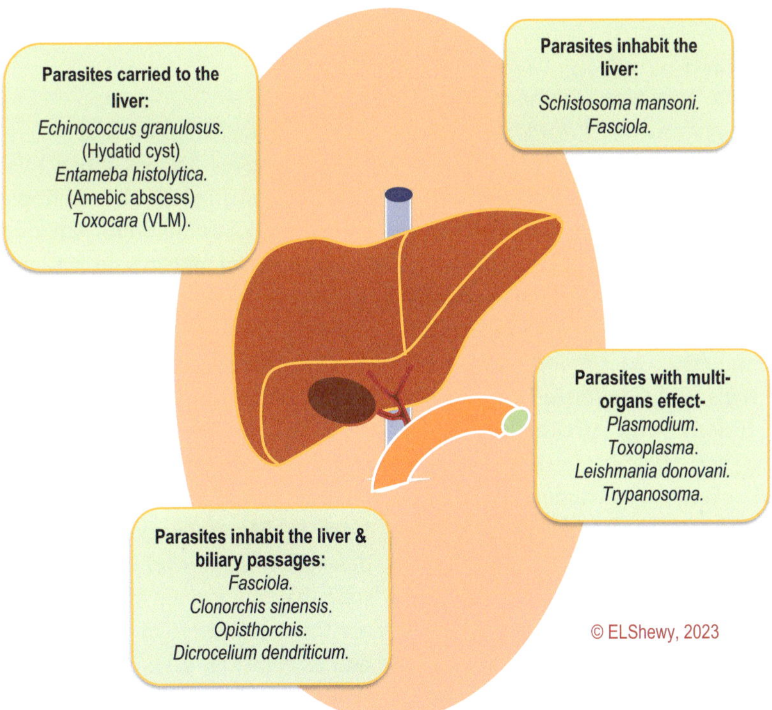

Liver Parasites

K. ElShewy, *Medical Parasitology*, https://doi.org/10.1007/978-3-031-64736-9_8

Test your learning and check your understanding of this book's contents: use the "Springer Nature Flashcards" app to access questions using ▶ https://sn.pub/oyFLIx. To use the app, please follow the instructions in Chap. 1.

Introductory

- The biggest organ in the gastrointestinal tract is the liver. It is considered an important target for many parasites that could be responsible for very serious or even life-threatening diseases. Liver parasites may cause the following:
  1. **Hepatomegaly:** mostly caused by *Schistosoma mansoni, Leishmania donovani, Trypanosoma, Plasmodium,* and *Toxoplasma.*
  2. **Liver cirrhosis, fibrosis, and liver cell failure:** e.g., *Schistosoma mansoni, Fasciola,* and *Capillaria hepatica.*
  3. **Cholangitis, cholecystitis, metabolic disorders, jaundice, and cholangiocarcinoma:** e.g., *Fasciola, Clonorchis sinensis, Opisthorchis,* and *Capillaria hepatica.*
  4. **Cystic lesions with pressure necrotic changes:** e.g., *Echinococcus granulosus* (hydatid cysts), *Entamoeba histolytica* (amebic liver abscess), and *Toxocara.*
- The liver parasites could be one of the following:
  1. **Parasites carried to the liver leading to its destruction:** e.g., *Echinococcus granulosus* (hydatid cysts), *Entamoeba histolytica* (Amebic liver abscess) **(Chap. 6),** and *Toxocara* (visceral larva migrans (VLM)) **(Chap. 11).**
  2. **Parasites inhabiting the liver vascularity and biliary passages:** e.g., *Schistosoma mansoni, Fasciola, Clonorchis sinensis, Opisthorchis,* and *Capillaria hepatica.*
  3. **Parasites with multiorgan effects:** e.g., *Plasmodium* **(Chap. 12),** *Toxoplasma, Leishmania donovani* (visceral leishmaniasis) **(Chap. 12),** and *Trypanosoma* **(Chap. 10).**

In this chapter, we will discuss the most common parasites infecting the liver; other parasites are mentioned in other chapters.

Schistosoma mansoni (*S. mansoni*)

Intestinal Bilharziasis

Schistosoma mansoni (*Schisto* = divided; *soma* = body) is a blood trematode inhabiting the intestinal mesenteric plexuses of veins. It is the causative parasite of intestinal bilharziasis, which affects the liver, spleen, intestine, and portal circulation.

Geographical Distribution

- *Schistosoma* species infect over 200 million people in tropical and subtropical regions of Africa, Asia, and South America.
- *S. mansoni* prevails in vast areas of Africa (Egypt, Sudan, and Uganda), Brazil, and parts of the Caribbean.
- Other species of intestinal schistosome include *S. japonicum*, S. *intercalatum,* and S. *mekongi,* which prevail in the southeast of Asia and Japan.

Parasite Morphology

Schistosoma mansoni is a heterophroditic trematode that has a male and female. The following are the morphological criteria of the adult worm and its egg (Fig. 8.1):

- Adult worm (♂):
  - It has a cylindrical body with a gynecological canal, 10 mm in length × 1.2 mm in width.
  - Two suckers (oral and ventral); its body is covered with coarse tubercles.
  - 6–9 small testes behind the ventral sucker.
  - Intestinal ceca reunite in the anterior 1/3 of the body.
- Adult worm (♀):

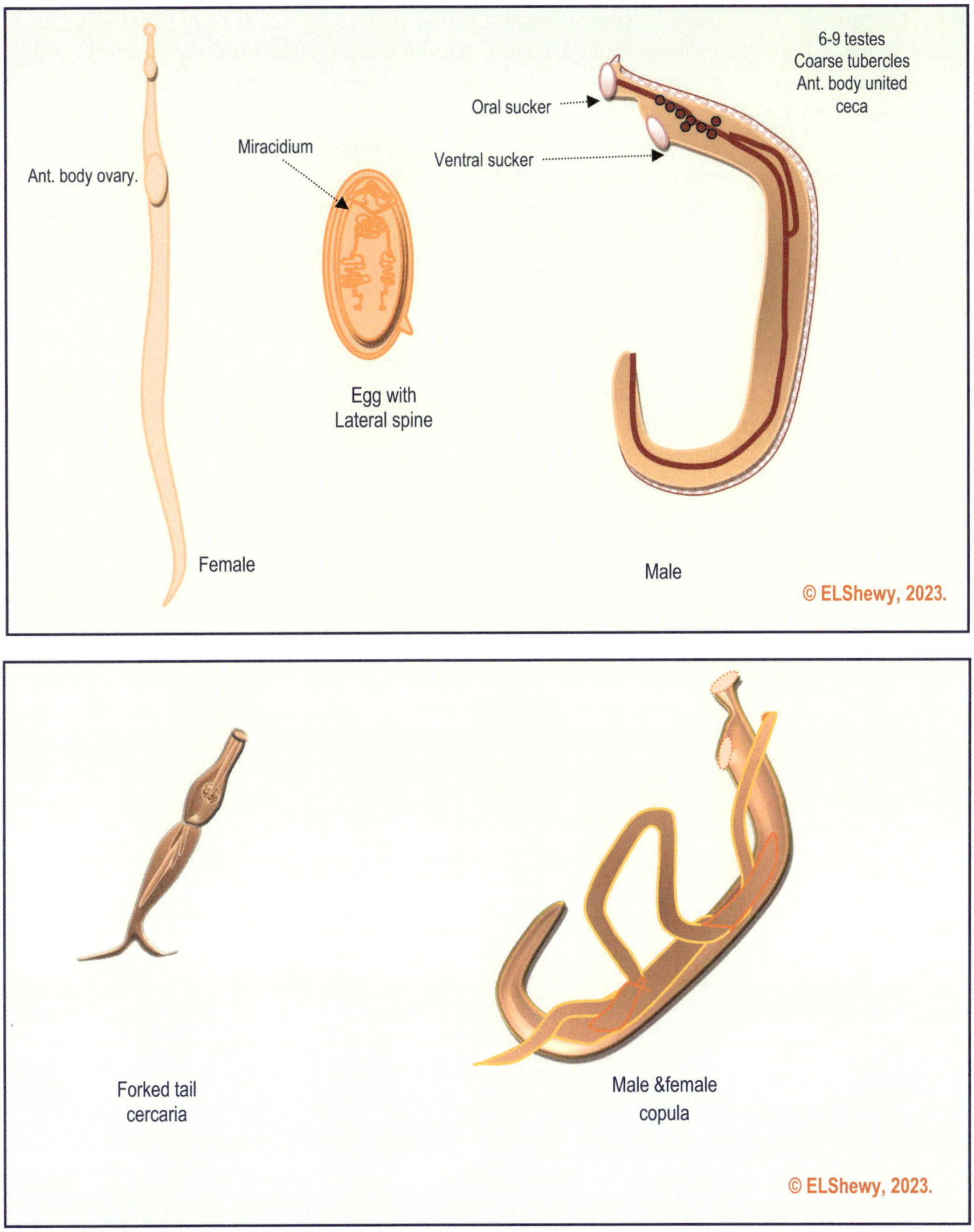

Fig. 8.1 *Schistosoma mansoni:* morphology

- It has a rounded, elongated body with pointed ends, 15 mm in length × 0.25 mm in width.
- Intestinal ceca reunite in the anterior 1/3 of the body.
- It has a short uterus with 1–4 eggs inside.

- Egg:
  - It is an oval, rounded at both ends, with a lateral spine.
  - 150–60 μm.
  - It contains a ciliated miracidium.

N.B. Cercaria is the infective stage for all *Schistosomes*. It has a head with a small body and a forked tail.

Habitat

S. mansoni inhabits the smaller branches of the mesenteric venous plexus of veins in the region of the lower colon. Occasionally, it could be found in the portal system or in the venous plexus of the intestinal tract and liver.

Transmission

Schistosomes are transmitted through penetration of the host skin by the infective forked tail cercariae.

Life Cycle

The life cycle of all *Schistosoma* species is almost similar. It includes three main stations:

1. **Penetration** of the host skin by cercaria.
2. **Migration** by young schistosomes.
3. **Eggs deposition** by adult worms.

The life cycle could be postulated as follows (Fig. 8.2):

In freshwater:

- When the patient defecates near a freshwater source, eggs pass with his stool onto the water to hatch and release miracidia.

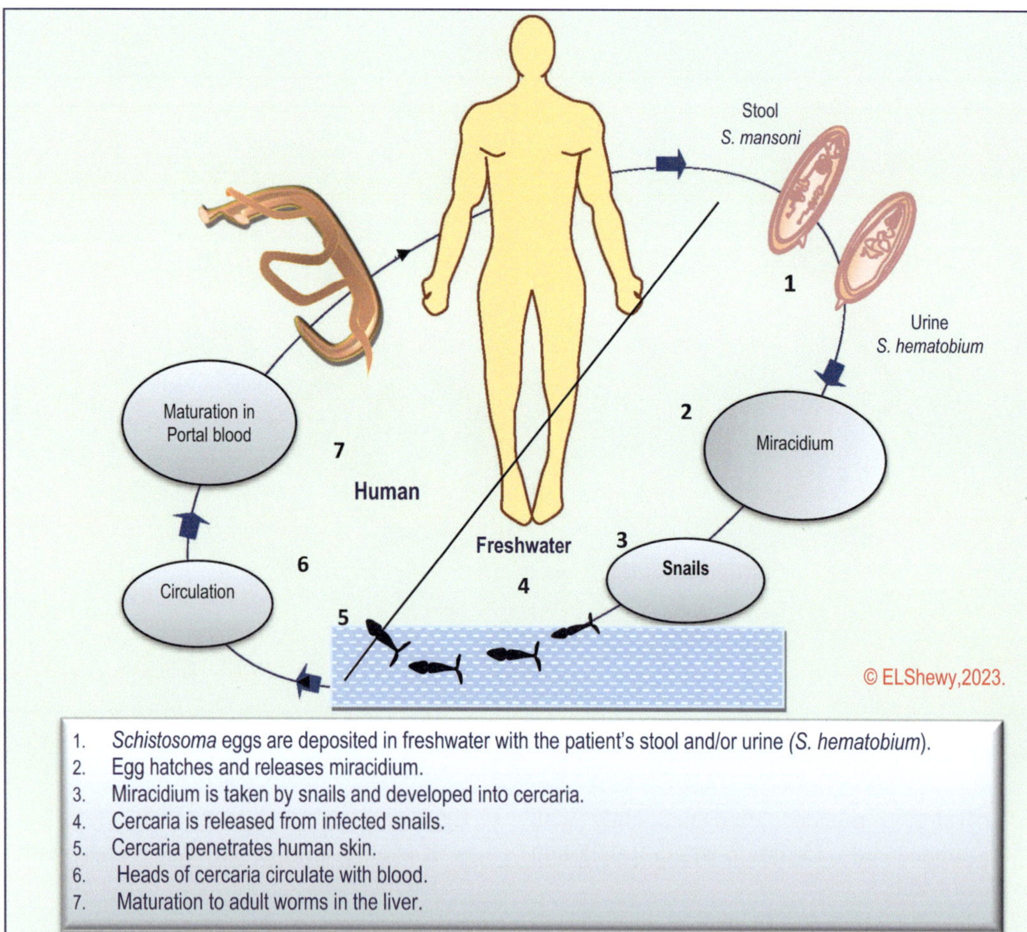

Fig. 8.2 *Schistosoma* spp.: life cycle

- Miracidia swim for about 48 h in water, trying to find and penetrate a freshwater snail (*Biomphalaria*).
- Inside the snail, miracidium transforms into a sporocyst, then into cercaria.
- Cercariae are released from snails and swim for a few hours to find their final host (man) penetrating his bared skin by their heads, leaving tails outside.

In human body:

- Cercaria heads circulate in blood as young schistosomes (schistosomulae) and mature into adults, male and female, in the portal vessels of the liver.
- Males and females start mating and migrate as copula to the narrower blood vessels.
- At the very narrow vessels, male *Schistosoma* leaves its female to continue and lays eggs (300 eggs/day) at the intestinal submucosal blood vessels (4 weeks to 2 months from the start of infection).
- On colonic contraction, deposited eggs in the intestinal mucosa are squeezed to escape the blood vessels and pass with the patient's stool.

Final hosts:	Humans.
Infective stage:	Cercaria.
Diagnostic stage:	Eggs in the patient's stool.
Intermediate hosts:	Freshwater snails: *Biomphalaria* spp. for *S. Mansoni*.

Pathogenesis

The pathogenesis of schistosomiasis is mainly immunologic in origin. The immunopathologic changes associated with intestinal schistosomiasis could be postulated as follows (Figs. 8.3 and 8.4):

1. **Penetration** of the host skin by cercaria stimulates a local inflammation with the formation of an itchy nodule (ground itch).
2. **Migration** of the schistosomulae in the circulation stimulates minimal host response.
3. **Egg deposition** stimulates a local cell-mediated reaction (granuloma formation) with damage to the affected intestinal mucosa (**the most important stage**).
4. Multiple granulomas lead to hypertrophy of the affected mucosa with multiple papillomas and polyps formation. Subsequently, multiple ulcerations appear, giving rise to dysenteric manifestations (blood-flecked stools).
5. Later, **fibrosis** supervenes, leading to a decrease in the egg passage as infection progresses (closed infection).
6. Sweeping of some eggs back to the liver through portal blood flow and their lodgment in the finer branches result in the formation of liver granulomas around the trapped eggs.
7. In heavy infection, **periportal fibrosis** may develop. The portal branches surrounded with thick white fibrous layers give a picture of **clay pipe stem** fibrosis (**Symmers' fibrosis**).

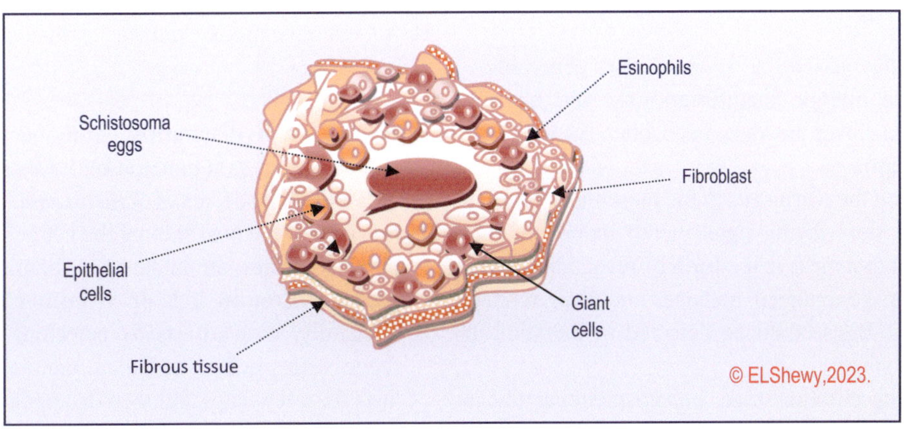

Fig. 8.3 Granuloma formation in intestinal schistosomiasis

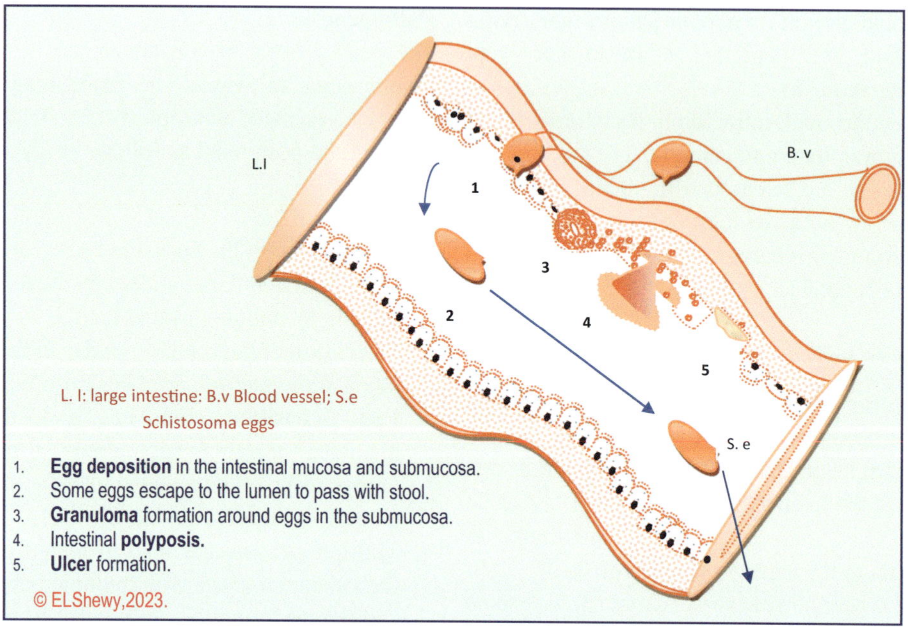

Fig. 8.4 Pathogenesis of intestinal schistosomiasis

8. Granuloma formation with pipe stem fibrosis leads to the formation of **liver cirrhosis**.
9. Liver cirrhosis leads to **portal hypertension** with backpressure reflected as splenomegaly, esophageal varices, hemorrhoids, ascites, and finally, liver cell failure.

Clinical Picture

Intestinal Bilharziasis:

1. Penetration of the host skin with the infective cercaria results in a petechial hemorrhage with pruritus and local edema lasting a few days.
2. In the succeeding 3 weeks after penetration, some allergic manifestations could be suffered (fever, malaise, headache, dyspnea, and cough).
3. When the worms reach the mesenteric venules and start laying eggs, acute manifestations start to appear in the form of fever, abdominal pain, generalized malaise, and liver tenderness. Eggs could be detected in the stool at this stage.
4. In the chronic stage, hepatosplenomegaly is the main presentation. Portal hypertension, esophageal varices, hematemesis, bleeding piles, and massive ascites are common sequelae of hepatosplenomegaly. Eggs sometimes cannot be detected during this stage because of fibrosis and chronicity (closed infection).
5. The intestine is greatly affected by chronic schistosomiasis. Intestinal polyps, ulcers, and intermittent bloody mucoid stools are characteristics of colonic schistosomiasis.

Schistosomal Skin Diseases:

Skin manifestations during schistosomiasis include:

1. **Schistosomal dermatitis** could be demonstrated during skin penetration with cercariae of nonhuman species of *Schistosoma*.
2. Invasion of the host skin by the cercaria of the human species stimulates an inflammatory reaction (**ground itch or swimmer's itch**). Clinically, it manifests as petechial hemorrhage with pruritus and local edema, which lasts for a few days and then disappears.

Other clinical manifestations:

- Occasionally, *Schistosoma* eggs reach the pulmonary artery and its branches as egg emboli, causing pulmonary hypertension with right ventricular hypertrophy (**bilharzial cor pulmonale**). It is characterized by hemoptysis, exertional dyspnea, right ventricular hypertrophy, and subsequent heart failure.

– Spinal cord involvement is frequently seen with *S. mansoni* infection in the form of acute myelopathy.

Immune Response

Host immune reactions against *Schistosoma* infection are summarized as follows:

1. During penetration of the skin by cercaria, a mild localized reaction with the release of histamine from the mast cells is noticed on the initial infection **(type I immune reaction).**
2. Transformation of the cercariae to schistosomulae stimulates an antibody-dependent cell-mediated cytotoxic effect. IgG and eosinophils and stimulated macrophages help damage the schistosomulae **(type II immune reaction)**.
3. After a few days in the host, the surviving schistosomulae are covered with either host antigens or produce self-antigens similar to that of the host. This enables the schistosomulae and, later, the adult worms to evade the immune response.
4. Circulating soluble egg antigens (SEA) could be deposited with their specific antibodies to form antigen–antibodies complex on the membrane of the renal tissue, resulting in nephritic manifestations **(type III immune reaction).**
5. Deposition of the eggs is associated with the release of many immune mediators, particularly IL13. These stimulate cell-mediated immune reactions with monocellular infiltration, fibroblasts, and granuloma formation **(type IV immune reaction)**.

Diagnostic Procedures

Diagnosis of bilharziasis depends on the clinical picture, especially, in endemic localities, together with direct and indirect laboratory investigations:

- **Direct:**
  1. Microscopic stool examination. It detects the characteristic eggs during acute infection.
  2. Rectal swabs are taken to be examined microscopically for the characteristic eggs.
  3. Rectal biopsy by endoscopy, followed by microscopic examination, is useful during chronic **closed infection**.
  4. An egg hatching test is used to detect the **viability** of the egg and, hence, **active infection** (useful as a **test of cure**).
  5. Kato technique is used to estimate the **intensity of infection** in intestinal bilharziasis.
- **Indirect Immunological:**
  1. Detection of specific antibodies is commonly used during surveys in endemic areas, e.g., ELISA, IHA, IFA, and RIA.
  2. Detection of specific schistosomal antigens is more reliable in the diagnosis of bilharzial infection, e.g., Dipsticks Dye Tests are used for the detection of specific soluble cercaria and egg antigens (SCA and SEA) especially, in survey studies.
  3. Circumoval precipitin test, using the suspected patient's serum and viable eggs.
  4. Detection of interleukins and other immune mediators.
- **Other investigations:**
  Abdominal sonography, barium enema, CAT, MIR, and plain X-ray are used for the detection of the pathological effects and complications of schistosomiasis.

N.B.

– Detection of the eggs in stool must be done after maturity of the worms (6–8 weeks from the start of infection).
– The viability of the eggs could also be detected by monitoring the active motility of the flame cells in the miracidium and the presence of RBCs around viable eggs.

– Occasionally, *S. mansoni* eggs could be detected in the urine.

Treatment

Drug treatment is effective in the early stages of bilharziasis. However, once chronic infection is established, drugs have no role in relieving clinical manifestations. There are many drugs commonly used in the treatment of schistosomiasis among these drugs are:

1. **Praziquantel**:
   - The drug of choice for treatment of schistosomiasis.
   - Single oral dose of 40 mg/kg body weight, with a cure rate of 60%.
   - Three doses of 20 mg/kg body weight for a single day, with a cure rate of 100%.
   - It is only effective against the mature adult worms.
   - **S.E.** headache, dizziness, nausea, and abdominal pain.
2. **Oxamniquine**:
   - It is a better drug for the treatment of S. mansoni infection.
   - It is effective in mass treatment.
   - 20 mg/kg/body weight daily for 3 days is recommended.
3. **Metrifonate**:
   - The drug of choice for the treatment of S. haematobium infection.
   - 10 mg/kg/body weight is recommended once every other week for three doses.
   - **S.E.** Abdominal pain, vomiting, headache, and dizziness.
4. **Merazid:**
   - Herbal extraction from an Egyptian plant.
   - It is a promising anti-bilharzial drug.

Prevention and Control

1. Health education and increasing awareness.
2. Mass treatment programs.
3. Avoidance of contact with infected water sources.
4. Control of the snail host by the use of molluscicides, as well as biological control with snail-eating fish, birds, or other competing snail species.
5. The use of Niclosamide as skin lotion to prevent penetration of the cercariae.
6. Schistosomal vaccines:

   Although there is no effective vaccine for mass application so far, available schistosomal vaccines include:

   (a) Cercarial cobalt attenuated cercaria vaccines.
   (b) SEG antigen vaccines.
   (c) Sh28GST (Bilhalvax) and Sm-p28.

- **Parasitology:** *S. mansoni* is a blood trematode fluke that needs a snail as an intermediate host. It is transmitted through the penetration of human skin by cercaria. The worms mature in the liver, and eggs with spines are deposited in the mucosa of the intestine to pass with the patient's stool.
- **Clinical:** Intestinal bilharziasis: Blood in stool with dysenteric manifestations, hepatosplenomegaly ascites, portal hypertension, and liver cirrhosis.

Fasciola hepatica **and** ***gigantica*** **(*****F. hepatica, gigantica*****) (Liver worm)**

Fasciola is a zoonotic liver trematode worm that infects both humans and animals. In humans, it causes hepatobiliary disease with gastrointestinal manifestations.

A massive infection with *Fasciola* worms could cause portal cirrhosis with all its manifestations. **Halzoun** is a pharyngeal disease caused by adult worms. In animals, the disease is called liver rot.

Geographical Distribution

Human and animal fascioliasis occurs worldwide. Fascioliasis prevails mainly in the cattle and sheep-raising areas. There are two species of *Fasciola*:

1. *F. hepatica* is the most widespread species.
2. *F. gigantica* is less prevalent and restricted to some African and Asian countries.

Parasite Morphology

Fasciola worm is hermaphroditic with both male and female genitalia. The following are morphological criteria of *Fasciola hepatica* (Fig. 8.5)**:**

- Adult worm (*F. hepatica*):
  - Flat leaf-like worm with two suckers (anterior and ventral).
  - 3 cm length × 1.5 cm width.
  - It has a cephalic cone and prominent shoulders.
  - The body is covered with spines with converging lateral sides.
  - Ceca close together, with rudimentary lateral branches.
- Egg:
  - Oval in shape, with an operculum and a slight thickening at the posterior end of the shell.
  - 140–80 µm.
  - It is immature when excreted with an unsegmented ovum and granules in yolk sacs.
- Cercaria:
  - It has a head and a single simple tail.

N.B. *F. gigantica* is larger (5–6 cm) than *F. hepatica*, with parallel body outlines. Shoulders are less prominent, and intestinal ceca have distinctive T- or Y-shaped branches.

Transmission

- Eating vegetation (freshwater aquatic plants) containing the infective stage (encysted metacercaria).
- The pharyngeal form of the disease (**Halzoun**) could appear after eating raw liver infected with adult *Fasciola* worms.

Habitat

Adult worms inhabit the biliary passages of the liver of humans, sheep, goats, cattle, and many other herbivorous animals.

Life Cycle

The life cycle of both *F. hepatica* and *F. gigantica* is similar; however, the intermediate host is different. The life cycle is postulated in (Fig. 8.6).

Final hosts:	Humans and other herbivorous animals.
Infective stage:	Encysted metacercaria on vegetation.
Diagnostic stage:	Characteristic eggs in the stool.
Intermediate host:	The first is freshwater snail (*Lymnaea* spp.); the second is freshwater vegetation.
Reservoir hosts:	Herbivorous animals.

N.B. The first intermediate host for *F. hepatica* is *Lymnaea truncatula*, while for *F gigantica* is *Lymnaea cailliaudi.*

Pathogenesis

The following pathological events are reported during fascioliasis:

1. During the migration of the worms to reach the liver, they produce local irritation with minimal inflammatory reactions.
2. In the biliary passages, adult worms cause mechanical and toxic irritation resulting from their large sizes and secretory/excretory metabolites.
3. Both mechanical and toxic irritation stimulate hyperplasia of the biliary epithelium and the proliferation of connective tissue around the duct with biliary obstruction.
4. Adult worms may erode the biliary passage to reach the liver parenchyma.
5. A secondary bacterial infection could occur, with manifestations of hepatitis.
6. Liver cirrhosis could be detected with severe *Fasciola* infection.

Clinical Picture

Light infection with *Fasciola* could pass unnoticed; however, in severe heavy infection, the following manifestations could be demonstrated:

1. Fever, chills, right hypochondrial pain, tenderness, jaundice, and eosinophilia are common signs and symptoms of fascioliasis.
2. Chronic cholangitis with biliary obstruction could be suffered in long-standing fascioliasis with marked eosinophilia. Anemia could be detected in chronic fascioliasis.
3. The pharyngeal disease of fascioliasis could be seen when the patients eat raw infected

Fig. 8.5 *Fasciola hepatica*: adult and egg morphology

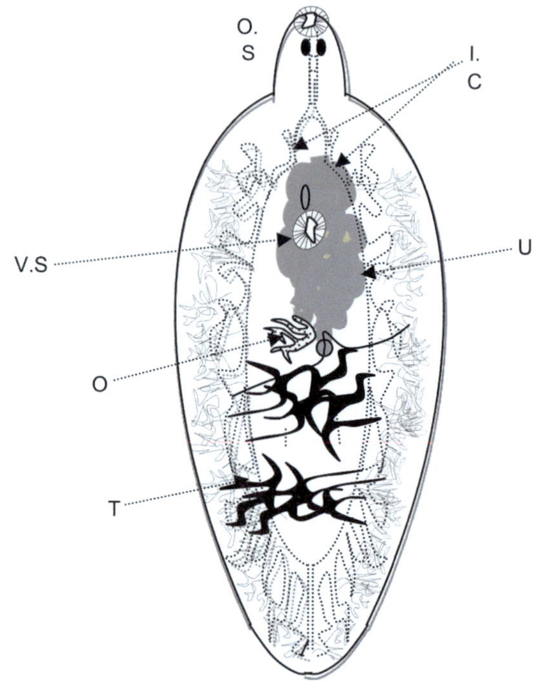

O.S= oral sucker, V.S= ventral sucker; G.S = genital sucker, T= testes, O= ovary, U= uterus, S= seminal receptacle, I.C= intestinal ceca.

Adult

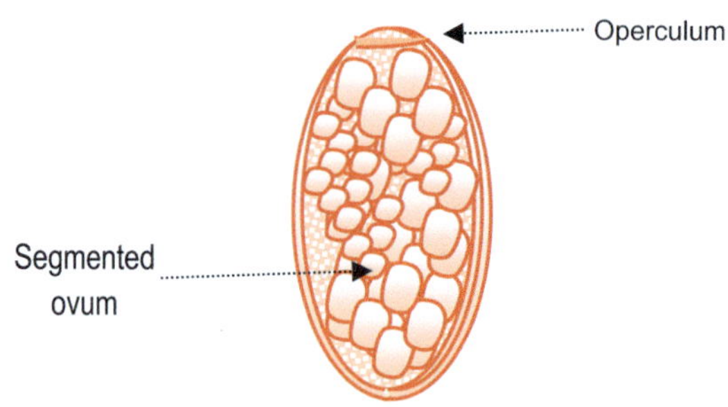

Egg

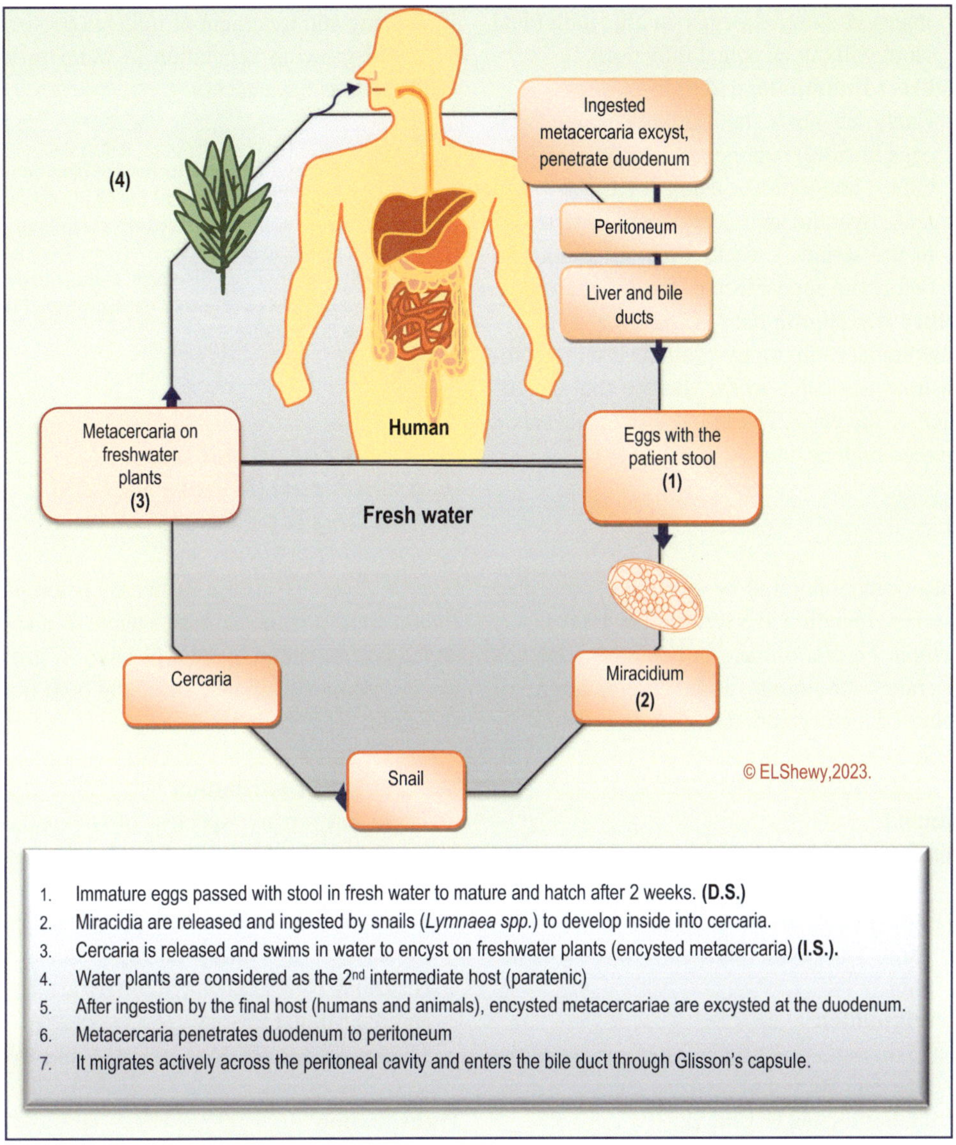

Fig. 8.6 *Fasciola* worm: life cycle

liver. This is called **Halzoun**; young adult worms attach themselves to the pharyngeal mucosa by their suckers, causing pain, bleeding, and angioneurotic edema and may be suffocation.

Diagnostic Procedures

- **Direct:**

  Diagnosis of fascioliasis is made by the detection of the characteristic eggs in the stool of the patient (eggs appear after 6–8 weeks from the start of infection):

  1. **Microscopic stool** examination of wet or stained smears is usually used.
  2. An **enterotest** (entero-capsule test) could be used to bring eggs stained with bile.
  3. Individuals are instructed not to eat liver at least 1 week before examination to exclude **spurious** (false) infection (passage of

ingested *Fasciola* eggs in the individual stool without an actual infection).

- **Indirect Immunological:**
  - **Early** diagnosis (before the appearance of eggs in stool) is done by detecting the specific antibodies using ELISA and IHA. Specific antibodies could be detected in the serum 3 weeks after infection and before the appearance of the *Fasciola* eggs.
- **Other Investigations:**
  Abdominal sonography could be used to demonstrate the adult worms before the appearance of the eggs. The blood picture reveals a marked high eosinophilia.

N.B.

- False infection could be ensured by repeated stool examination and serological assays.
- Ectopic *Fasciola* infection could be observed in rare conditions after implantation of encysted metacercaria in tissues other than the liver.

Treatment

1. **Bithionol** (Bitin):
   - 30–50 mg/kg/ every other day for 10 doses.
   - **S.E.** photosensitivity skin reaction, urticaria, vomiting, diarrhea, and abdominal pain.
2. **Praziquantel:**
   - 25 mg/kg/3 times daily/7 days is less effective than Bithionol.
3. **Triclabendazole** (Fasinex):
   - Single oral dose of 10 mg/kg/b.w. is effective.

Prevention and Control

1. Snail control.
2. Following standards of health and good sanitary practices.
3. Proper veterinary inspection.
4. Proper washing of green leafy vegetables (soaking in acetic acid or K. permanganate for 15 m, and washing with running water before consumption).
5. Tracing and treatment of infected persons.
6. Avoid growing vegetation in water to which herbivores have access.

- **Parasitology:** *Fasciola hepatica* is a 3–6 cm, leaf-like worm; it inhabits the liver of both humans and herbivores; it needs a snail for its life cycle. It is transmitted through eating plants containing encysted metacercaria.
- **Clinical:** Hepatobiliary disorders, chronic hepatitis with liver cirrhosis at the end. Eating infected raw liver causes Halzoun.

Echinococcus granulosus (*E. granulosus*) (Minute Hydatid Worm, Dog Tapeworm)

E. granulosus is a cestode worm belonging to *Taenia* spp.; it is the causative agent of a cestode larval disease called hydatid disease. *E. granulosus* is a **zoonotic** parasite infecting both humans and canines (dogs, foxes …).

Geographical Distribution

- There are many species of *Echinococcus* worms, and each has its clinical and geographical pattern.
- *E. granulosus* is the most prevalent species, especially in sheep-raising areas, e.g., Australia, New Zealand, South Africa, Europe, and the Middle East.
- The final hosts are dogs, coyotes, and wolves.

Transmission

- Human infection (**hydatidosis**) is transmitted through accidental ingestion of *E. granulosus* eggs with foods or during dealing with dogs.
- Dogs and other canine animals (foxes, wolves) acquire infection (**echinococcosis**) by ingestion of dead animal viscera and carcasses containing hydatid cysts.

Parasite Morphology

E. granulosus has three morphological forms (Fig. 8.7)**:**

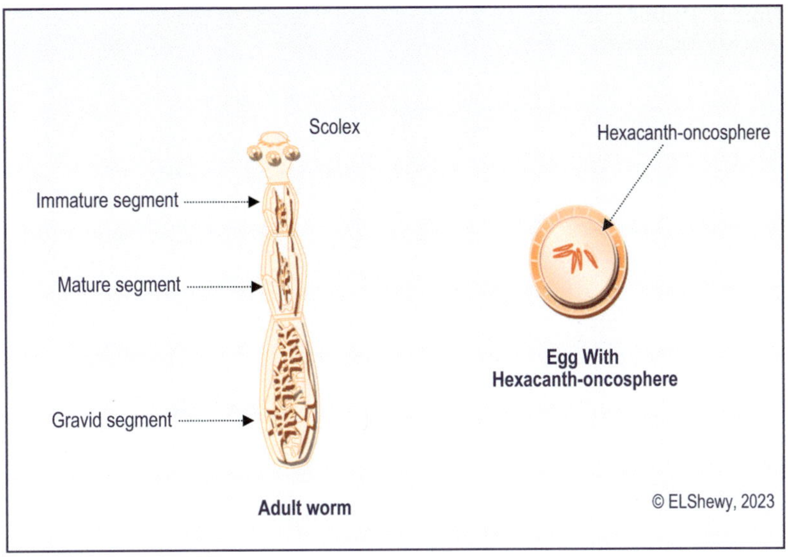

Fig. 8.7 *Echinococcus granulosus*: adult and egg morphology

- Adult worm:
  - A small tapeworm of 3–6 mm long.
  - It has a scolex and three proglottids (immature, mature, and gravid).
  - Scolex has four suckers with a hooked rostellum.
  - The gravid proglottid is longer than wide.
- Egg:
  - Similar to *Taenia* egg.
- Hydatid cyst:
  - It ranges in size from a few millimeters to the size of grapefruit.
  - The wall of the cyst is consisting of:
    (a) Outer rigid fibrous layer (host origin).
    (b) Non-nucleated laminated layer (parasite origin).
    (c) Inner nucleated germinal epithelium.
  - Budding scolecies and brood capsules from the germinal layer.
  - Fluid inside the cyst (Hydatid fluid) contains hooklets, daughter cysts, and degenerating tissues (Hydatid sand) (Fig. 8.8).

Habitat

1. Adult worms of *E. granulosus* inhabit the small intestine of canine animals, e.g., dogs, foxes, and wolves.
2. Hydatid cyst is localized mainly in the liver, lungs, and CNS of humans and in herbivorous animals (cattle, sheep, and others).

Life Cycle

The life cycle of *E. granulosus* is postulated in Fig. 8.9.

Final hosts:	Dogs and other canines.
Infective stage:	Hydatid cysts for dogs and the eggs of the worm for humans.
Diagnostic stage:	Hydatid cysts in human tissues.
Intermediate hosts:	Humans and herbivorous animals, e.g., sheep and cattle.
Reservoir hosts:	Dogs and canine animals.

Pathogenesis

Hydatid cyst is commonly detected in the right lobe of the liver, lungs and brain. Other tissues could be infected:

1. The existence of a hydatid cyst results in pressure necrosis of the surrounding tissues.
2. A growing cyst in the liver stimulates an inflammatory reaction around it, ending in the formation of a fibrous capsule.

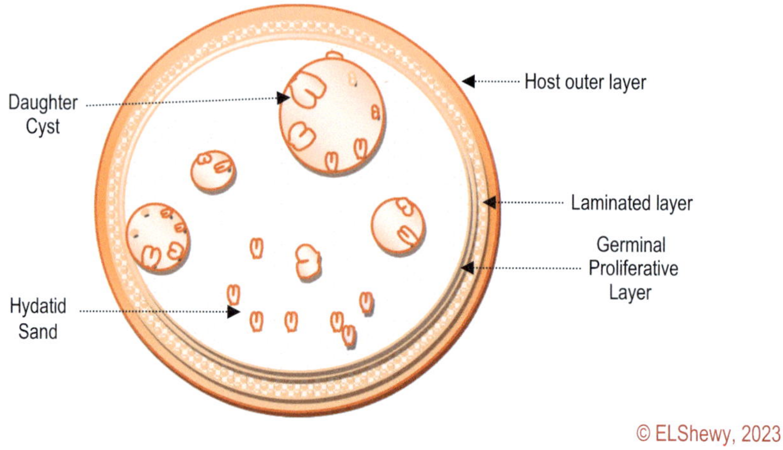

Fig. 8.8 Hydatid cyst: structure and contents

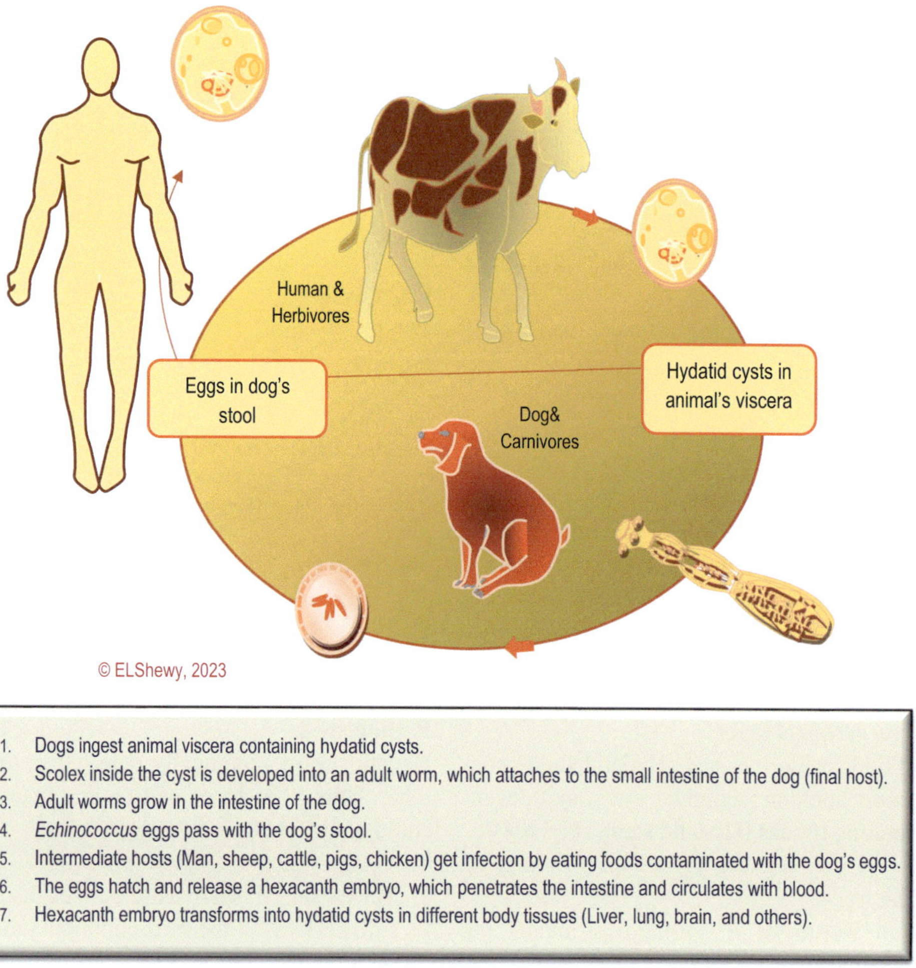

Fig. 8.9 *Echinococcus granulosus*: life cycle

3. Slow leakage of the hydatid fluid (if it happens) usually sensitizes the patient and produces marked eosinophilia.
4. Rupture of hydatid cyst during surgery or trauma produces anaphylactic shock with the possibility of growing and spreading of the cyst germinal epithelium to new hydatid cysts.
5. Rupture of pulmonary hydatid cyst results in severe allergic manifestations with blood-flecked fluid.

Clinical Picture

Clinical presentations of hydatid cysts vary according to the location of the cyst:

- Cysts in the liver
  Most of the hydatid cysts are localized at the right lobe of the liver. A slowly growing cyst does not produce so many symptoms, but when size increases, encroachment on a blood vessel, bile duct or intraciliary rupture occurs, severe hepatic symptoms are observed, e.g., fever, liver tenderness, jaundice, hepatitis, and hepatomegaly.
- Cysts in the lung
  They are asymptomatic until they become large enough to produce cough, dyspnea, and chest pain.
- Cysts in the CNS
  They produce severe tissue damage with many nervous manifestations, e.g., convulsion, epilepsy, diplopia, aphasia, paralysis and other nervous manifestations.
- **Hydatid cyst in bones** is termed **an osseous** hydatid cyst. It is small, contains gelatinous material, has no scolecies, and grows very slowly in the direction of less resistance.

Diagnostic Procedures

1. Clinical pictures and history were taken.
2. Demonstration of the cyst inside the human body using the following:
   - Ultrasonography (U) and radiographic examination (X-ray).
   - Computerized tomographic scan (CTS) and magnetic resonance imaging (MRI).
3. Immunological assay:
   This is the **main diagnostic tool** for the diagnosis of hydatidosis.
   - Many serologic tests are commonly used for the detection of hydatid diseases, e.g., ELISA, CF, IFA, and IHA.
   - Skin hypersensitivity test (**Casoni test**): 0.2 ml of sterile hydatid fluid is injected into the intradermally and checked for skin reaction after 24–48 h.
   - Cross-reactions have been noted between cysticercosis and hydatid disease. So, a combination of assays, including immunoblot and ELISA, has been suggested.
4. Aspiration of the hydatid fluid and microscopic detection of hydatid sand could be used in diagnosis, but this carries a high risk of rupture.

Treatment

Surgical removal of operable cases, while inoperable cases or cases with multiple cysts, chemotherapy, and/or **PAIR** (puncture-aspiration-injection-re-aspiration) techniques are applied. In both cases, the patient should receive the following drugs before and after the intervention:

1. **Albendazole (Albenza):**
   - It is the drug of choice.
   - 5 mg/kg/body weight/ twice daily for 8–12 weeks.
   - Should be given to the inoperable cases.
   - Could be given pre-and post-operatively for the operable cases.
   - Repeated courses are recommended.
2. **Mebendazole (Vermox).**
3. **Benzimidazole.**
4. **Praziquantel (Biltricide):**
   - It is effective as scolicidal drug in high doses.
   - 120–200 mg/kg/body weight for 5 days is effective preoperatively.
   - **S.E.:** headache, dizziness, abdominal pain, and nausea.

Prevention and Control

1. Good sanitary practices with pet dogs are essential.
2. Anti-helminthic treatment and veterinary care of dogs.
3. Avoidance of ingestion of sheep viscera by keeping dogs away from slaughterhouses.
4. Avoidance of direct and close contact with dogs.

- **Parasitology:** *E. Granulosus* is a minute worm of the canines, humans are an intermediate host. Humans are infected through accidental ingestion of contaminated foods with the worm's eggs.
- **Clinical:** Hydatid cysts scattered in different tissues and organs with different clinical manifestations, diagnosed mainly through serological methods and imaging.

Capillaria hepatica (*C. hepatica*)

C. hepatica is a nematode animal parasite that occasionally infects the human liver.

Habitat

- Adult worms of *C. hepatica* inhabit liver tissues of carnivorous animals and occasionally human liver.

Life Cycle

- Adult worm inhabits the liver of carnivorous animals.
- Worm deposits eggs in the liver tissues and remain there for the rest of the host's life.
- When the infected animal is eaten by carnivorous, the eggs are released in its stool and disseminated in the environment,
- Man can get the infection by an accidental ingestion of these eggs in his food.
- Eggs hatch and release larvae that penetrate the human intestine. Larvae pass through the portal circulation to the liver where they mature into adult *C. hepatica* and repeat the cycle.

Pathogenesis and Clinical Picture

- Acute hepatitis due to the presence of the worm and its eggs.
- Chronic inflammatory reaction with granuloma formation.
- Fibrosis and eventually liver cirrhosis.
- The clinical picture includes:
  1. Hepatitis, high fever, chills, and jaundice.
  2. Right hypochondrial pain and tenderness.
  3. Weight loss and decreased appetite.
  4. Hepatomegaly, ascites, and gallstones in the bile ducts.

Diagnosis

A liver biopsy is done to detect the characteristic eggs.

Treatment

Albendazole:

- It is the drug of choice.
- 5 mg/kg/body weight/twice daily with meal for 8–12 weeks.

Clonorchis sinensis (C. Sinensis) (Chinese Liver Fluke)

C. sinensis is a trematode fluke that could infect fish-eating animals and humans. It is responsible for clonorchiasis, a hepatobiliary disease characterized by severe liver damage, jaundice, cholangitis, and cholelithiasis.

Geographical Distribution

C. sinensis prevails in many areas of China, Taiwan, Japan, Korea, and Vietnam.

Parasite Morphology

The following are the morphological criteria of the adult worm and its egg (Fig. 8.10).

Fig. 8.10 *Clonorchis sinensis:* adult and egg morphology

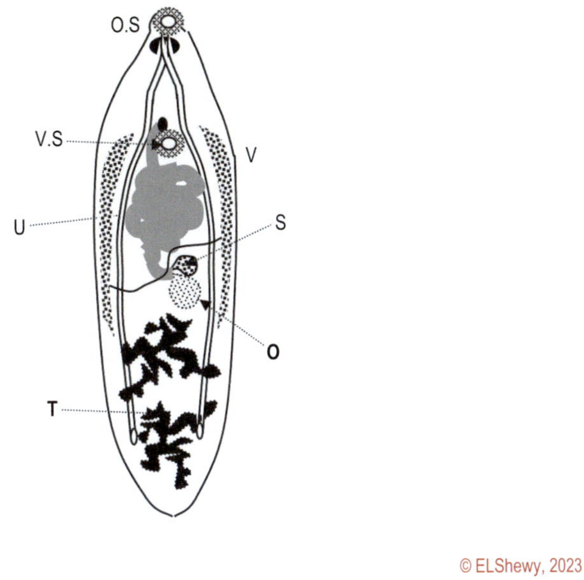

O. S= oral sucker, V. S= ventral sucker; G.S = genital sucker, T= testes, O= ovary, U= uterus, V= vitellaria, S= seminal receptacle

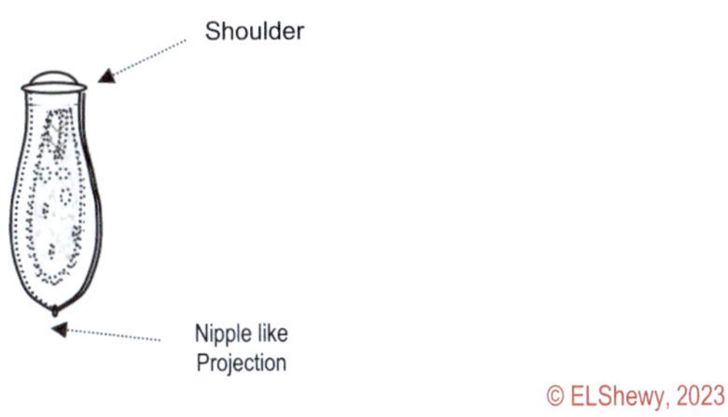

- Adult worm:
  - It is a flat, oblong-shaped fluke.
  - Size: 1–2.5 cm × 0.5 cm.
  - It has a small ventral sucker.
  - The body has no spines.
  - Testes located at the posterior 1/3 of the body.
- Egg: (similar to *H. heterophyes*).
  - It is an oval, operculated small egg.
  - It has thick, prominent shoulders with a small nipple-like projection opposite to the operculum.
  - 30 μm × 15 μm.
  - Yellowish brown.
  - It is embryonated (containing miracidium) when first passed.

Transmission

Infection with *C. sinensis* is transmitted through the consumption of raw or undercooked freshwater fish containing encysted metacercaria.

Habitat

C. sinensis inhabits the biliary passages of humans, dogs, and cats.

Life Cycle

The life cycle of *C. sinensis* is similar to that of *H. heterophyes* and *M. yokogawai*. When the embryonated eggs pass in freshwater, they are ingested by a freshwater snail (first intermediate host). Eggs hatch to release miracidium. Miracidium develops in the snail tissue to cercaria. Cercaria leaves snails to penetrate and encyst in the freshwater fish (second intermediate host). Humans are infected by eating raw or undercooked fish. Metacercariae start to excyst in the duodenum and pass through the common bile duct to the biliary passages. Adult worms mature and lay eggs in the bile duct to pass with bile to the feces (Fig. 8.11).

Final hosts:	Fish-eating animals and humans.
Infective stage:	Encysted metacercaria.
Diagnostic stage:	Eggs containing fully mature miracidium.
Intermediate hosts:	The first is freshwater snail (*Parafossarulus*); the second is freshwater fish.
Reservoir hosts:	Fish-eating animals.

Pathogenesis

1. In early infection, there is an acute inflammatory reaction with leucocytes and eosinophils infiltration in the infected biliary passage.
2. A localized dilatation of the bile ducts is seen in heavy infection.
3. Hyperplasia and thickening of the biliary mucosal lining are usually detected in long-standing infections.
4. There is a significant correlation between adenocarcinomas and clonorchiasis.

Clinical Picture

1. Light infection is usually passed unnoticed.
2. Heavy infection with *C. sinensis* is associated with tender liver, epigastric pain, nausea, vomiting, and vague abdominal pain.
3. Early detection of high leukocytosis and eosinophilia is remarkable.
4. Eggs are detected in the stool after 1 month from infection.
5. Chronic and repeated infection with *C. sinensis* is usually associated with signs and symptoms of cholecystitis and cholelithiasis.
6. Hepatitis, gallstones, pancreatitis, and carcinoma are complications of chronic infection.

Diagnostic Procedures

Diagnosis of *C. sinensis* is done by detection of the characteristic eggs in the patient's stool. A routine stool examination is usually used. Care must be taken to distinguish between *C. sinensis* and the intestinal fluke *Heterophyids*. Adult worms are rarely seen in the stool.

Treatment

1. **Praziquantel (Biltricide):**
   - It is a drug of choice.
   - It is given 25 mg/kg/body weight three times for 2 days.
2. **Albendazole (Albenza):**
   - It is the drug of choice.
   - 5 mg/kg/body weight/ twice daily for 8–12 weeks.

Prevention and Control

1. Proper fish cooking and salting.
2. Proper disposal of human excreta.
3. Snail eradication.
4. The use of human excreta as fertilizers should be prohibited.

- **Parasitology:** *C. Sinensis* is a flat, oblong worm, 1–2.5 cm, mainly inhabiting the biliary tract, and it needs snails during its life cycle. It is transmitted through eating undercooked or raw freshwater containing encysted metacercaria.
- **Clinical:** Hepatobiliary disorders, jaundice, cholangitis, cholecystitis, and cholelithiasis with an association with adenocarcinoma.

Opisthorchis felineus & viverrini (*O. felineus & viverrini*) (Southeast Asian Liver Flukes) /(Siberian Fluke)

O. felineus and *O. viverrini* are both trematodes infecting the biliary passages of humans and animals. They, together with *C. sinensis,* are biological causes of cholangiocarcinoma. They have a similar life cycle as *C. sinensis*.

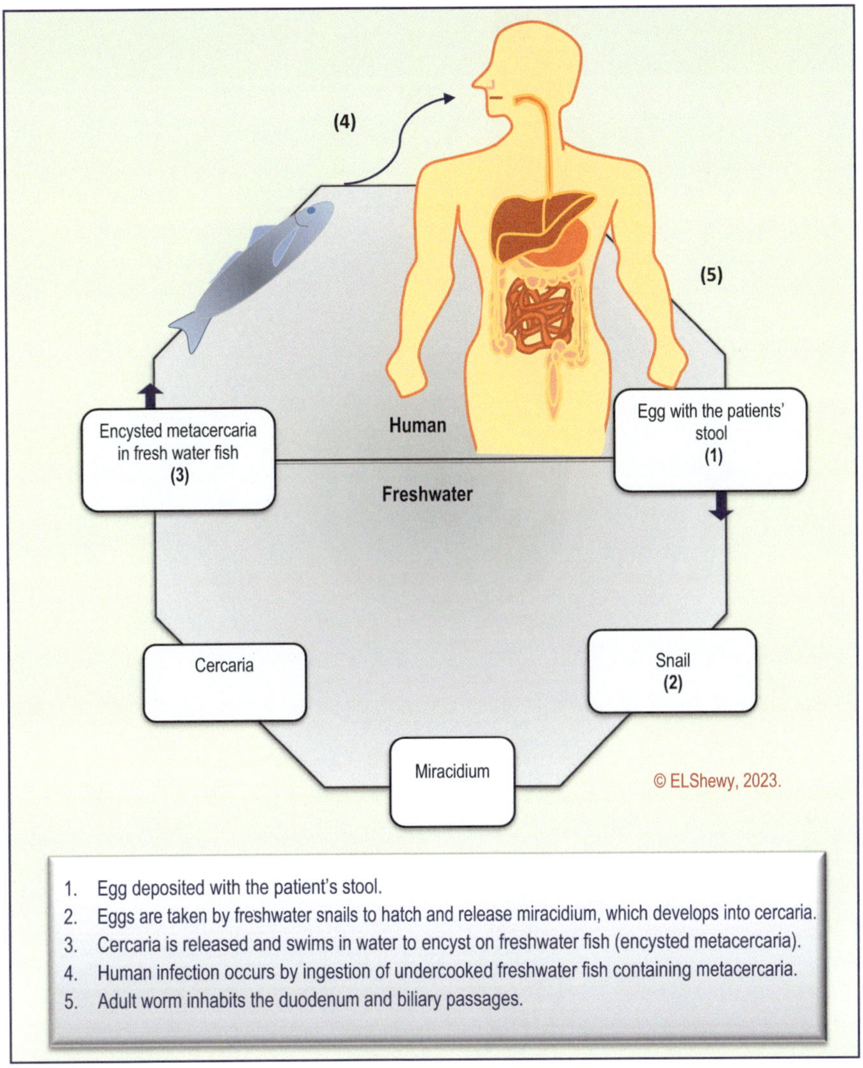

Fig. 8.11 *Clonorchis sinensis*: life cycle

Geographical Distribution

O. felineus prevails mainly in Poland, Russia, Siberia, and some areas of Vietnam and India. *O. viverrini* mainly prevails in Thailand.

Parasite Morphology

Adult worms of *Opisthorchis* species are similar in morphology to *C. sinensis*. Their eggs are similar to *C. sinensis* but with smaller and narrow bodies (Fig. 8.12).

Transmission

Infection with *O. felineus* and *O. viverrini* results from consumption of raw, undercooked, and incompletely salted fish freshwater fish containing encysted metacercaria.

Habitat

O. felineus and *O. viverrini* inhabit the biliary passages of humans, dogs, and cats.

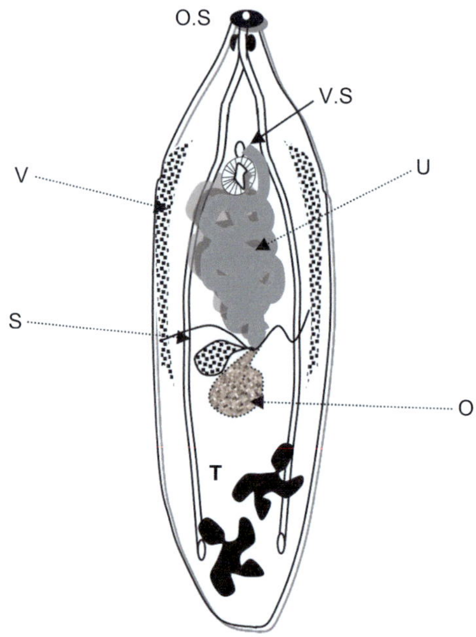

O. S= oral sucker, V. S= ventral sucker, T= testes, O= ovary, U= uterus, V= vitellaria, S= seminal receptacle

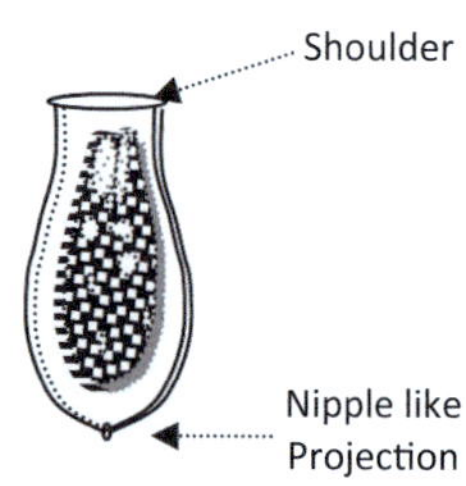

Fig. 8.12 *Opisthorchis felineus and viverrini:* adult and egg morphology

Life Cycle

The life cycle of *O. felineus* and *O. viverrini* is similar to *C. sinensis*.

Clinical Picture

Similar to *C. sinensis,* however, there is a strong correlation between cholangiocarcinoma and infection with *O. viverrini.*

N.B. Diagnosis, treatment, and prevention as in *C. sinensis* infection.

Dicrocoelium dendriticum (D. dendriticum) (Lancet Liver Flukes)

D. dendriticum is a common parasite of ruminants. Humans are accidentally infected with it. It causes the same clinical manifestations as *C. sinensis.*

Geographical Distribution

D. dendriticum could be detected in Europe, Asia, North Africa, and the Middle East.

D. dendriticum is a liver trematode that causes the same clinical manifestation as *C. sinensis.*

Parasite Morphology

The following are the morphological criteria of the adult worm of *D. dendriticum* with its egg (Fig. 8.13).

- Adult worm:
  - It is an oval, small fluke with a smooth cuticle.
  - 0.8–1 cm × 1.5–2.5 mm.
  - It has a small ventral sucker that lies just behind the cecal bifurcation.
  - Testes located in the anterior 1/3 of the body.
- Egg:
  - It is an oval, operculated small egg.
  - It has a thick shell.
  - 45 μm × 30 μm.
  - Golden brown.
  - Embryonated when freshly passed.

Transmission

D. dendriticum is transmitted by ingestion of water contaminated with ants. However, a spurious (false) infection with *Dicrocoelium* could occur through ingestion of a liver infected with adult worms and the appearance of the worm eggs in the patient's stool.

Habitat

Adult worms inhabit the biliary passages of ruminants and humans.

Life Cycle

The life cycle of *D. dendriticum* is similar to that of *C. sin*ensis, except that the first intermediate host is a land snail, which ingests the egg and releases cercaria during rainy days. The second intermediate host is an **ant** (*Formica* ants). The liberated cercariae from the land snail are encysted in ants. Humans can be accidentally infected when ingesting bottled water, drinks, or foods contaminated with ants (Fig. 8.14).

Clinical Picture

1. Mostly, infection with *D. dendriticum* is asymptomatic.
2. Biliary colic and digestive disturbance could be seen in light infection.
3. In heavy infection, marked enlargement of the biliary passages, mucosal hyperplasia, periductal fibrosis, and eventually portal cirrhosis are usually demonstrated.

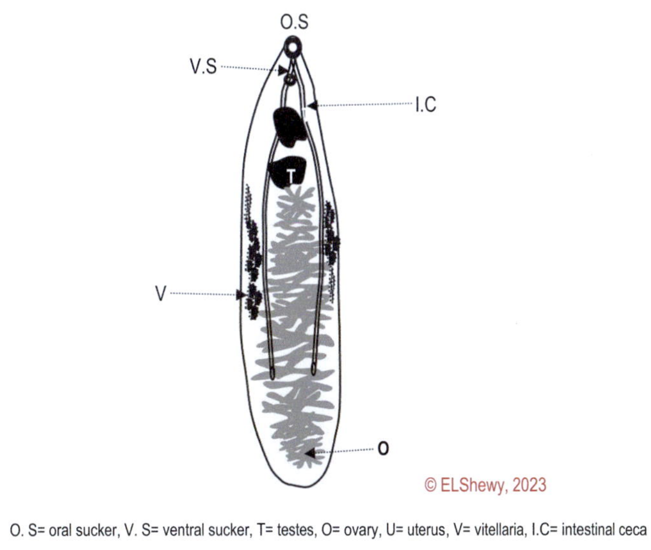

Fig. 8.13 *Dicrocoelium dendriticum:* adult morphology

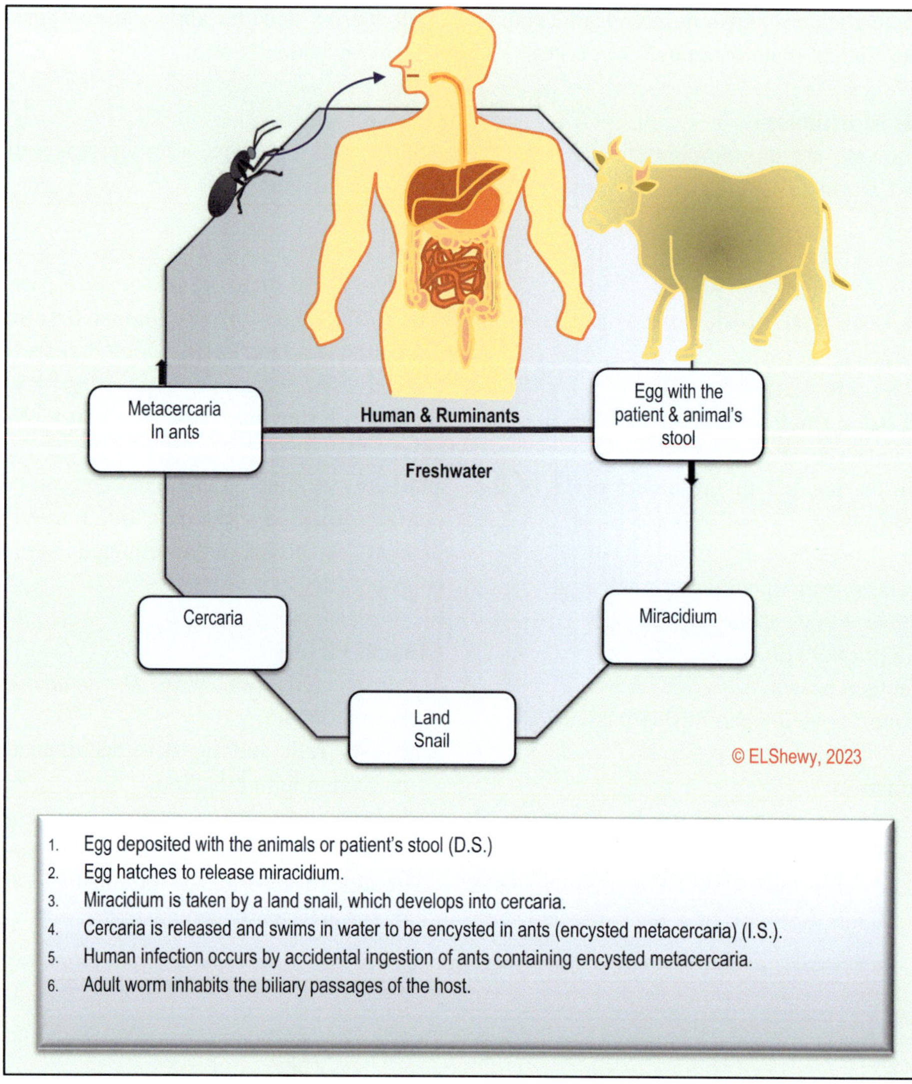

Fig. 8.14 Dicrocoelium dendriticum: life cycle

Diagnostic Procedures

Diagnosis of *D. dendriticum* is done by the detection of the characteristic eggs in the stool of the patients. Care must be taken to distinguish between different eggs of the intestinal flukes: *Heterophyids, C. sinensis,* and *D. dendriticum*. Adult worms are rarely seen in the stool.

Treatment

1. **Praziquantel:**
   - It is a drug of choice.
   - 20 mg/kg/3 times/for one day is recommended.
2. **Triclabendazole (Fasinex).**
3. **Merazid.**

Prevention and Control

1. Proper and clean water supply.
2. Proper disposal of human excreta.
3. Avoid eating raw animal liver.

- **Arasitology:** *D. Dendriticum* is an oval worm, 1–2.5 cm, mainly inhabits the biliary passages; it needs land snail for its life cycle; it is transmitted through accidental ingestion of ants containing encysted metacercaria.
- **Clinical:** Infection is usually asymptomatic, but heavy infection is presented with cholangitis, cholecystitis, cholelithiasis, and portal cirrhosis.

Review Questions

1. **Rectal biopsy could be used to detect infection with.**
   (a) *Taenia solium.*
   (b) *Heterophyes heterophyes.*
   (c) *Schistosoma mansoni.*
   (d) *Fasciola hepatica.*
2. **Which of the following techniques is not commonly used for diagnosis of *Schistosoma mansoni* infection?**
   (a) Stool examination.
   (b) Kato technique.
   (c) Skin snip.
   (d) Enzyme-linked immunosorbent assay (ELISA).
3. **Eating liver should be prohibited 1 week before the stool examination required for diagnosis of.**
   (a) heterophyiasis,
   (b) fascioliasis,
   (c) schistosomiasis,
   (d) taeniasis.
4. **Which of the following worms needs two intermediate hosts to complete their life cycle?**
   (a) *Schistosoma mansoni and Heterophyes heterophyes.*
   (b) *Fasciola hepatica and Schistosoma haematobium.*
   (c) *Heterophyes heterophyes and Fasciola hepatica.*
   (d) *Schistosoma haematobium and Schistosoma mansoni.*
5. **During intestinal bilharziasis, the following pathological changes could be observed except.**
   (a) granuloma formation,
   (b) ureteric fibrosis,
   (c) liver cirrhosis,
   (d) portal hypertension.
6. **The egg hatching test helps to demonstrate the viability of the egg of.**
   (a) *Taenia* species.
   (b) hookworm,
   (c) *Trichuris trichiura.*
   (d) *Schistosoma* species.
7. **Hematemesis could be detected during infection with.**
   (a) *Hymenolepis diminuta.*
   (b) *Schistosoma mansoni.*
   (c) *Taenia saginata.*
   (d) *Hymenolepis nana.*
8. **One of the following eggs is immature when immediately passed outside the human body.**
   (a) *Taenia solium.*
   (b) *Heterophyes heterophyes.*
   (c) *Schistosoma mansoni.*
   (d) *Fasciola hepatica.*
9. **Man can act as an intermediate host for which of the following parasites.**
   (a) *Trichuris trichiura.*
   (b) *Echinococcus granulosus.*
   (c) *Strongyloides stercoralis.*
   (d) *Wuchereria bancrofti.*

9 Parasites Infecting the Urinary Tract

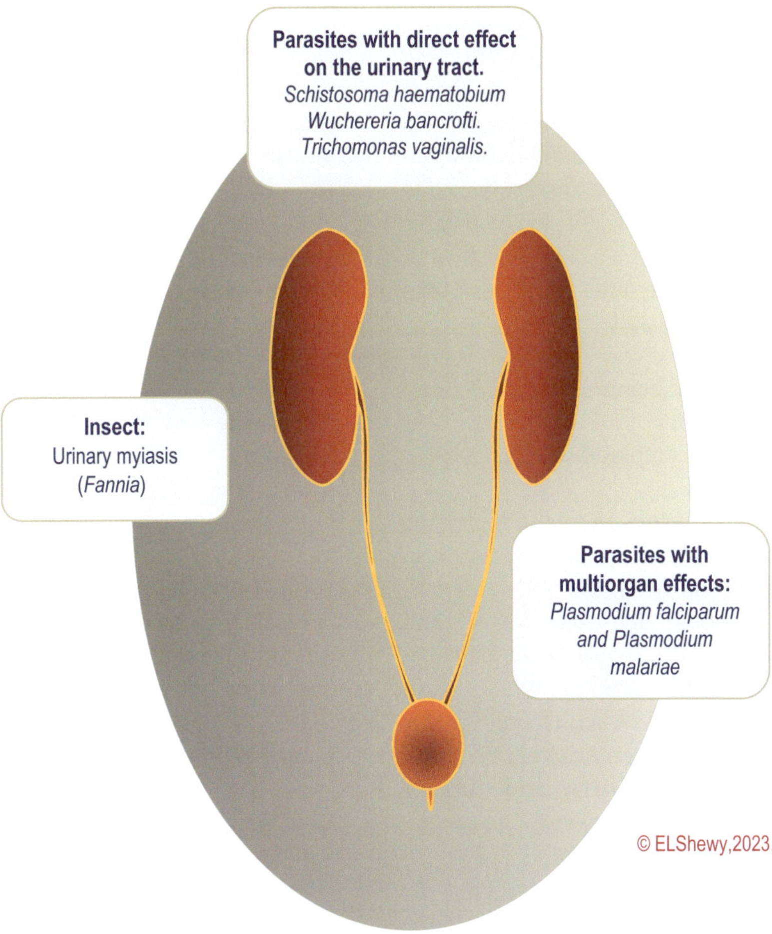

Parasites Infecting the Urinary Tract

K. ElShewy, *Medical Parasitology*, https://doi.org/10.1007/978-3-031-64736-9_9

Test your learning and check your understanding of this book's contents: use the "Springer Nature Flashcards" app to access questions using ▶ https://sn.pub/oyFLIx. To use the app, please follow the instructions in Chap. 1.

Introductory

Helminths and, to a lesser extent, protozoan parasites could affect different organs of the urinary tract. These parasites could reach the body through different modes of infection, e.g., penetration of the skin (*Schistosoma haematobium),* a bite of vector (*Wuchereria bancrofti and Plasmodium),* or through direct contact with an infected person (*Trichomonas vaginalis*). In many parts of the world, particularly underdeveloped countries, these parasites are recognized to be endemic with severe health and economic problems.

Parasites of the urinary tract could be responsible for the following manifestations:

1. **Terminal and total hematuria**: e.g., *Schistosoma haematobium.*
2. **Cystitis and cancer bladder**: e.g., *Schistosoma haematobium.*
3. **Dysuria and frequency of micturition**: e.g., *Schistosoma haematobium, Trichomonas vaginalis,* and urogenital myiasis.
4. **Ureteric stenosis, hydronephrosis, and renal failure**: e.g., *Schistosoma haematobium.*
5. **Urethritis**: e.g., *Trichomonas vaginalis.*
6. **Chyluria**: e.g., *Wuchereria bancrofti.*
7. **Nephritis, hemoglobinuria, and tubular necrosis:** e.g., *P. falciparum and P. malariae.*
8. **Urinary myiasis:** e.g., *Fannia.*

Parasites of the urinary tract could be one of the following:

1. **Parasites directly infect the urinary tract:** e.g., *Schistosoma haematobium* and *Trichomonas vaginalis.*
2. **Parasites with multisystem effects:** e.g., *Wuchereria bancrofti, P. falciparum,* and *P. malariae.*
3. **Insects:** e.g., *Fannia.*

In this chapter, we discuss the most common parasites infecting the urinary tract; other related parasites are mentioned in other chapters.

Schistosoma haematobium (*S. haematobium*)

Urinary Bilharziasis

S. haematobium is a blood trematode inhabiting the vesical plexuses of veins. *S. haematobium* infection results in urinary bilharziasis involving the urinary bladder, ureter, kidney, and genital organs.

Geographical Distribution

S. haematobium is mainly endemic in:

1. Africa (Egypt, Sudan, Zambia, Nigeria, and South Africa).
2. The Middle East (Saudi Arabia, Iraq, Jordan, and Iran).
3. Asia, in some areas of India.

Parasite Morphology

Schistosoma is a heterophroditic (separate sex) trematode. The following are the morphological criteria of the adult worm and its characteristic egg (Fig. 9.1):

S. haematobium:

- Adult worm (♂):
  - It has a cylindrical body with a gynecological canal.
  - 10–15 mm length × 1 mm width.
  - Its body has small tubercles with fine spines.
  - Intestinal ceca reunite at the posterior 1/3 of the body.
  - It has 4–5 large testes behind the ventral sucker.

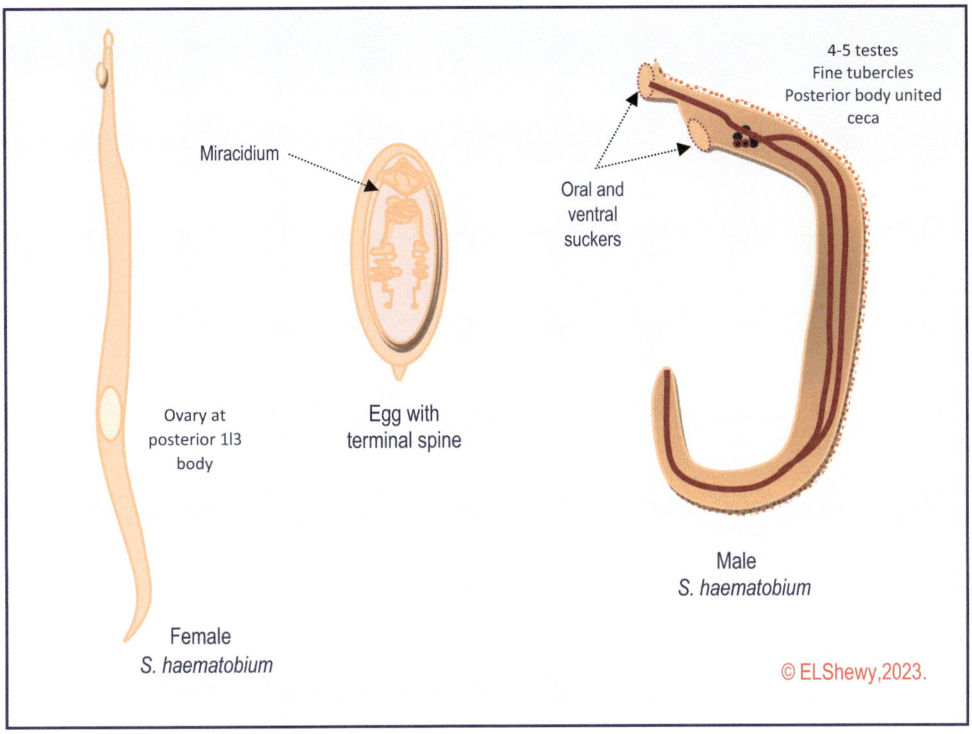

Fig. 9.1 Morphology of *Schistosoma haematobium*

- Adult worm (♀):
  - It has a rounded body with pointed ends.
  - 20 mm length × 0.5 mm width.
  - Intestinal ceca reunite in the posterior 1/3.
  - It has a short uterus with 20–100 eggs inside.
- Egg:
  - Oval to spindle in shape.
  - 140–50 μm.
  - It has a terminal spine.
  - It contains miracidium.

N.B. Cercaria is the infective stage for all schistosomes. It has a head with a body and a forked tail.

Habitat

S. haematobium inhabits the vesical (urinary) and pelvic venous plexus of veins.

Transmission

Schistosomes are transmitted through penetration of the host skin by the infective forked tail cercariae.

Life Cycle

The life cycle of the *Schistosoma haematobium* is almost similar to that mentioned with *S. mansoni* (see Fig. 8.2). However, the following are some differences:

- The snail's intermediate host is *Bulinus truncatus*.
- At the very narrow vessels, the male *Schistosoma haematobium* leaves its female to continue and to lay eggs at submucosal blood vessels of the urinary bladder.

Final hosts:	Humans.
Infective stage:	Cercaria.
Diagnostic stage:	Eggs with terminal spine in urine.
Intermediate hosts:	Freshwater snails: *Bulinus truncatus*.

Pathogenesis

In urinary bilharziasis, the pathological changes are mainly immunologic in origin. These changes could be postulated as follows (Figs. 9.2 and 9.3):

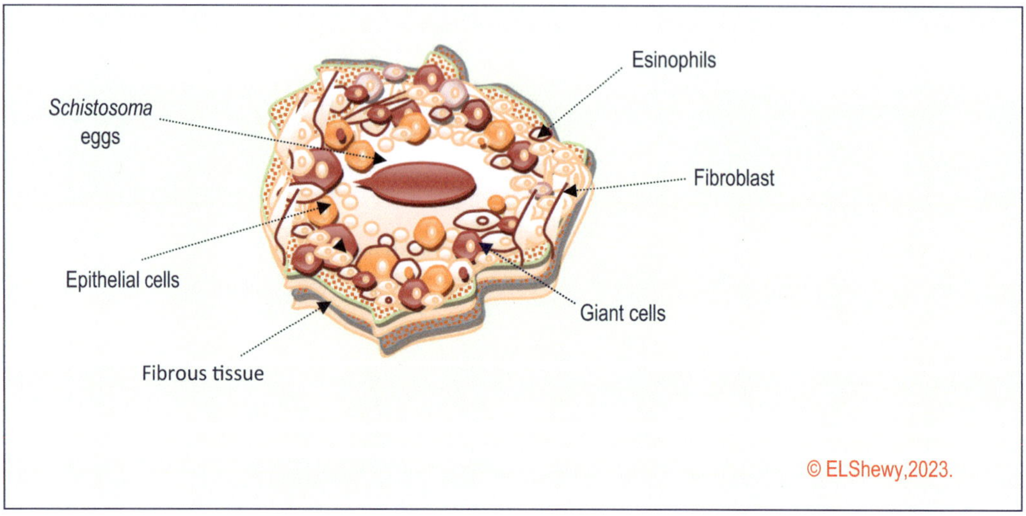

Fig. 9.2 Granuloma formation in urinary schistosomiasis

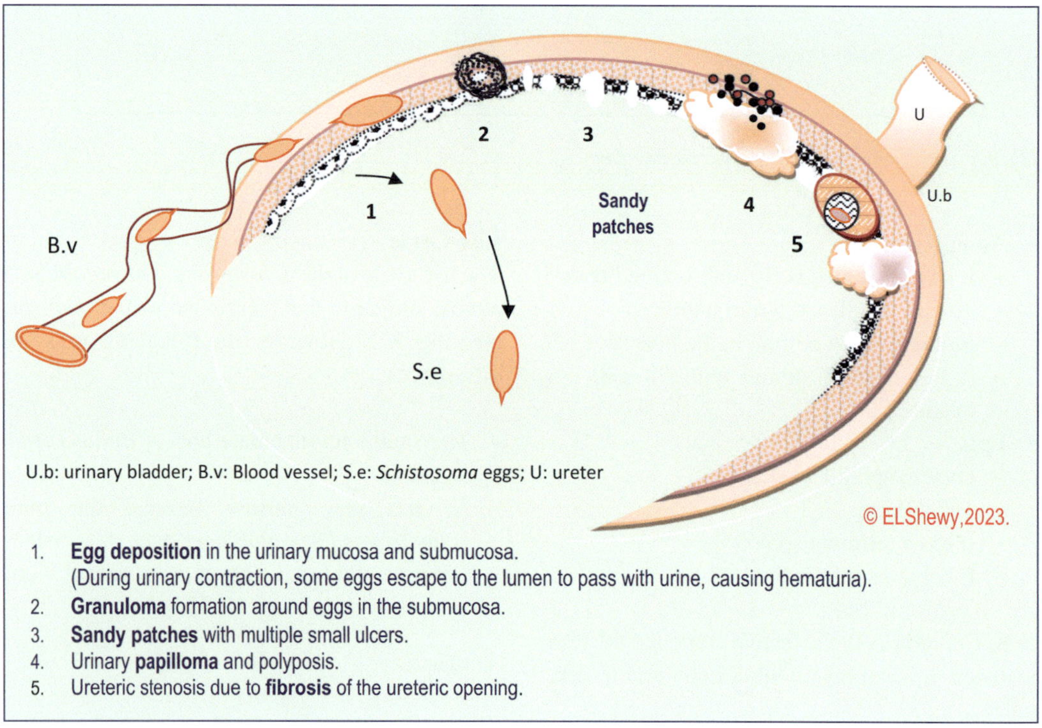

Fig. 9.3 Pathogenesis of urinary schistosomiasis

1. **Penetration** of the host skin by cercaria stimulates local inflammation with the formation of an itchy nodule (ground itch).
2. **Migration** of the schistosomulae in the circulation stimulates minimal host response.
3. **Egg deposition** in the wall of the urinary bladder stimulates local inflammatory reaction with a laceration that gives rise to hematuria at the early stages of infection. Later on, a granuloma will be formed with more inflam-

matory cell infiltration of the mucosa, leading to its hypertrophy, papilloma, and **polyps' formation**, but without calcification.
4. **Sandy patches** with tiny tubercles, thickened walls, and repeated infections could be seen in chronic urinary schistosomiasis, which has a high risk of malignancy.
5. Fibrosis of the ureteric opening leads to obstruction with backpressure on the kidneys (**hydronephrosis**).
6. The urethra, prostate, seminal vesicles, or even the spermatic cord may be involved in heavy *Schistosoma* infection.

Clinical Picture

Urinary Bilharziasis:

1. Penetration of the host skin with the infective cercaria results in petechial hemorrhage with pruritus and local edema, which last a few days.
2. In the succeeding 3 weeks after penetration, some allergic manifestations could be suffered (fever, malaise, headache, dyspnea, and cough).
3. Dysuria, frequency, and terminal hematuria (drops of blood at the end of micturition) are early manifestations.
4. Terminal hematuria is usually associated with eosinophiluria, together with chronic bacteriuria.
5. Eggs are detected in the patient's urine in this acute stage.
6. Recurrent urinary tract infections.
7. In chronic heavy infection and with fibrosis, there is an obstructive uropathy.
8. Eggs could not be detected during this stage because of the fibrosis and chronicity (closed infection).
9. Urinary schistosomiasis could be finally associated with cancer bladder.

N.B. Hepatosplenomegaly could be seen in urinary schistosomiasis but less frequently than in *S. mansoni.*

Diagnostic Procedures

Diagnosis of bilharziasis depends on the clinical picture, especially in endemic localities, together with the following laboratory investigations:

- **Direct:**
  1. Microscopic urine examination using direct and concentration techniques. These methods help detect the characteristic eggs during acute infection.
  2. Cystoscopy and vesical biopsy followed by microscopic examination are useful during chronic **closed infection**.
  3. Egg hatching tests are used to detect the **viability** of the egg and hence **active infection**; also, they are useful as a **test of cure**.
  4. Nucleopore for urine samples helps in the diagnosis and estimation of the intensity of urinary bilharziasis.
- **Indirect immunological:**
  1. Detection of specific antibodies for diagnosis of schistosomiasis could be used; however, serology could not differentiate between recent and old infections. It is usually used in the screening of individuals in endemic areas. ELISA, IHA, IFA, and RIA are commonly used in practice.
  2. Detection of the specific schistosomal antigens is more reliable in the diagnosis of bilharzial infection. For example, dipsticks are developed for the detection of specific antigens in the urine samples, particularly in survey studies.
  3. Circumoval precipitin test, using the suspected patient's serum and viable eggs.
  4. Cercarial Hulin reaction, using the suspected patient's serum and live cercaria.
  5. Detection of interleukins and other immune mediators is useful in the estimation of the immunopathologic changes during infection.
- **Other investigations:**
  Abdominal sonography, barium enema, and plain X-ray are used for detection of the pathological effects and complications of schistosomiasis.

N.B.

- Detection of the eggs in the urine and stool must be done after maturity of the worms (6–8 weeks from the start of infection).
- The viability of the eggs could also be detected by monitoring the active motility of the flame

cells in the miracidium, the presence of RBCs around viable eggs, and their translucency.
- Occasionally, eggs of *S. mansoni* could be detected as a contaminant in the urine.

Treatment
The same as in *S. mansoni.*

Prevention and Control
The same as in *S. mansoni*.

- **Parasitology:** *S. Mansoni* is a blood trematode fluke that needs a snail as an intermediate host. It is transmitted through penetration of human skin by the cercaria, maturation of the worms in the liver, and eggs with spines deposited in the mucosa of urinary bladder to pass with the patient's urine.
- **Clinical:** Urinary bilharziasis: Hematuria, sandy patches, polyposis, and fibrosis. Urinary carcinoma could be a complication of chronic infection.

Urogenital Myiasis

- Urogenital myiasis involves infestation of the urinary bladder and genital organs such as vaginal or penile orifices.
- Larvae of *Fannia* are the most common fly larvae infecting the urogenital system.
- Urogenital myiasis is considered an accidental infestation.
- The condition is rare in developed countries but relatively more common in individuals with low socioeconomic and poor hygienic conditions in developing countries.
- Factors leading to urogenital myiasis include:
  (a) Psychiatric disorders.
  (b) Use of urinary catheters under septic conditions.
  (c) Infected uterine prolapse.
  (d) Abortion and during menstruation with bad sanitary conditions.
- Prevention of urogenital myiasis involves dressing and covering open wounds, maintaining proper personal hygiene, and control of fly populations.

Renal Involvement in Malaria

Plasmodium parasite is a well-known parasite that affects kidney and renal functions. All *Plasmodia* species could share in these effects; however, *P. malariae* and *falciparum* are particularly involved **(see chap. 12).**

A. **Factors Contributing to Renal Complications During Malaria:**
   1. Severe pyrexia, vomiting, and dehydration.
   2. Cytoadherence of parasitized RBCs with other normal cells, platelets, and capillary endothelium. Formation of clumps with decreased microcirculation.
   3. Disturbance of hemodynamics, including hypovolemia and shock.
   4. Endothelial irritation by clumps leads to the production of many cytokines and inflammatory mediators.
   5. Immune complex precipitation on glomerular membranes during *P. malariae.*
   6. Hemolysis and hemoglobinuria.
   7. Liver involvement, including hyperbilirubinemia and cast formation (hepatorenal syndrome).

B. **Renal Complications During Malaria:**
   1. Acute renal injury.
   2. Acute tubular necrosis.
   3. Glomerulonephritis.
   4. Interstitial nephritis.
   5. Chronic renal insufficiency.
   6. Irreversible renal failure.

C. **Management of Malaria Induced Renal Complications:**
   1. Antimalaria treatment.
   2. Fluids and electrolytes replacement.
   3. Renal dialysis (should be done early).

Parasites Affecting Human Urine

A. **Parasites Changing Color of Urine:**
   1. **Red urine:**
      Schistosoma haematobium **(see above).**

2. **Black urine:**
   Plasmodium falciparum causes massive RBC destruction with the appearance of hemoglobin in the blood (hemoglobinemia), which passes through the kidney, forming hemoglobinuria and giving black urine on micturition. It is commonly seen during paroxysm of the *P. falciparum* (black water fever).
3. **White urine(chyluria):**
   Infection with *Wuchereria bancrofti* results in urinary lymphatics varices that could be ruptured. Ruptured lymphatic varices in the urinary bladder result in the passage of white urine **(chyluria).**

B. **Parasites Commonly Found in Urine:**
1. ***Schistosoma haematobium:***
   Ova of *S. haematobium* are frequently detected during active urinary bilharziasis. They are characterized by terminal spines. However, *S. mansoni* eggs could also be found in cases of mixed urinary and intestinal infections.
2. ***Trichomonas vaginalis:***
   - Trophozoites of *T. vaginalis* are usually contaminants from vaginal infection, and they are accompanied by an increased number of WBCs.
   - Trophozoites of *T. vaginalis* are highly motile due to an undulating membrane and multiple anterior flagellae; however, movement may deteriorate after 30 min from collection.
   - Confirmation of motility is essential as nonmotile organisms can resemble WBCs, transitional or renal epithelial cells. For this reason, it is essential to examine a freshly collected urine sample.
3. ***Enterobius vermicularis:***
   - Although *E. vermicularis* (pinworm) is not a primary urine parasite; however, its eggs are usually present in the urine as a fecal **contaminant**.
   - Eggs are ovoid with one side flattened. Embryos (larvae) may be visible within the ova.
4. ***Wuchereria bancrofti:***
   - Larvae of *W. bancrofti* could occasionally be detected in the urine. Ruptured lymphatic varices in the urinary bladder result in the passage of white urine as well as larvae.

C. **Parasites Affecting Micturition:**
1. ***Trichomonas vaginalis:***
   - Frequency and itching during micturition in both male and female patients.
2. ***Enterobius vermicularis:***
   - Nocturnal enuresis, especially in children, due to perianal itching and insomnia.
3. ***Schistosoma haematobium:***
   - Frequency dysuria and anuria in late and chronic cases.

Review Questions

1. **Bilharziasis is transmitted through**
   (a) adult worm
   (b) cercaria
   (c) snail
   (d) eggs
2. **Red urine could be due to infection with**
   (a) *Trichuris trichiura*
   (b) *Ascaris lumbricoides*
   (c) *Schistosoma haematobium*
   (d) *Echinococcus granulosus*
3. **White urine (chyluria) is caused by**
   (a) *Wuchereria bancrofti*
   (b) *Schistosoma mansoni*
   (c) *Schistosoma haematobium*
   (d) *Plasmodium falciparum*
4. **Which of the following is not true about *Schistosoma haematobium*?**
   (a) Infection is transmitted through cercaria.
   (b) *Biomphalaria* is the intermediate snail host.
   (c) Terminal hematuria is commonly seen.
   (d) Hydroureter and hydronephrosis could be seen in the chronic stage.
5. **An identifying characteristic of the egg of *Schistosoma* species is the presence of.**
   (a) a hexacanth embryo
   (b) polar filaments
   (c) bipolar plugs
   (d) spines

6. **Cancer bladder is a possible complication of**
   (a) *Heterophyes heterophyes*
   (b) *Schistosoma mansoni*
   (c) *Schistosoma haematobium*
   (d) *Fasciola hepatica*
7. **Adult worms of *Schistosoma haematobium* could be found in the**
   (a) superior mesenteric veins
   (b) renal artery
   (c) inferior mesenteric veins
   (d) vesical venous circulation
8. **All of the following parasites could be detected in the urine except**
   (a) *Trichomonas vaginalis*
   (b) *Plasmodium malariae*
   (c) *Schistosoma haematobium*
   (d) *Enterobius vermicularis*
9. **Urogenital myiasis is commonly caused by larvae of**
   (a) *Sarcophaga*
   (b) *Fannia*
   (c) *Phlebotomus*
   (d) *Simulium*

10 Neuroparasites

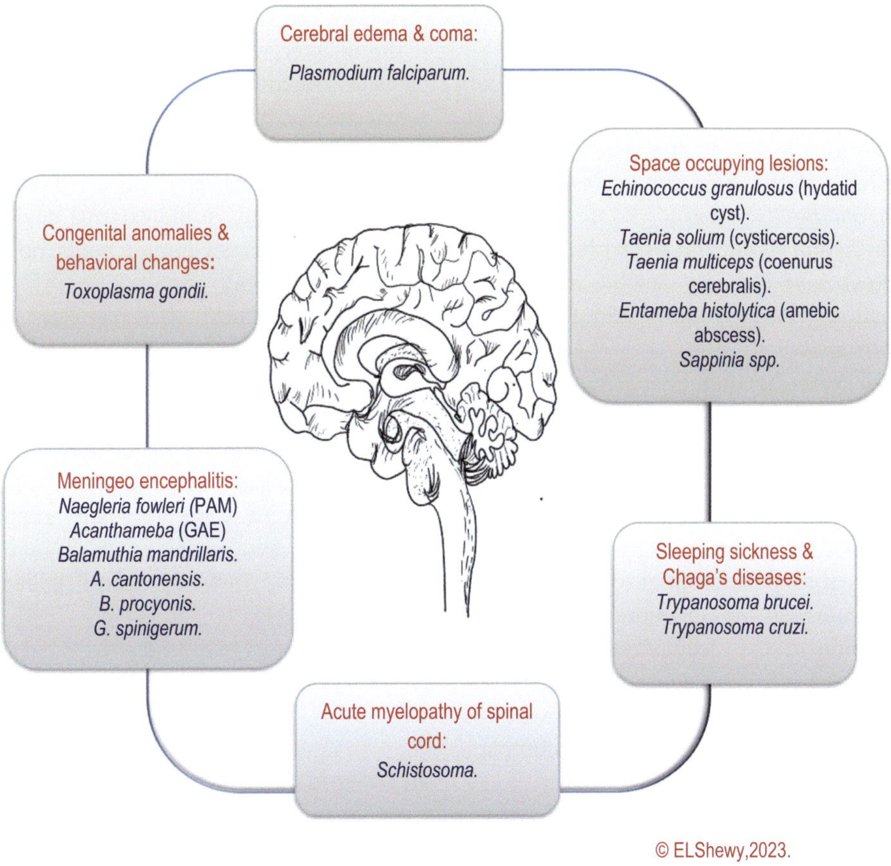

Neuroparasites

K. ElShewy, *Medical Parasitology*, https://doi.org/10.1007/978-3-031-64736-9_10

Test your learning and check your understanding of this book's contents: use the "Springer Nature Flashcards" app to access questions using ► https://sn.pub/oyFLIx. To use the app, please follow the instructions in Chap. 1.

Introductory

Neuroparasites are parasites that affect the nervous system either directly or indirectly. They could be responsible for very serious or even life-threatening diseases. Neuroparasites can produce a variety of symptoms and signs:

1. **Meningitis and meningoencephalitis:** e.g., *Naegleria fowleri, Acanthamoeba, Balamuthia mandrillaris, Angiostrongylus, Gnathostoma, and Baylisascaris.*
2. **Brain hemorrhage and necrosis:** e.g., *Naegleria fowleri* (primary amebic meningoencephalitis), *Acanthamoeba, and Balamuthia mandrillaris.*
3. **Spinal cord affection**: e.g., *Schistosoma.*
4. **Cerebral edema and coma**: e.g., *Plasmodium falciparum* (cerebral malaria).
5. **Space occupying lesion and pressure atrophy**: e.g., *Taenia solium* (neurocysticercosis), *Echinococcus granulosus* (hydatid cysts), *E. histolytica* (amebic brain abscess), *Sappinia,* and *Taenia multiceps* (coenurus cerebralis).
6. **Alteration of the sleep rhythm**: e.g., *Trypanosoma* (sleeping sickness).
7. **Sensory changes, paresthesia, anxiety, and behavioral changes:** e.g., most of neuroparasites.
8. **Motor changes:** paresis, tremors, and convulsions, e.g., most neuroparasites.

- Neuroparasites could be one of the following:
  1. **Parasites inhabiting CNS:** e.g., free-living amebae (*Naegleria fowleri, Acanthamoeba, and Balamuthia mandrillaris)* **(Chaps. 10 and 17)**.
  2. **Parasites carried to CNS:** e.g., *Taenia solium* (neurocysticercosis) **(Chap. 3), ***Echinococcus granulosus* (hydatid cysts) **(Chap. 8),** *E. histolytica* (Amebic brain abscess) **(Chap. 6),** *Taenia multiceps* (coenurus cerebralis) and *Schistosoma.*
  3. **Parasites with multiorgan effects:** e.g., *Trypanosoma, Plasmodium, Toxoplasma* **(Chap. 14),** *Toxocara* (VL migrans) **(Chap. 13)**, *Angiostrongylus, Gnathostoma,* and *Baylisascaris* **(Chap. 17).**

Free-Living Amebae (FLA)

(*Terramebae*, Soil *Amebae*, Coprozoic *Amebae*)

In this chapter, we will discuss the most common parasites affecting the nervous system; other neuroparasites are mentioned in other chapters.

- Free-living amebae (FLA) are those amebae that exist freely in the external environment, e.g., in bounds, swimming pools, hot springs, and in moist soil.
- FLA are also known as *Terramebae*, soil amebae, or coprozoic amebae.
- FLA could be transmitted from the external environment to infect humans; in some instances, they act as opportunistic dangerous parasites causing serious and fatal diseases.
- Four main FLA are clinically important to humans:
  1. *Naegleria fowleri* (amebo-flagellated).
  2. *Acanthamoeba.*
  3. *Balamuthia mandrillaris.*
  4. *Sappinia* spp. (Chap. 17).

A. *Naegleria fowleri* (*N. fowleri*)

(Brain-Eating Ameba)

N. fowleri is the causative agent of primary amebic meningoencephalitis (PAM). It is an acute fatal disease transmitted through nasal inhalation of contaminated water.

Geographical Distribution

- There are more than 30 *Naegleria* species, and only *fowleri is* pathogenic.

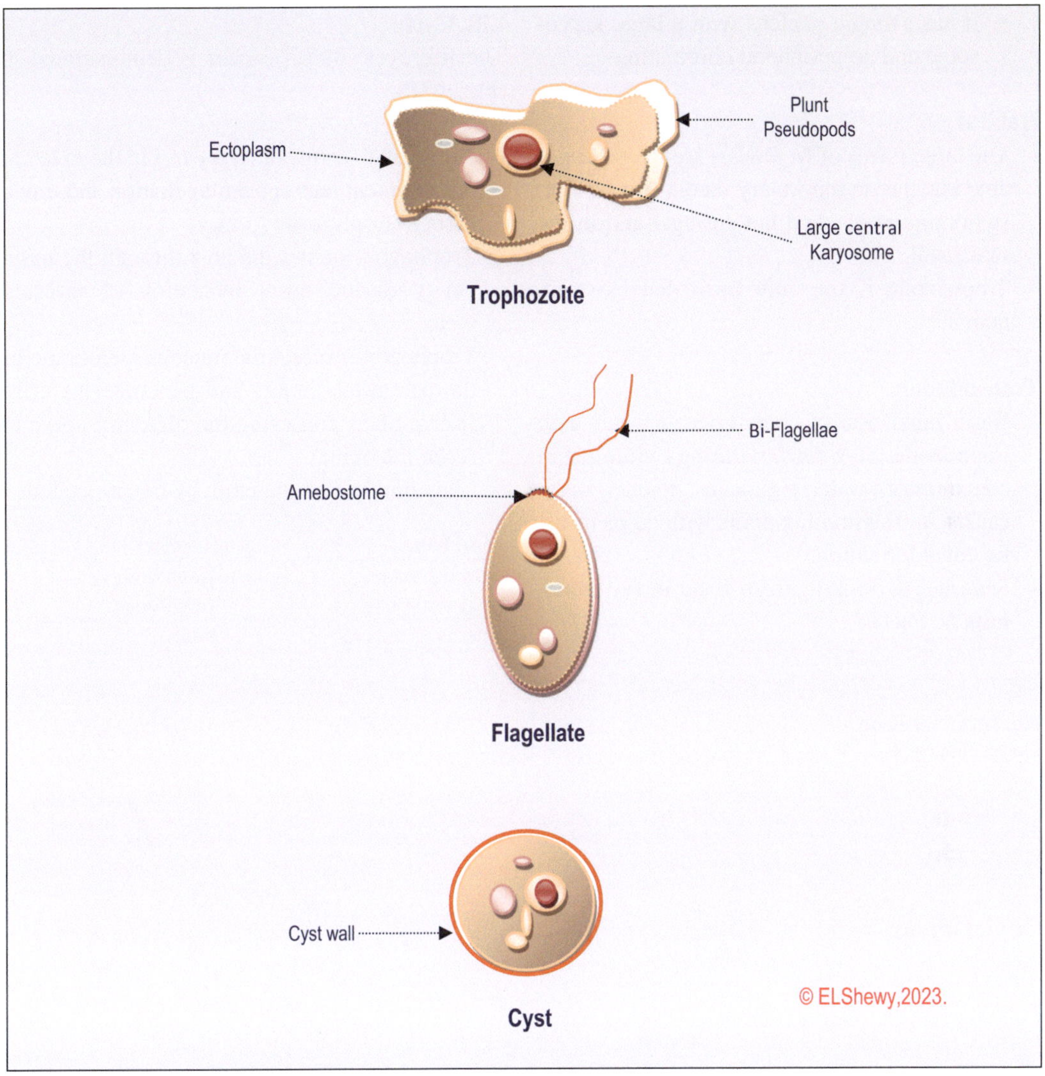

Fig. 10.1 *Naegleria fowleri*: morphology

- *N. fowleri* infection is uncommon and sporadic cases have been reported (may be due to unawareness of the parasite), with a high prevalence in warmer months.
- *N. Fowleri* is thermophilic (up to 45°C).
- It is found in swimming pools, soil, sewage, hot springs, irrigation canals, water tanks, lakes, rivers, and water wells. However, it is never found in seawater.

Parasite Morphology

N. fowleri could be detected in three interchangeable morphological forms: trophozoite flagellate form and the cyst form (Fig. 10.1):

- Trophozoite*:*
  - 10–35 µm.
  - Ameboid shape with blunted pseudopods.
  - It has a single nucleus with a large karyosome and no peripheral chromatin.
- Flagellate form:
  - 7–15 µm.
  - Pear-shaped, biflagellate.
  - It has amebostome.
  - It has a single nucleus with a large karyosome and no peripheral chromatin.
- Cyst form:
  - 7–10 µm.
  - Rounded shape with a smooth surface.

- It has a single nucleus with a large karyosome and no peripheral chromatin.

Habitat

- The three forms of *N. fowleri* are interchangeable together in the watery media (freshwater, swimming pools, and hot springs) and in the moist soil.
- Trophozoite is the only form that exists in humans.

Transmission

- Nasal inhalation of contaminated water with trophozoites of *N. fowleri* during swimming in contaminated water, e.g., lakes, bounds, water canals, and swimming pools with no or insufficient chlorination.
- Seawater or drinking fresh water **never transmits** *N. fowleri*.

Life Cycle

The life cycle of *N. fowleri* is demonstrated in (Fig. 10.2).

- Biflagellate forms of *N. fowleri* in the external environment undergo enflagellation and transform to trophozoite form.
- Trophozoites enter the host through the nasal cavity during nasal inhalation of infected water.
- Trophozoites infect the mucous membrane of the paranasal sinuses and penetrate the cribriform plate, following the olfactory nerve to reach the brain.
- They multiply in the brain by binary fission.

Final hosts:	Humans.
Infective stage:	Trophozoite.
Diagnostic stage:	Trophozoite.
Intermediate hosts:	None.

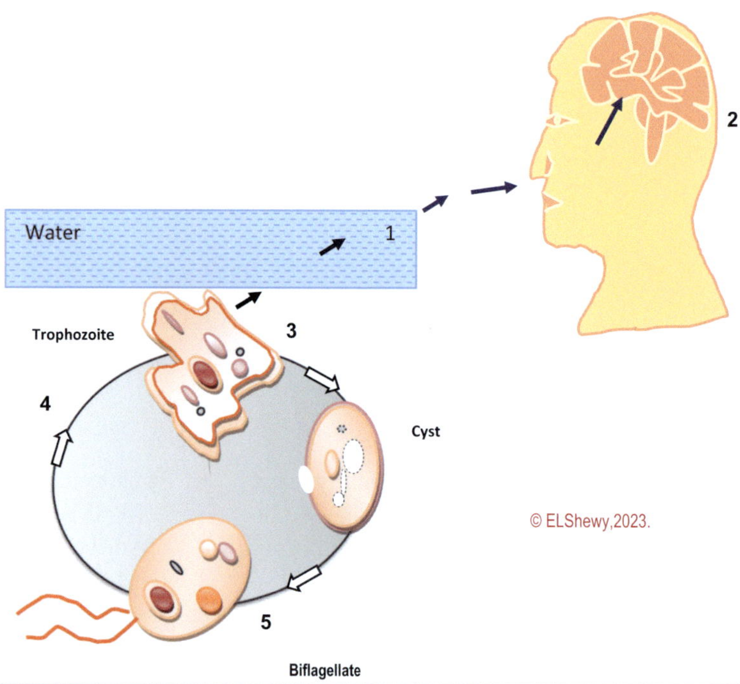

Fig. 10.2 *Naegleria fowleri*: life cycle

1. Intranasal infiltration of *Naegleria* trophozoite from the water to the cribriform plate, then to the brain.
2. Only trophozoite inhabits the brain and CSF.
3. Trophozoite encysts in water.
4. Trophozoite enflagellates in water.
5. Both cyst and flagellate forms transform into trophozoite.

Pathogenesis

Pathological changes leading to primary amebic meningoencephalitis (PAM) are postulated as follows:

1. Once trophozoites reach the meninges, they start an acute inflammatory process leading to acute meningitis.
2. From meninges towards brain tissues, trophozoites stimulate severe inflammatory cellular reactions with marked exudates extending to the grey matter.
3. Many monocytes and neutrophils with hemorrhage and necrosis are usually supervening.
4. Focal demyelination of the white matter by enzyme-like substances or phospholytic enzymes secreted by trophozoites leads to acute encephalitis.

Clinical Picture

Primary amebic meningoencephalitis (PAM) is characterized by:

1. A history of contact with water shortly before the start of manifestations.
2. Acute onset with a day of prodromal symptoms of high fever and headache, followed by a rapid onset of nausea and vomiting.
3. Signs of meningeal irritation (neck stiffness, seizures with **Kernig's** sign).
4. Taste and/or smell changes (affection of the olfactory lobe).
5. Death occurs early, within 3–6 days.

Diagnostic Procedures

CSF sample is taken by lumbar puncture to be examined as follows:

1. **Microscopic** examination of the wet mount and stained CSF smears for detection of the characteristic moving trophozoites (single nucleus, large pig karyosome, and no peripheral chromatin).
2. **Culture** of the CSF on low/non-nutrient agar seeded with *E. coli* at 37°C–42 °C, under aerobic conditions, and examined daily for the presence of trophozoite.
3. **Flagellation test:** demonstration of the trophozoite transformation to flagellate form (adding water to the CSF or the tissue specimen for 2–20 h at 37 °C, trophozoite of *N. fowleri* starts to flagellate).
4. Identification of the **specific antibodies** using IFA, counter-immunoelectrophoresis (CIE), and enzyme-linked immunosorbent assay (ELISA).
5. **PCR** is usually performed for research works.

N.B

- CSF of PAM is cloudy, purulent or sanguinopurulent with increased WBCs, increased pressure and low bacterial cells.
- Demonstration of the trophozoite transformation to flagellate form is used to differentiate between the *Naegleria* and *Acanthamoeba*, which never flagellate.

Treatment

There is no effective treatment for *N. fowleri* because of the following:

- The rarity of the disease (lack of awareness).
- The short fulminate clinical course of PAM.
- Difficulties in making a rapid diagnosis.

However, the best drug regimen should include:

- Amebicidal (a combination of drugs capable of crossing the blood–brain barrier).
- Antibiotics and steroids.

1. **Amphotericin-B**:
   - It is given as 1 mg/kg/body weight, once daily intravenously for 3 days.
   - It is followed by another 6-day course, at 1 mg/kg/daily by intrathecal administration.
2. **Miconazole:**
   - 117 mg/m^2/body surface × 3 times for 10 days intravenously.
   - It could be given intrathecally, as 10 mg daily for 2 days, and then 10 mg every other day through the lumbar catheter.

3. **Rifampicin:**
   - 3.3 mg/kg/body weight × 3 times daily × 9 days.
4. **Miltefosine (oral):**
   - An anticancer drug was found to be effective against N. fowleri.
   - Up to 45 kg body weight: 100 mg daily (i.e., one 50 mg cap./45 kg body weight and higher: 150 mg daily.

Prevention and Control
Controlling *N. fowleri* is not an easy matter because of its extensive distribution:

1. Conducting safety campaigns to educate the public.
2. Adequate chlorination of heavily used swimming pools.
3. Monitoring of recreational water.
4. *N. fowleri* should be considered by local public health authorities.

- **Parasitology**. *N. Fowleri* is a FLA that has trophozoite, cyst, and flagellated forms. All morphological forms are characterized by having a single nucleus with a large karyosome without peripheral chromatin dots—Infection with N. fowleri is transmitted from water by nasal inhalation to reach the human brain.
- **Clinical:** It causes acute and fatal primary amebic meningitis (PAM).

B. *Acanthamoeba* Species

Acanthamoeba species are free-living amebae that occasionally infect humans. Most of these species are sources of chronic granulomatous amebic encephalitis (**GAE**) and corneal infection (**amebic keratitis**).

Geographical Distribution
Acanthamoeba has a worldwide distribution in the soil (moist sands, flowerpot soil), air, sewage, humidifiers, air conditioners, dental and dialysis units, and it can be found in contact lenses.

Parasite Morphology
Acanthamoeba exhibits two morphological forms: trophozoite and cyst (Fig. 10.3).

- Trophozoite:
  - 15–45 μm.
  - The body has spiny, thorn-like projections.
  - It has a single nucleus with a large karyosome and no peripheral chromatin.
- Cyst:
  - 10–20 μm.
  - The body has a wrinkled outer wall with a polygonal inner wall.
  - It has a single nucleus with a large karyosome and no peripheral chromatin.

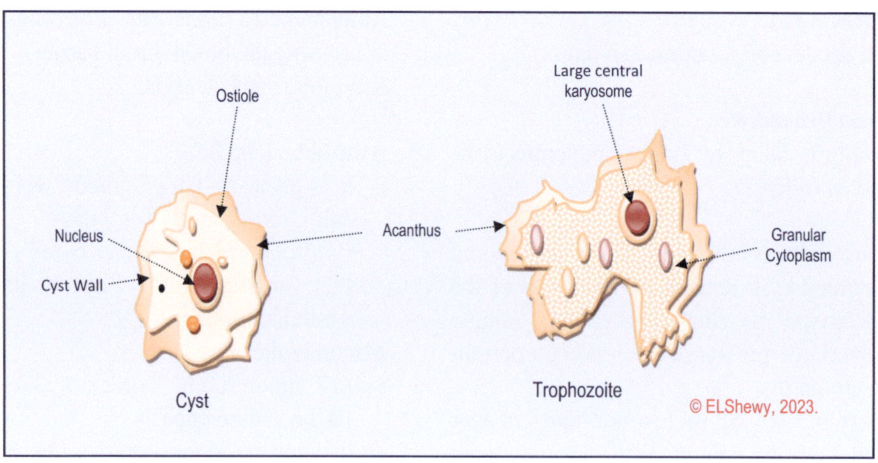

Fig. 10.3 *Acanthamoeba:* morphology

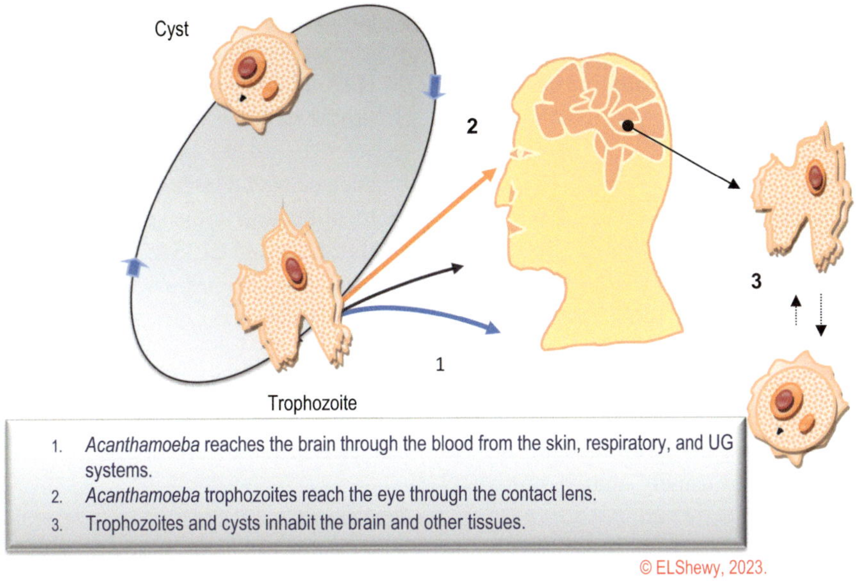

Fig. 10.4 *Acanthamoeba:* life cycle

Transmission

Acanthamoeba enters the body through the respiratory tract, skin abrasions, and/or the genitourinary tract. They reach the CNS from these tissues through the blood. Eye infection occurs through contact lenses, sterilizing solutions, and corneal abrasions.

Habitat

Trophozoite and cyst forms can be detected inside the infected tissues, e.g., respiratory, CNS, skin, and cornea.

Life Cycle

Trophozoites and cysts are interchangeable together in the external environment. Trophozoites reach the human body through skin abrasion/ulcers and the respiratory tract. Trophozoites reach the CNS through the bloodstream. In the human tissues, both trophozoite and cyst could be detected (Fig. 10.4).

Final hosts:	Humans.
Infective stage:	Trophozoite.
Diagnostic stage:	Trophozoite and cyst in infected fluids and tissues.

Pathogenesis

Pathogenesis of granulomatous amebic encephalomyelitis (GAE) and amebic keratitis:

1. *Acanthamoeba* reaches the brain through the blood from a skin wound and/or the respiratory tract.
2. Both trophozoites and cysts are found in the deeper tissues of the brain; then, they spread to the periphery to initiate an inflammatory reaction.
3. Slowly growing subacute or chronic granulomatous encephalitis develops at the brain tissues with many lymphocytes and eosinophils infiltration.
4. *Acanthamoeba,* when reaching the cornea, initiates subacute to chronic progressive ulcerative keratitis with scattered opacities followed by corneal ulceration.
5. Finally, a corneal ulceration that may progress to corneal perforation is established.

Clinical Picture

Granulomatous amebic encephalitis (GAE) and keratitis are the most common manifestations of *Acanthamoeba* infection. However, it may cause other tissue infections, particularly in immunocompromised individuals:

Chronic granulomatous amebic encephalomyelitis (GAE):

1. An irregular and generally low fever.
2. Headache sets in early and is insidious.
3. Nausea and vomiting.
4. Focal neurologic changes, hemiparesis, drowsiness, personality changes, and seizures are common early symptoms.
5. Diplopia, paresis, lethargy, and cerebellar ataxia.
6. The disease progresses throughout one to several weeks and usually ends in coma and death.

***Acanthamoeba* keratitis:**

1. Severe ocular pain with inflammation of the cornea and red eye.
2. Rapid deterioration of the visual acuity.
3. Deep tissue affection with the appearance of a ring shape.
4. Ulcerative keratitis is followed by perforation of the cornea and eventually blindness.
(See eye parasites for more details)

Diagnostic Procedures

1. **Microscopic** examination of CSF for detection of trophozoites and cysts.
2. **Culture** of CSF, as in *N. fowleri*, but with a low recovery rate.
3. **Histological examination** of the corneal or skin swabs (for demonstration of the characteristic thorny trophozoite that has a single nucleus with a large central karyosome).
4. Identification of the **specific antibodies** using IFA, CIE, and ELISA.
5. **PCR** and IFA are usually performed for the research works.

N.B. *N. fowleri* should be differentiated from *Acanthamoeba*; differences between them are tabulated in Table 10.1.

Treatment

As in *N. fowleri,* there is no satisfactory treatment for *Acanthamoeba*; however, the following are commonly used in treatment:

GAE:

1. Ketoconazole.
2. Penicillin and chloramphenicol.
3. Sulfamethazine.

Acanthamoeba keratitis:

1. Itraconazole (orally).
2. Topical application with otramidryl (ointment).
3. Brolene (drops).
4. Neomycin (drops).

Table 10.1 Differences between *N. fowleri* and *Acanthamoeba* spp

	Naegleria fowleri	*Acanthamoeba* spp.
Disease	Primary amebic meningoencephalitis (PAM).	Granulomatous amebic encephalitis (GAE) and amebic keratitis.
Entry	Nares, swimming, and diving.	Skin lesions, eye, and respiratory tract. Reach the brain through blood.
Pathology	Acute suppurative inflammation.	Granulomatous chronic inflammatory reaction.
Tissue form	Trophozoite.	Cyst mainly and with little trophozoite.
Trophozoite	10–35 μm blunted pseudopodia.	10–45 μm spiny, thorn-like projections.
Nucleus	Large big karyosome with no peripheral chromatin.	
Cyst	7–10 μm. Smooth surface.	10–20 μm. Wrinkled outer wall and polygonal inner wall.
Flagellate form	Present with pear shape biflagellate and amebostome.	Not present.
Culture	Frequently positive.	Usually, negative.
CSF	Contains many neutrophils.	Contains many lymphocytes.

Prevention and Control

1. Contact lens wearers should not use nonsterile homemade saline when cleaning their lenses.
2. Contact lens wearers should not wear the lenses while swimming.
3. When there are any skin lesions, do not swim in common pools.

- **Parasitology**: *Acanthamoeba* is FLA characterized by having a single nucleus with a large karyosome without peripheral chromatin dots. Trophozoite has thorny projections. The infection spreads from water, urogenital tract, skin, or from respiratory, through the blood to the brain. Contact lenses transmit it to the eye.
- **Clinical:** It causes subacute or chronic GAE and amebic keratitis.

C. *Balamuthia mandrillaris*

(*B. mandrillaris*)

Characteristics

- *B. mandrillaris* is one of the free-living amebae that could infect humans.
- It is an opportunistic parasite.
- Both *Acanthamoeba* and *Balamuthia mandrillaris* have trophozoite and cyst forms, but they never flagellate like *N. fowleri*.
- *B. mandrillaris* causes subacute or chronic amebic encephalomeningitis (GAE).
- *B. mandrillaris* has the same diagnosis, treatment, and preventive measures as in *Acanthamoeba*.
- *B. mandrillaris* enters the body when contaminated soil comes in contact with skin wounds and cuts or when dust-containing *Balamuthia* is breathed in or gets in the mouth.
- It reaches the brain through blood.
- Trophozoites of *B. mandrillaris* are recognized by the presence of long, slender, spine-like processes.

Parasite Morphology

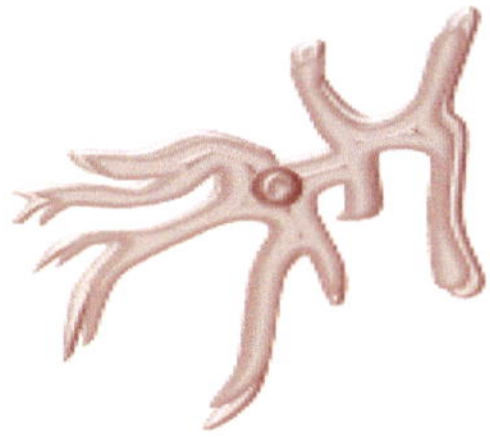

B. mandrillaris trophozoite

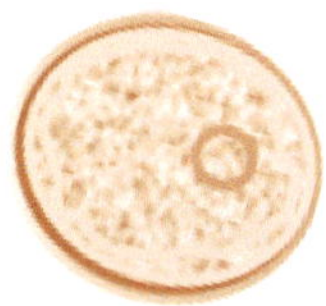

B. mandrillaris Cyst

- Trophozoite:
  - 25–40 µm, with many slender projections.
  - Finely granular cytoplasm.
  - It has a nucleus with a large central nucleolus.
- Cyst:
  - Three layers-walled spherical cysts.
  - It has no pores.
- 15–20 µm in diameter, a nucleus with a large central nucleolus.

Trypanosoma

T. brucei (*brucei, gambiense, rhodesiense*) & *T. cruzi*

- *Trypanosoma* is a flagellated protozoan parasite with a single flagellum.
- Two distinctly different species of *Trypanosoma* could infect humans.
- Each one of these two species has a different mode of transmission, geographical distribution, and different clinical presentations.

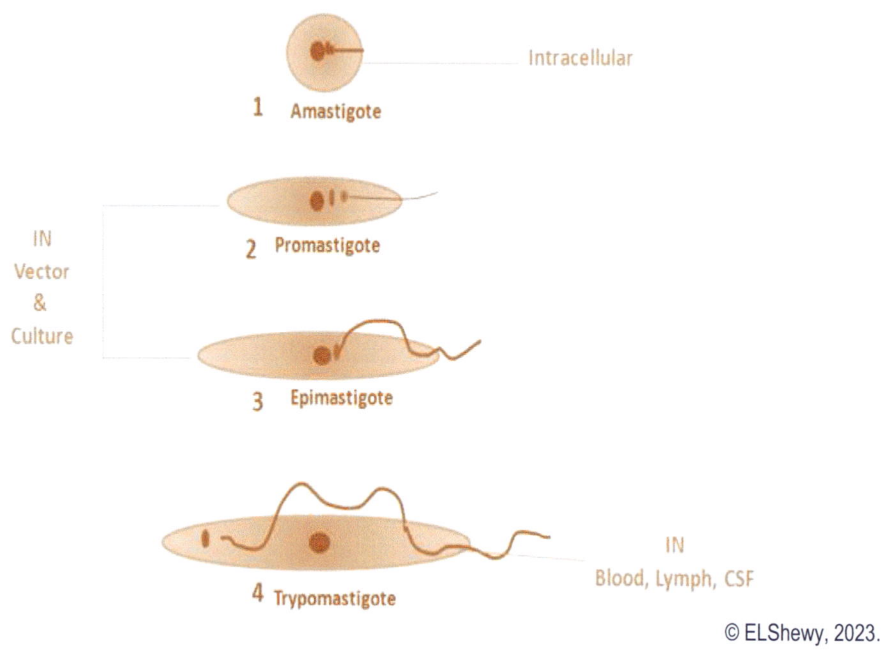

Fig. 10.5 Morphological forms of blood and tissue flagellates

Table 10.2 Differences between human trypanosomes

	Trypanosoma b. gambiense	*Trypanosoma b. rhodesiense*	*Trypanosoma cruzi*
Disease	Chronic African sleeping sickness. Late CNS affection.	Acute African sleeping sickness. Early CNS affection.	American sleeping sickness (Chagas disease). Late CNS.
Vector	*Glossina palpalis & tachinoides.*	*Glossina morsitans pallidipes.*	*Reduviid* bugs.
Transmission	Anterior station (saliva).	Anterior station (saliva)	Posterior station (feces).
Reservoir	Man and domestic animals.	Wild animals (true zoonotic).	Domestic and wild animals.
Distribution	West African regions.	East and central African regions.	Brazil, Argentina, Mexico, and South America.
Parasitic forms	Wavy trypomastigotes with more *pleomorphism in the blood.	Wavy trypomastigotes with less pleomorphism in the blood	Amastigotes in the tissues. S- and C-shaped trypomastigotes in the blood

[a]Pleomorphism: the trypomastigotes demonstrate different sizes and shapes in the blood (short, stumpy, slender, metacyclic)

- Both species could take one or more than one of the four morphological forms of the blood flagellates (amastigote, promastigote, epimastigote, and trypomastigote) (Fig. 10.5).
- Human trypanosomes include:
  1. ***T. brucei* (*T. brucei gambiense and rhodesiense*):** causes African type of sleeping sickness. It presents in the form of trypomastigote in the blood, lymph, and CSF.
  2. ***T. cruzi***: is responsible for American trypanosomiasis (**Chagas disease**). It presents in the blood, lymph, and CSF in the form of trypomastigote and in the tissues in the amastigote form. The main differences between trypanosomes are tabulated in Table 10.2.

A. *Trypanosoma brucei*

(*T. b. gambiense and T. b. rhodesiense*)

Human African Trypanosomiasis (HAT)

T. brucei includes two species *rhodesiense* and *gambiense*, both are responsible for the Human African Trypanosomiasis (HAT) or African sleeping sickness. The Gambian type is a chronic disease, while the Rhodesian is an acute fatal disease.

Geographical Distribution

1. *T. gambiense* prevails in West African areas (Congo, Zaire).
2. *T. rhodesiense* prevails in the East and Central African countries (Zimbabwe, Zambia, and Angola).

Parasite Morphology

1. Both *T. b. rhodesiense* and *T. b. gambiense* are present in the blood in wavy trypomastigotes form (Fig. 10.5 and 10.6).
2. *T. b. gambiense* is a highly pleomorphic organism.
3. Both species exhibit the epimastigote form in culture media and in vectors.
   - Epimastigote:
     - Long, slightly wide body.
     - 9–12 μm.
     - It has one nucleus situated at the posterior end.
     - Kinetoplast is located anterior to the nucleus, with half body undulating membrane.
     - It has one free anterior flagellum.
   - Trypomastigotes:
     - **Wavy shape found** in the blood, CSF, and lymph.
     - 12–35 μm.
     - One single nucleus anterior to the kinetoplast.
     - Kinetoplast gives rise to full-body length undulating membrane.
     - It has one free anterior flagellum.

Transmission

The following are the modes of transmission of African trypanosomiasis:

1. Vector bites, both male and female *Glossina* (Tsetse fly), can transmit infection.
2. Blood transfusion.
3. Organ transplant.
4. Congenital.

Habitat

- Wavy shape trypomastigotes inhabit blood, CSF, and lymph.
- Epimastigote form is detected in culture and the vector.

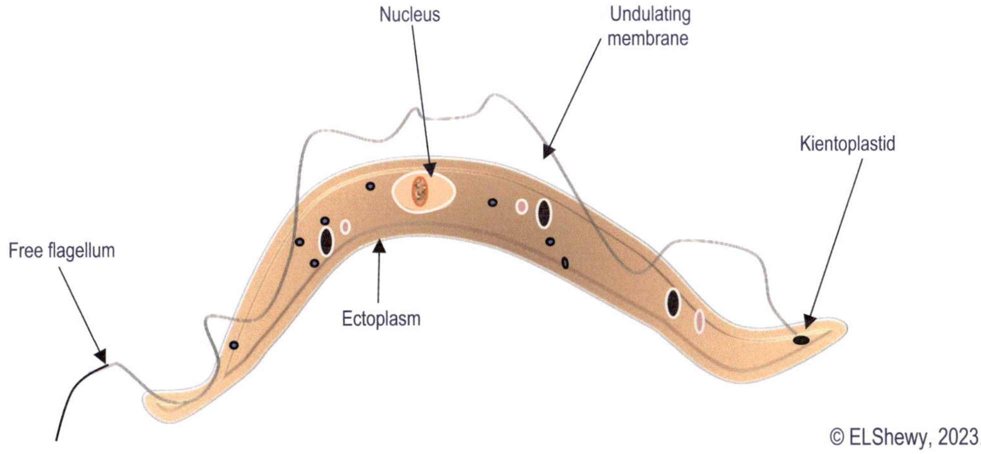

Fig. 10.6 Trypomastigote: morphology

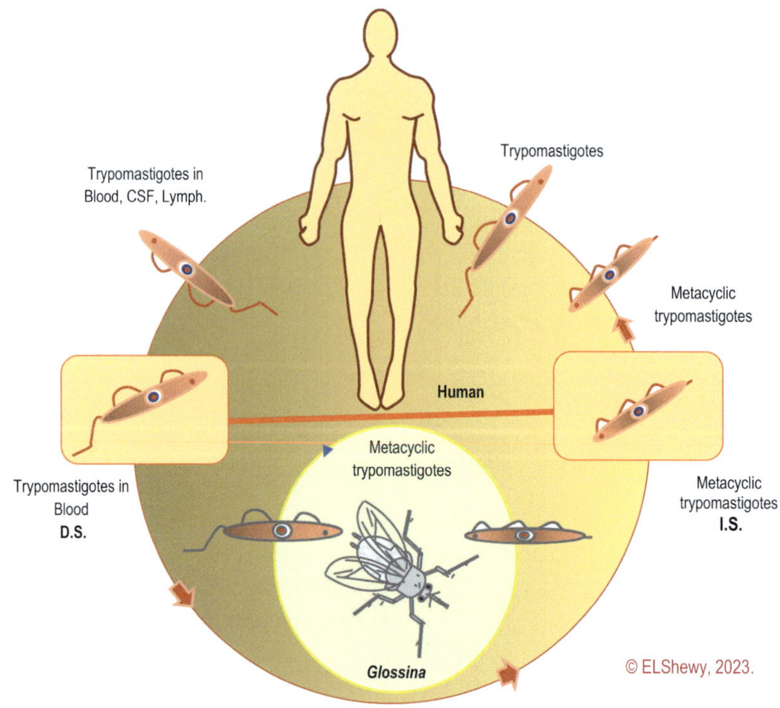

1. Metacyclic trypomastigote is injected into humans through *Glossina* bite.
2. In human metacyclic form, it transforms into a trypomastigote.
3. Trypomastigotes multiply in the blood, lymph, and CSF.
4. Trypomastigotes ingested by *Glossina*.
5. Trypomastigote transforms into epimastigote in *Glossina*.
6. Epimastigote transforms into infective metacyclic form in *Glossina* salivary gland.

I.S.= infective stage D.S. = diagnostic stage

Fig. 10.7 *Trypanosoma brucei*: life cycle

Life Cycle

Both species of *T. brucei* have the same life cycle (Fig. 10.7):

Transmission of the *T. brucei* from host to host occurs through the bite of an infected *Glossina* (Tsetse) fly.

In humans, *Glossina* (tsetse fly) bites and injects the infective stage (metacyclic trypomastigotes) onto human skin.

- The parasites multiply locally at the bite site, then migrate through the bloodstream and lymphatic to reach the adjacent lymph nodes (L.N.).
- Finally, they reach the central nervous system (CNS) through cerebrospinal fluid (CSF).

Glossina takes a blood meal from infected humans containing the trypomastigotes (D.S.), which move to the insect gut, where they undergo multiple developmental cycles. Then they migrate to the salivary gland, where they become infective metacyclic trypomastigotes (I.S.) (it takes 2–3 weeks).

Final hosts:	Humans.
Infective stage:	Metacyclic trypomastigotes.
Diagnostic stage:	Wavy shape trypomastigotes.
Intermediate host:	*Glossina* (*tsetse*) flies.
Reservoir hosts	– Wild animals (especially *T. Rhodesiense*). – Domestic animals (especially *T. Gambiense*).

Pathogenesis

– Pathologic changes of both *T. b. gambiense* and *T. b. rhodesiense* are the same. However, Gambian trypanosomiasis demonstrates a late CNS involvement (Fig. 10.8).

1. **At an early stage:** trypomastigotes are detected in large numbers in blood before they invade lymph nodes. The following blood changes are demonstrated during this stage:
   - Destruction of the RBCs with marked anemia.
   - Normal WBCs.
   - Increased lymphocytes.
   - Thrombocytopenia.
   - Disseminated intravascular coagulopathy (DI).
3. Invasion of lymph nodes causes inflammation and hypertrophy.
4. **At late stage:** invasion of the CNS with the parasites results in leptomeningitis with marked lymphocytes and plasma cell infiltration.
5. Increase in the protein level and cellular infiltration in the CSF.
6. Increase in the microglia and astrocyte cells with brain edema.
7. **Decreased immunity** with a possibility of secondary infection.

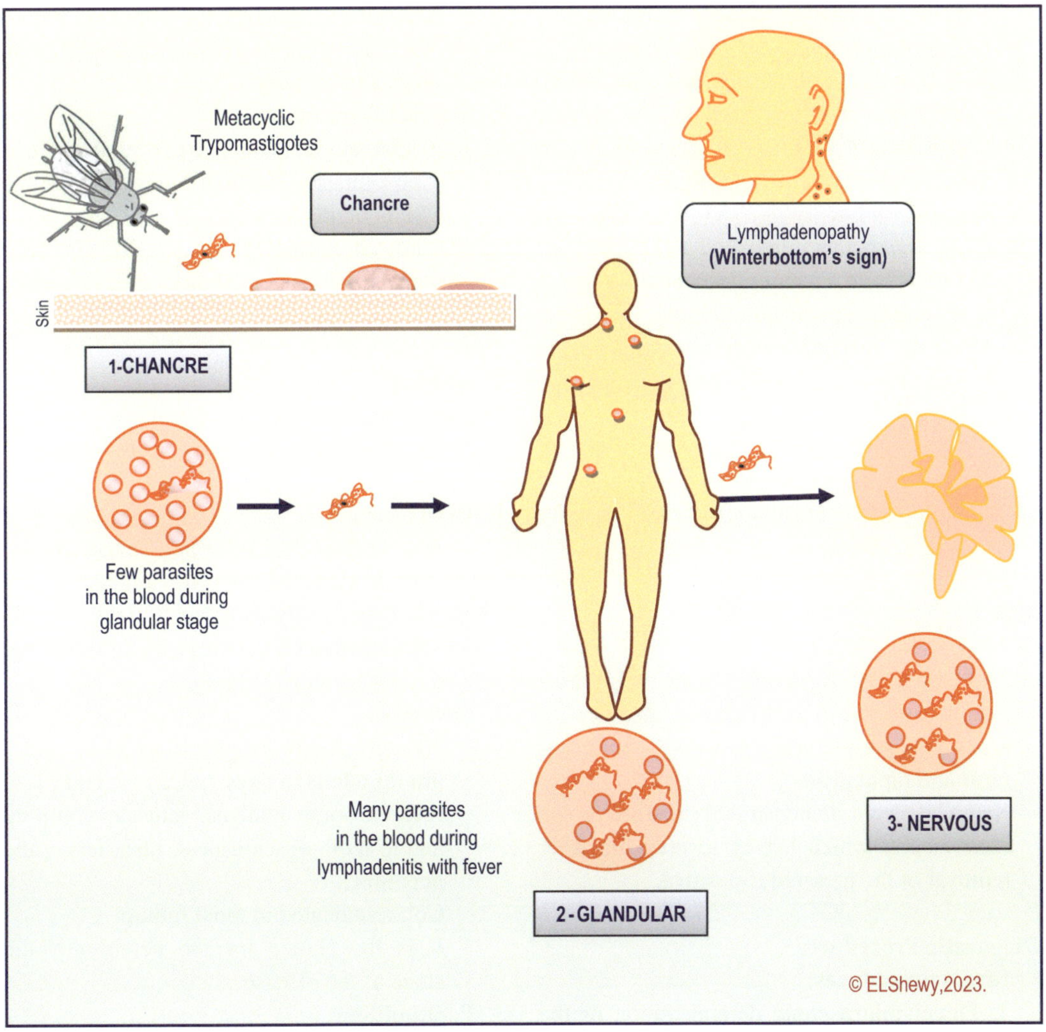

Fig. 10.8 Pathogenesis of human African Trypanosomiasis (HAT)

Clinical Picture

Clinical manifestations during Human African Trypanosomiasis (HAT) pass through the following successive stages:

(a) **Incubation Period:** It ranges from a few days to weeks after the tsetse bite.
(b) **Chancre Formation:**
   1. **Chancre** (a painful tender skin area) appears at the site of tsetse fly bite.
   2. Chancre slowly disappears.
   3. At the end of this stage, either the infection is aborted, or invasion of the lymphatic tissues supervenes.
(c) **Glandular Phase***:*
   1. Invasion of the L.N. is associated with fever, lymphadenitis, and increased trypanosomes in the blood (parasitemia).
   2. A few days later, fever and parasitemia disappear and lymphadenopathy appears, particularly in posterior cervical lymph nodes **(Winterbottom's sign).**
(d) **Nervous Phase:**
   1. Apathy and lassitude.
   2. Progressive meningoencephalitis with increasing apathy and fatigue.
   3. Lastly, marked muscle twitches, urine incontinence, convulsion, and somnolence.
   4. Sensory changes in the form of paresthesia, positive **Kerandel's sign** together with severe emaciation, which usually ends fatally.

N.B.

- The Rhodesian form of African trypanosomiasis is similar to the Gambian one but with a severe acute picture, early CNS manifestations and rapid death.
- Kerandel's sign demonstrated pressure on the ulnar nerve, which leaves severe pain after removal of the pressure if positive.

Diagnostic Procedures

- **Direct procedures:**
   1. Direct **microscopic** demonstration of the wavy **trypomastigotes** form of the parasite in samples stained with Leishman or Giemsa stain. Samples could be taken from blood, CSF, or lymph node aspirate.

      Blood samples are concentrated using one of the following techniques:
      - Buffy microcapillary centrifugation: trypomastigotes are concentrated at the buffy layer of the hematocrit samples.
      - Anion exchange column.
      - Triple centrifugation (for blood and CSF).
   2. **Culture** of the samples for demonstration of **epimastigotes** form of the parasite in:
      - Conventional culture media: Weinman's medium.
      - Tissue culture: Insect cell lines (Schneider's).
   3. **Animal inoculation**: the sample is inoculated intraperitoneally in guinea pigs, rats, dogs, and monkeys.
- **Indirect Immunological:**
   - ELISA and IFA could be used to detect trypanosomiasis. The high titer of IgM in the CSF is highly specific.
   - Card agglutination Test (CAT) is used especially in the epidemiological screening of the disease.

Treatment

Treatment of African trypanosomiasis should go along the progress of the disease as follows:

1. **Fexinidazole:**
   - It is the first choice for non-severe disease with or without CNS involvement.
   - It is taken orally daily for 10 days.
   - 1800 mg/day for patients ≥35 kg, 1200 mg/day for patients 2–34 kg.
2. **Suramin:**
   - 10 mg/kg/body weight for children and 1 gm for adults in days 1, 3, 7, 17, and 21.
   - **S.E.** in some patients, nausea, vomiting and loss of consciousness, rash, fever, and hepatitis.
   - Contraindicated in renal failure.
   - It is the choice for the hemolymphatic stage of the disease.
3. **Pentamidine:**
   - 4 mg/kg/body weight once daily for 10 days.

- **S.E.** in some patients, nausea, and vomiting.

4. **Mel B + Dimercaprol** (BAL):
   - 6 mg /kg/body weight, I.V. injection on 3 days course.
   - The course could be repeated after 2 weeks.
   - **S.E.** in some patients, nausea and vomiting.
5. **DL alpha-difluoromethylornithine** (DFMO, Eflornithine):
   - 100 mg/kg/body weight/6 h/ I.V. injection for 3 weeks course.

   S.E. diarrhea, abdominal pain, and anemia.
6. **Ivermectin:**
   - It could be used (needs more investigations and research).

Prevention and Control

1. Bush clearing along the breeding sites of tsetse fly.
2. Application of insect control measures **(see Chap. 18).**
3. Surveillance for case detection and treatment.
4. Visitors to the endemic areas should wear protective clothing together with screening, bed netting, and insect repellents.

- **Parasitology**: *T. Brucei* takes the wavy trypomastigotes in the blood, lymph, and CSF. Epimastigote forms in culture and in the vector. Metacyclic trypomastigote is the infective stage. *Glossina* is the vector.
- **Clinical:** It causes HAT (African sleeping sickness); Gambian type is chronic with late CNS affection. Rhodesian type is an acute with early CNS affection. Suramin is used for treatment.

B. *Trypanosoma cruzi* (*T. cruzi*)

(American Trypanosomiasis or Chagas Disease)

T. cruzi is the causative agent of American trypanosomiasis **(Chagas disease)**. It is a blood and tissue flagellate that exists as an amastigote in organs and tissues; meanwhile, it acquires the trypomastigote form in blood, CSF, and lymph.

Geographical Distribution

T. cruzi prevails in Mexico, Argentina, and South and Central America.

Parasite Morphology

T cruzi presents as a C or S shape trypomastigote in the blood and an amastigote forms in the tissues. In culture and vector, it takes the epimastigote form (Fig. 10.5 and 10.6).

- Amastigote:
  - It is an oval body, 3–5 μm.
  - It has one peripheral nucleus.
  - A rod shape Kinetoplast (dot like in *Leishmania* amastigotes).
  - Usually present in a nest within the tissues.
  - Multiply within cells (macrophages, histocytes).
- Epimastigote:
  - Long, slightly wide body, 9–12 μm.
  - It has one nucleus situated at the posterior end.
  - Kinetoplast is located anterior to the nucleus, with half body undulating membrane.
  - It has one free anterior flagellum.
- Trypomastigotes:
  - **C or S shape** in the blood (wavy and pleomorphic in *T. brucei*).
  - 12–35 μm.
  - One single nucleus anterior to the kinetoplast.
  - Kinetoplast gives rise to full-body length undulating membrane.
  - It has one free anterior flagellum.

Transmission

T. cruzi is transmitted through the following ways:

1. Winged bugs (*Triatoma, Reduviid, and kissing* bugs): through contamination of the bite site on human skin with the bug feces (posterior station infection).
2. Accidental ingestion of infected bugs.
3. Blood transfusion and organ transplant.
4. Congenital.

Habitat

Different morphological forms of *T. cruzi* could inhabit humans:

- The amastigote form of *T. cruzi* is present in the human tissues and organs.
- C or S shape trypomastigotes are present in the blood, CSF, and lymph.
- Epimastigote form is detected in the culture and the vector.

Life Cycle

In human: Winged bugs excrete the metacyclic trypomastigotes (I.S.) with their feces to contaminate the bite sites. Trypomastigotes actively enter the histocytes and pass to other tissue cells when these sites are scratched or rubbed (especially at the lips and eye). Trypomastigotes transform into amastigotes, which multiply intracellularly and are then released from the infected cells and become trypomastigotes in the blood.

Winged bug: it gets the infection when it takes a blood meal from an infected human containing the trypomastigotes (D.S.). Trypomastigotes pass to the insect mid-gut and migrate to the hindgut, where they become metacyclic trypomastigotes (I.S) that pass with the bug feces to contaminate the bite wound (Fig. 10.9).

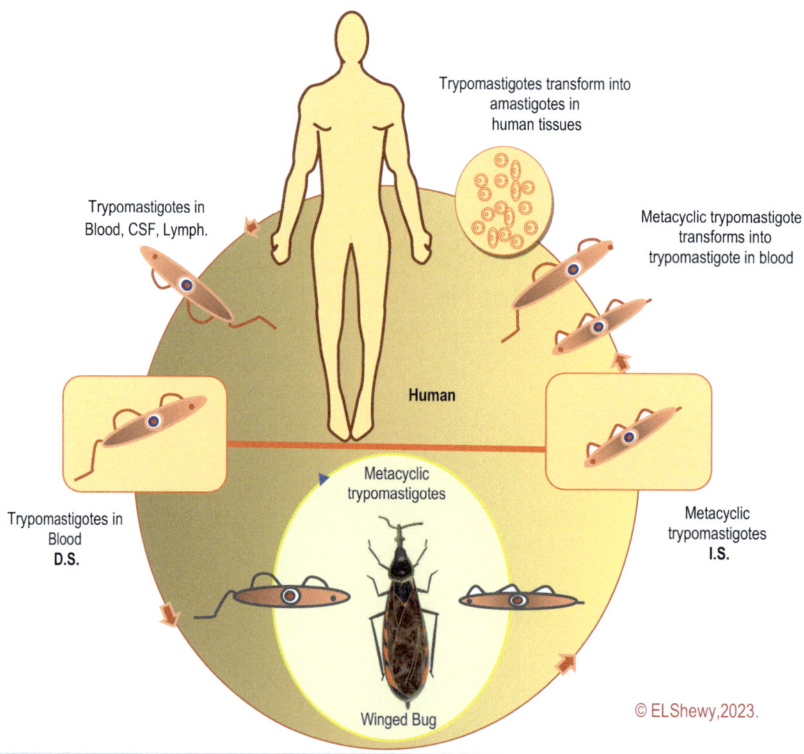

1. Metacyclic trypomastigote contaminates human skin with *Triatoma* stool. They enter the body through skin abrasions.
2. In human tissues and cells, metacyclic trypomastigotes transform into amastigotes, while in the blood, lymph, and CSF, amastigotes transform into trypomastigotes.
3. Trypomastigotes in blood ingested by *Triatoma*.
4. Trypomastigotes transform into epimastigotes in *Triatoma*.
5. Epimastigote transforms into an infective metacyclic form in the *Triatoma* hindgut to contaminate the insect bite.

Fig. 10.9 *Trypanosoma cruzi*: life cycle

Final hosts:	Humans.
Infective stage:	Metacyclic trypomastigotes.
Diagnostic stage:	C or S shape trypomastigotes.
Intermediate host:	*Reduviid* bugs (winged bugs).
Reservoir hosts:	Domestic and wild animals.

Pathogenesis

Infection with *T. cruzi* passes through the following sequences (Fig. 10.10):

1. At the bite site, a local inflammatory reaction is produced.
2. Increasing local inflammation results in a skin lesion called **Chagoma** (formed of adipose cells of the subcutaneous tissue, adjacent muscle cells, proliferating amastigotes, and many inflammatory cells).
3. Trypomastigotes then invade the blood or the lymph to reach tissue cells.
4. At the L.N. or other tissue cells, trypomastigotes transform into amastigotes, which multiply intracellularly and are then released from the infected cells to be trypomastigotes in the blood.
5. Trypomastigotes spread all over the body tissue cells (skeletal, cardiac, reticuloendothelial cells, and neuroglia cells are preferentially parasitized).
6. In the enlarged organs, a diffuse inflammation with fibrosis occurs, giving rise to a mega disease characterizing **Chagas** disease, which could be due to the following:
   (a) Fibrosis of the autonomic ganglion cells of the heart and gastrointestinal tract causes dilatation of the supplied organs.
   (b) Circulating antibodies react against the endocardium, blood vessels, and interstitium of the striated muscles.
   (c) Cytokines released from infected macrophages. These cytokines help in the production of antibodies against both the parasite and the host organs.

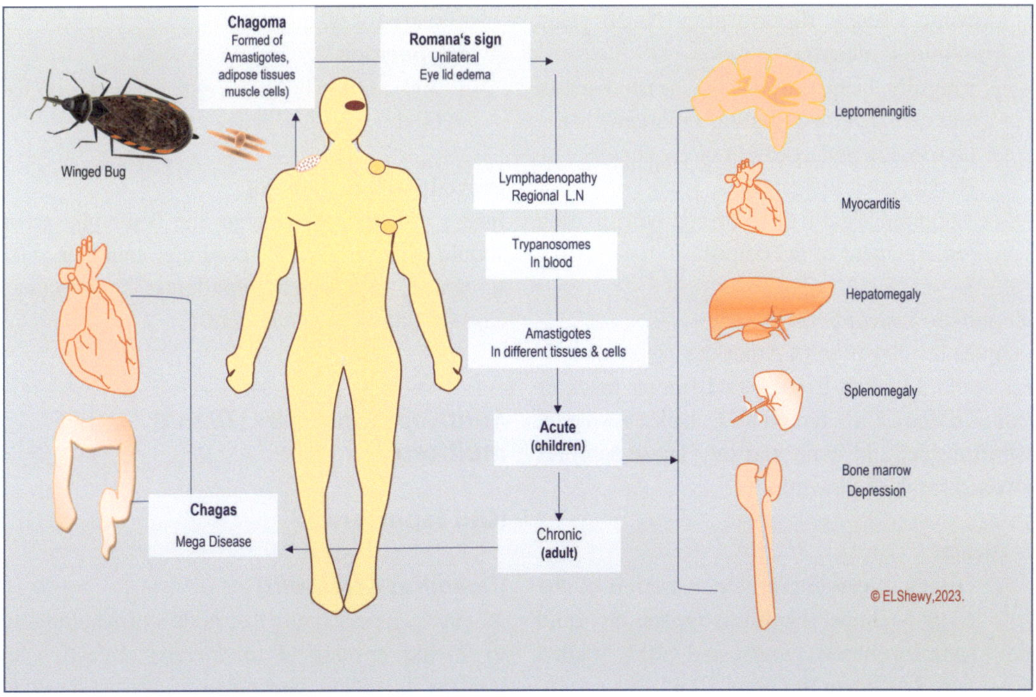

Fig. 10.10 Pathogenesis of Chagas disease (American Trypanosomiasis)

Clinical Picture

American trypanosomiasis is represented as an acute disease (especially in children), an indeterminate condition (no symptoms) and a chronic disease.

1. At the insect bite, swelling with an erythematous, indurated area called **Chagoma** appears and lasts for days, then disappears in 2–3 months.
2. **Acute Chagas** (mostly among children) the infection is severe in the form of:
   - Unilateral edema at the eyelid (**Romana's sign),** with facial edema.
   - Fever, chills, muscle ache, epistaxis with generalized well-defined red spots on the chest and abdomen.
   - Hepatosplenomegaly.
   - CNS affection (sensory and motor affection).
   - Myocarditis with ECG changes (prolongation of P-R and Q-T intervals, S-T depression and inverted T-wave).
3. **Chronic Chagas disease** after 10–20 years from initial infection:
   - Irregular heart, congestive heart disease, and cardiac arrest (cardiac enlargement).
   - Dysphagia (enlargement of esophagus).
   - Hepatosplenomegaly.
   - Abdominal colic and chronic constipation (enlargement of the colon).

Diagnostic Procedures

Samples for detection of American trypanosomiasis could be taken from: blood (finger brick or venous blood), CSF, lymph node aspirate, spleen, liver, muscle, and bone marrow. These samples are used for the following:

- **Direct:**
  1. **Direct microscopic** demonstration of the C or S shape **trypomastigotes** in liquid samples (blood, lymph, and CSF), stained with Leishman or Giemsa.
  2. Demonstration of **amastigote** form in tissue samples.
  3. **Culture** of the samples for demonstration of the **Epimastigotes**.
  4. **Animal inoculation**: the sample is inoculated intraperitoneally in guinea pigs, rats, dogs, and monkeys.
  5. **Xenodiagnosis**: Laboratory-reared bugs are allowed to feed on the patient's blood and are examined later for the parasite.
- **Indirect Immunological:**
  ELISA, IFA, RIA, and Card agglutination tests.

Treatment

The following drugs are commonly used for the treatment of American trypanosomiasis:

1. **Nifurtimox** (Lampit, Bayer 2502):
   - 25 mg/kg/body weight. 4 times daily, for 15 days orally.
2. **Benznidazole:**
   - 5 mg/kg body weight orally, once daily for 60 days for adults.
   - For children, 10 mg/kg body weight once daily for 60 days orally.
3. **Allopurinol:**
   - 300 mg/kg body weight twice daily for 60 days.

Prevention and Control

Insect control; visitors to the endemic areas should wear protective clothing, together with screening, and insect repellents: surveillance, case detection, and treatment.

Multiceps multiceps (*Taenia multiceps*)

(Gid Tapeworm)

(*Coenurus Cerebralis*)

M. multiceps is an animal cestode worm belonging to *Taenia* species (*T. multiceps*). Infection in humans is called coenurosis, a larval cestode infection affecting many body tissues, particularly the central nervous system (*coenurus cerebralis*).

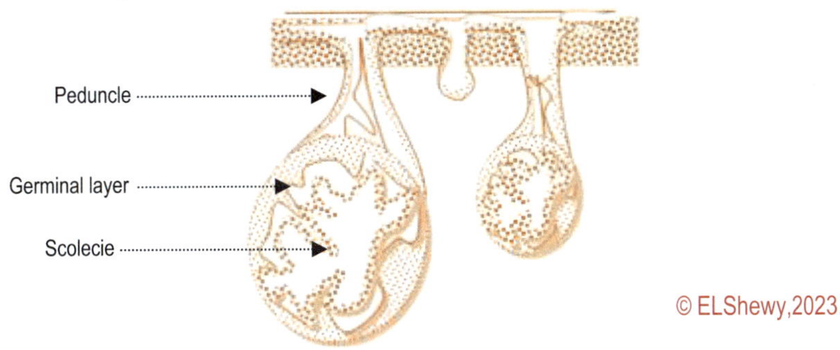

Fig. 10.11 Structure of *coenurus cerebralis*

Geographical Distribution

M. multiceps is detected in sheep-raising Arctic countries in Europe, South Africa, and Asia. However, human infection is uncommon or rare.

Parasite Morphology

M. multiceps has three morphological forms:

- Adult worm:
  - A small tapeworm of 30–60 cm long.
  - It is similar in morphology to *T. solium* (*T. multiceps*).
- Egg:
  - Similar to *Taenia* egg.
- Coenurus *cerebralis* cyst (Fig. 10.11).
  - It ranges in size from a few millimeters to the size of a hen's egg.
  - The cyst is characterized by:
    - (a) Lack of outer fibrous layer.
    - (b) Inner nucleated germinal epithelium.
    - (c) Neither broad capsules nor daughter cysts.
    - (d) Budding scolecies from the germinal layer.
    - (e) Clear translucent fluid inside the cyst.

Transmission

M. multiceps is transmitted through ingestion of the worm eggs with contaminated food or drinks.

Habitat

- Adult worms of *M. multiceps* inhabit the small intestine of the final hosts (canine animals: dogs, foxes, and wolves).
- Coenurus cysts located in intermediate hosts (human and herbivorous animals: cattle, sheep, and others) are localized mainly in the CNS.

Life Cycle

1. Dogs and other canine animals (fox, wolf ...) acquire infection by ingestion of viscera and carcasses of dead animals containing coenurus cysts.
2. Scolex inside the cyst is developed into an adult worm, which attaches to the small intestine of canines (final hosts).
3. Humans and other herbivorous animals (intermediate hosts) acquire infection through accidental ingestion of the eggs of *M. multiceps*.
4. The ingested eggs migrate to various body tissues, particularly the CNS, and transform into coenurus cysts (Fig. 10.12).

Final hosts:	Dogs and other canines.
Infective stage:	For canines, coenurus cysts; for humans, the worm eggs.
Diagnostic stage:	*Coenurus* cyst in human tissues.
Intermediate hosts:	Humans and other herbivores, e.g., sheep and cattle.
Reservoir hosts:	Carnivorous animals.

Clinical Picture

Clinical presentation of coenurosis varies according to the location of the cyst:

1. A slowly growing cyst does not produce so many symptoms; only symptoms appear when size increases, and pressure on adjacent tissue occurs.
2. Cysts in the CNS usually produce severe tissue damage depending on their location, e.g., convulsion, epilepsy, diplopia, and other nervous manifestations.

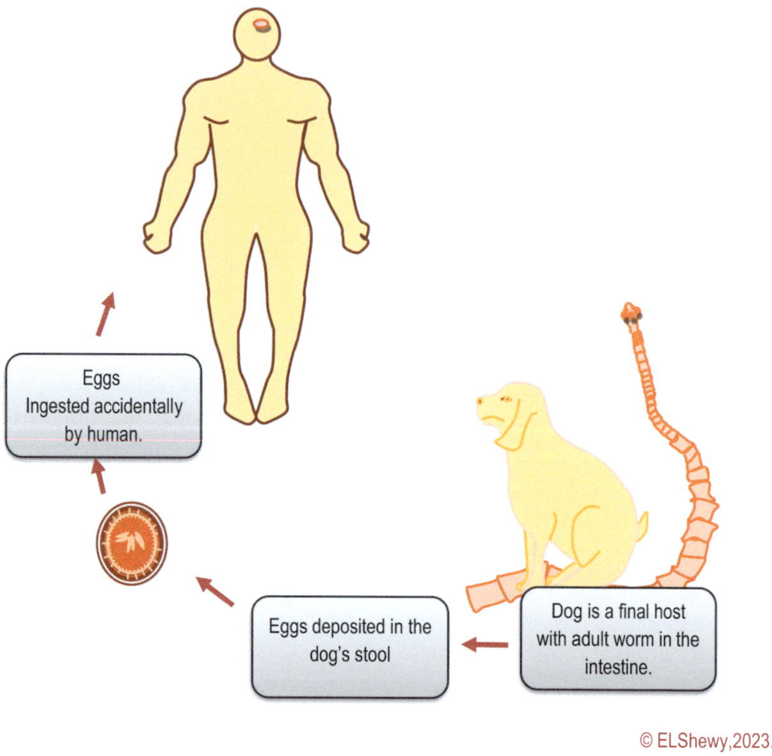

Fig. 10.12 *Taenia multiceps*: life cycle

3. Infection in cattle is called gid, which is characterized by nervous ataxia and blindness.

Diagnostic Procedures

1. Clinical picture and history taken.
2. Serology.
3. Imaging MRI and CATS.
4. Histological examination after surgical removal of the cysts is commonly helpful.

Treatment

Mainly through surgical removal of accessible and operable cysts together with the following drugs:

1. **Albendazole:**
   - It is the drug of choice.
   - 5 mg/kg/body weight/ twice daily for 8–12 weeks.
   - Should be given to the inoperable cases.
   - Could be given pre-and post-operatively for the operable cases.
   - Repeated courses should be recommended.
2. **Praziquantel:**
   - It is effective as scolicidal drug in high doses.
   - 120–200 mg/kg/body weight for 5 days is effective preoperatively.
   - **S.E:** headache, dizziness, abdominal pain, and nausea.

Prevention and Control

As mentioned with *E. granulosus*.

1. Good sanitary practices with pet dogs are essential in prevention.
2. Treatment of infected dogs with anthelminthic.
3. Healthy and clean food supply.
4. Avoidance of direct and close contact with dogs.

Review Questions

1. **The following parasitic stage(s) could be detected in a cerebrospinal fluid (C.S.F.) infected with *Acanthamoeba*.**
   (a) trophozoites and biflagellate forms,
   (b) trophozoites only,
   (c) cysts and trophozoites,
   (d) cysts and biflagellate forms.
2. **Xenodiagnosis could be used in the diagnosis of infection with.**
   (a) *Multiceps multiceps.*
   (b) *Naegleria fowleri.*
   (c) *T. cruzi.*
   (d) *Acanthamoeba.*
3. **A 25 male patient complained of severe headache, fever, and neck stiffness. His C.S.F. demonstrated a lot of neutrophils and trophozoites with a single nucleus and a big karyosome. This parasite is most probably.**
   (a) *Entamoeba histolytica.*
   (b) *Naegleria fowleri.*
   (c) *Balamuthia mandrillaris.*
   (d) *Acanthamoeba.*
4. **Suramin is the drug of choice for the treatment of.**
   (a) giardiasis,
   (b) leishmaniasis,
   (c) HAT.
   (d) amebiasis.
5. ***Naegleria fowleri* could reach the C.N.S through.**
   (a) blood,
   (b) lymph,
   (c) olfactory nerves,
   (d) skin abrasion.
6. **Primary amebic meningoencephalitis (PAM) is caused by.**
   (a) *Acanthamoeba.*
   (b) *Balamuthia mandrillaris.*
   (c) *Naegleria fowleri.*
   (d) *Multiceps multiceps.*
7. **Granulomatous amebic encephalitis (GAE) is caused by.**
   (a) *Acanthamoeba.*
   (b) *Entamoeba histolytica.*
   (c) *Naegleria fowleri.*
   (d) *Multiceps multiceps.*
8. **Which of the following is best used to diagnose *Naegleria fowleri* infection?**
   (a) Blood film.
   (b) Serological examination.
   (c) Blood culture.
   (d) Microscopic CSF examination.
9. **The following specimens could be used for diagnosis of American trypanosomiasis, except.**
   (a) splenic,
   (b) lymph node,
   (c) stool,
   (d) spinal fluid.

11 Parasites Infecting the Respiratory Tract

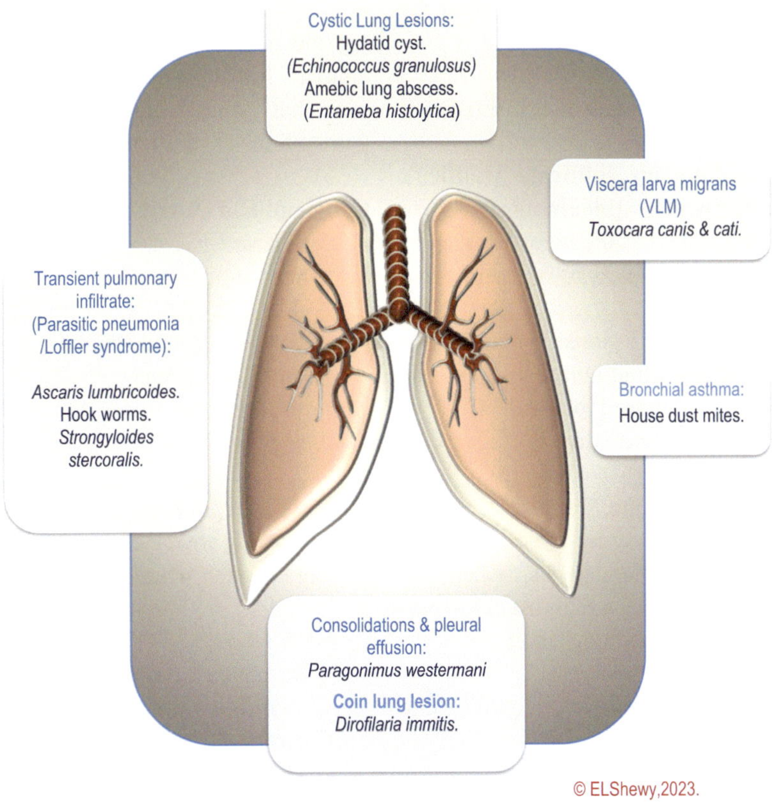

Parasites Infecting the Respiratory Tract

K. ElShewy, *Medical Parasitology*, https://doi.org/10.1007/978-3-031-64736-9_11

Test your learning and check your understanding of this book's contents: use the "Springer Nature Flashcards" app to access questions using ▶ https://sn.pub/oyFLIx. To use the app, please follow the instructions in Chap. 1.

Introductory

Parasitic infections of the respiratory tract occur worldwide among both immunocompetent and immunocompromised patients. Parasites of the respiratory system could be responsible for many chronic and debilitating diseases, including the following:

1. **Bronchial asthma and allergic rhinitis:** e.g., house dust mite (HDM).
2. **Transient pulmonary infiltrate (Loffler's syndrome/parasitic pneumonia):** e.g., *Ascaris,* hookworms *(Ancylostoma duodenale, Necator americanus), S. stercoralis and Toxocara.*
3. **Cystic and coin lung lesions:** Hydatid cysts, amebic lung abscess, and *Dirofilaria immitis.*
4. **Lung consolidation and pleural effusion:** e.g., *Paragonimus westermani.*
5. **Tropical pulmonary eosinophilia:** e.g., *Wuchereria bancrofti* and *Brugia* species.
6. **General signs and symptoms:** Dyspnea, chest pain, cough, hemoptysis, wheezing, fever, chills and malaise: e.g., most of the parasites affecting the respiratory tract.

Parasites of the respiratory tract could be one of the following:

1. **Parasites inhabiting the lung:** e.g., *Paragonimus westermani.*
2. **Parasites carried to the lung:** e.g., *Echinococcus granulosus* (hydatid cysts) **(Chap. 8),** *Entamoeba histolytica* (Amebic lung abscess) **(Chap. 6).**
3. **Parasites pass through the lung during their normal development:** e.g., *Ascaris,* hookworms, *S. stercoralis, Wuchereria bancrofti* **(Chap. 12),** and *Brugia malayi.*
4. **Ectopic parasitic stages infecting the lung:** e.g., *Toxocara cati, Toxocara canis,* and *Dirofilaria immitis.*

In this chapter, we discuss the most common parasites infecting the respiratory system; other parasites are mentioned in other chapters.

Paragonimus westermani (*P. westermani*) (Lung Fluke)

P. westermani is a lung trematode worm inhabiting the pulmonary architecture of humans and animals. Infection in humans causes a lung disease that typically mimics tuberculosis called paragonimiasis or **parasitic hemoptysis.**

Geographical Distribution

P. westermani has a wide distribution, particularly in:

(a) Japan, Korea, and China.
(b) South East of Asia.
(c) Some localities of Africa (Nigeria and Congo).

Parasite Morphology

The following are the characteristics morphology of both the adult worm and its egg (Fig. 11.1):

- Adult worm:
  - It is an oval flat fluke with tiny spines.
  - 10 mm × 4 mm, with an oral and a ventral sucker.
  - The ventral sucker lies anterior to the mid-body.
- Egg:
  - It is an oval, operculated small egg.
  - It has a thick shell with shoulders.
  - 50 μm × 90 μm, brown.
  - Unembryonated (immature) when freshly excreted.

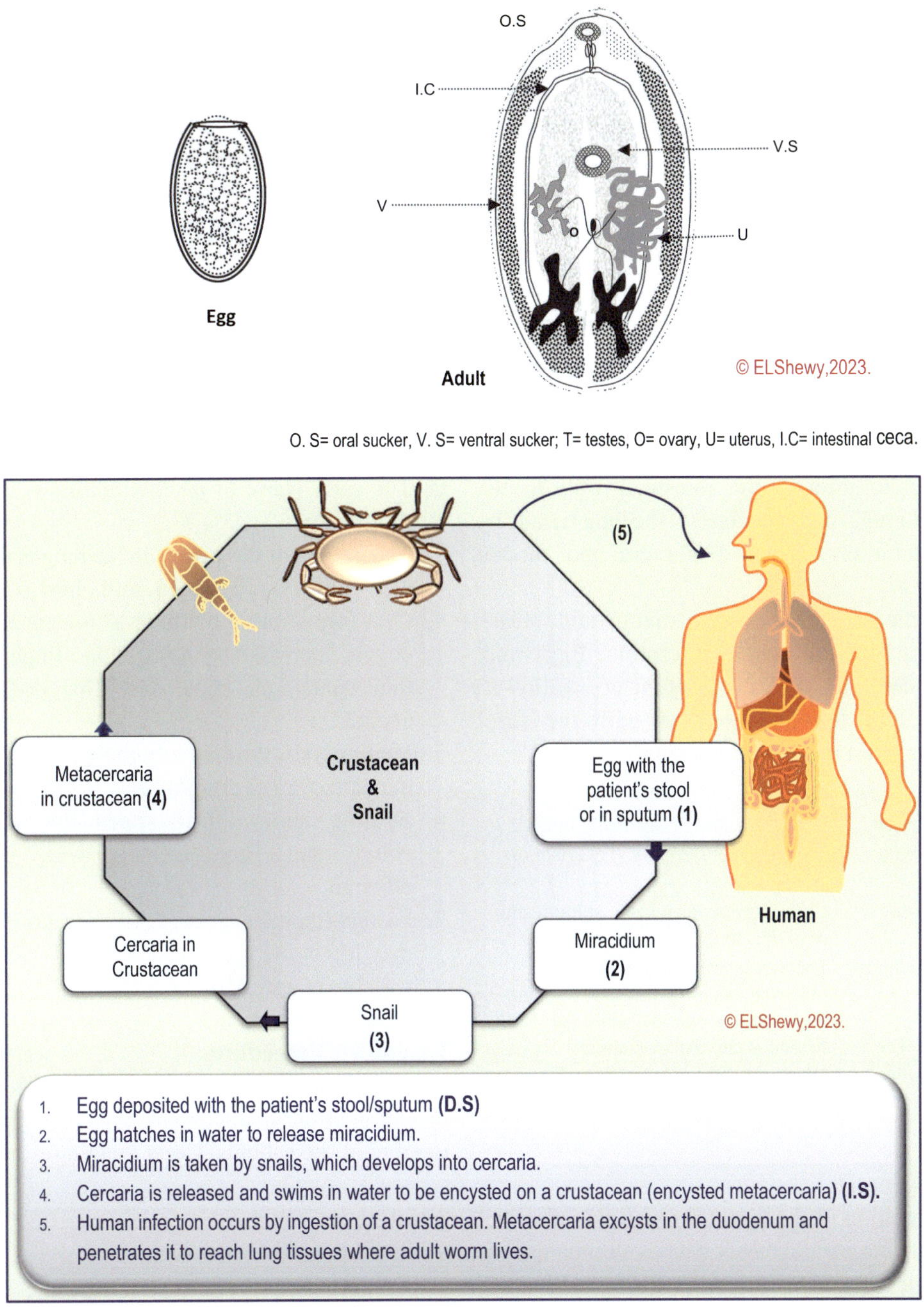

Fig. 11.1 *Paragonimus westermani*: morphology and life cycle

Transmission

Infection with *P. westermani* is transmitted through ingestion of raw or insufficiently cooked crayfish or freshwater crabs.

Habitat

Adult worm of *P. westermani* inhabits lung tissues, bronchi, and pleural cavities of dogs, cats and occasionally men.

Life Cycle

- The life cycle of *P. westermani* starts with the passage of unembryonated eggs in the freshwater, along with the patient's stool or sputum.
- After 2 weeks, eggs hatch to liberate miracidium that invades a freshwater snail (first intermediate host). Cercariae released from the snail to infect and encysted (encysted metacercaria) inside a crustacean (crayfish or crab) (second intermediate host).

  Human ingests infective metacercariae in undercooked crayfish or crabs.
- At the duodenum, metacercariae are excysted, migrate through the intestinal wall to the abdominal cavity and reach the lung tissues by penetration of the diaphragm and pleural cavity.
- At the lung tissue, worms mature and release eggs in the bronchial secretion. Eggs may be coughed with the sputum or swallowed back in the GIT to pass with the stool (Fig. 11.1).

Final hosts:	Dogs, cats, and occasionally humans.
Infective stage:	Encysted metacercaria in crayfish or crabs.
Diagnostic stage:	Characteristic eggs in the sputum and/or in the stool.
Intermediate hosts:	The first intermediate host is a freshwater snail (*Semisulcospira* and *Melania tuberculata*), while the second is crayfish or crabs.
Reservoir hosts:	Dogs and cats.

Pathogenesis

1. Migration of the worm through the intestinal wall and pleural cavity causes minimal pathological reactions in the form of local hemorrhage and leukocytic infiltration.
2. Adult worms in lung tissues cause leukocytic infiltration, chronic inflammation, and fibrous tissue surrounding the infiltrate, which causes the formation of many **cysts.**
3. Communication of the cysts with the bronchioles results in the spread of the eggs to the surrounding tissues or the bloodstream to give extrapulmonary extension.
4. Granulomatous reaction surrounds the eggs or the worms in the lung tissues, leading to lung consolidation with pleural effusion.
5. Extrapulmonary paragonimiasis, although rare, could affect the peritoneal cavity, brain, or the pericardium.

Clinical Picture

1. History of eating undercooked crabs or crayfish.
2. The early phase of paragonimiasis is usually asymptomatic.
3. The growth of the worms in the lung is associated with **fever, dyspnea, and chest pain.**
4. Cyst formation in the lung is associated with cough, increased sputum, blood-tinged sputum containing eggs, and Charcot-Leyden crystals.
5. **Hemoptysis,** with severe chest pain, follows the paroxysm of **coughing**.
6. By time, there will be **dyspnea**, chronic bronchitis, and bronchiectasis with pleural effusion.
7. Extrapulmonary spread of paragonimiasis may affect the brain, heart, or abdomen with variable manifestations.

Diagnostic Procedures

Diagnosis of *P. westermani* is usually done by detection of the characteristic eggs in the sputum or stool of the patients. Chest imaging is helpful.

Treatment

1. **Praziquantel** (Biltricide):
   - It is a drug of choice.
   - 25 mg/kg/3 times daily for 2 days.
2. **Bithionol:**
   - It is an alternative drug for pulmonary paragonimiasis.

– 25 mg /kg/2 times every other day for 15 days is recommended. Skin rash and contact dermatitis have been reported as side effects (S.E.)

Prevention and Control

1. Proper disposal of human excreta.
2. Snail eradication.
3. Eating raw crayfish and crabs should be prohibited.

- **Parasitology:** *P. westermani* is a 1–0.5 cm oval worm that mainly inhabits the lung in cysts; it needs a snail for its life cycle. It is transmitted through the ingestion of crabs and crayfish containing encysted metacercaria. Its eggs could be detected in sputum or stool.
- **Clinical:** Paragonimiasis (parasitic hemoptysis) is presented with cough, hemoptysis, chronic bronchitis, and bronchiectasis with the possibility of an extrapulmonary spread.

Visceral Larva Migrans (VLM)

Toxocara canis and Toxocara cati (T. canis and T. cati)

(Dog and Cat Roundworms)

T. canis and T. cati are nonhuman nematode worms; when they infect the human body, their developing larvae fail to complete their normal life cycle and implant in the human viscera, causing VLM.

Geographical Distribution

T. canis and *T. cati* have a worldwide distribution.

Parasite Morphology

Adult worms of *T. canis* and *T. cati* are similar to those of *Ascaris* worms but with smaller size and cervical alae at the anterior ends of both males and females. Their eggs are rounded, superficially pitted and unembryonated when passed (Fig. 11.2).

Transmission

Man could be infected with toxocariasis in the same ways as in *Ascaris* infection, i.e., through accidental ingestion of *Toxocara* eggs in his foods or drinks.

Habitat

Adult worms of *T. canis and T. cati* inhabit small intestine of dogs and cats.

Life Cycle

Toxocara is a nonhuman parasite when infect human they cannot complete its life cycle as follows:

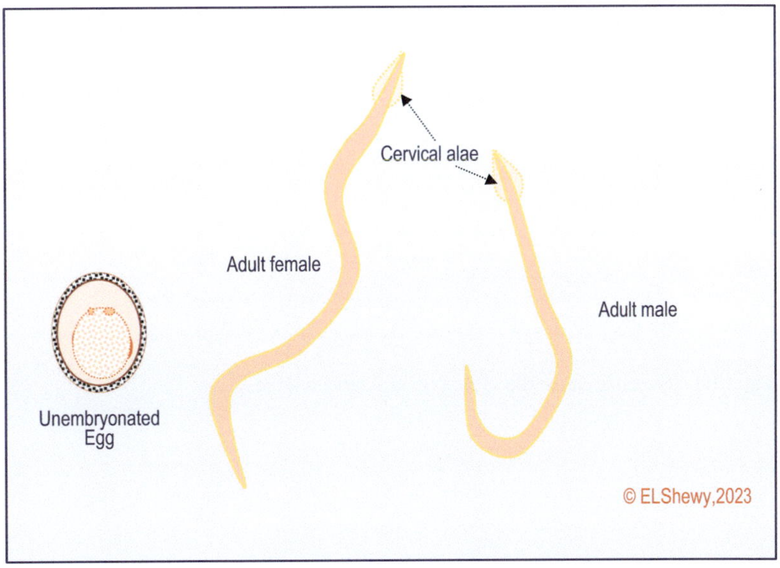

Fig. 11.2 *Toxocara canis*: morphology

1. Man accidentally ingests *Toxocara* eggs in his foods or drinks.
2. Eggs hatch and release larvae in the small intestine, which cannot complete the life cycle as normal animal hosts do.
3. Larva implants in human viscera (lung, heart, liver, and intestine), cause visceral larva migrans (VLM).
4. In some conditions, *T. canis* larva could reach the eye, forming ocular larva migrans (OLM) (Figs. 11.3 and 11.4).

Clinical Picture

- VLM usually causes fever, pneumonitis, myocarditis, and hepatomegaly. If CNS is infected by VLM encephalitis, convulsion and psychiatric changes are observed.
- OLM usually occurs as a separate entity that is not associated with VLM. It presents with blurring vision, decreased visual acuity and retinal detachment.

Diagnostic Procedures

Diagnosis of VLM infection depends on:

1. Clinical picture.
2. History of contact with dogs and cats.
3. Histological examination of the biopsy taken.
4. Serological test for detection of the specific antibodies.
5. Examination of the soil is done to detect the characteristics of the eggs.

Treatment

1. **Albendazole** is the drug of choice.
2. **Diethylcarbamazine.**
3. **Steroids**.

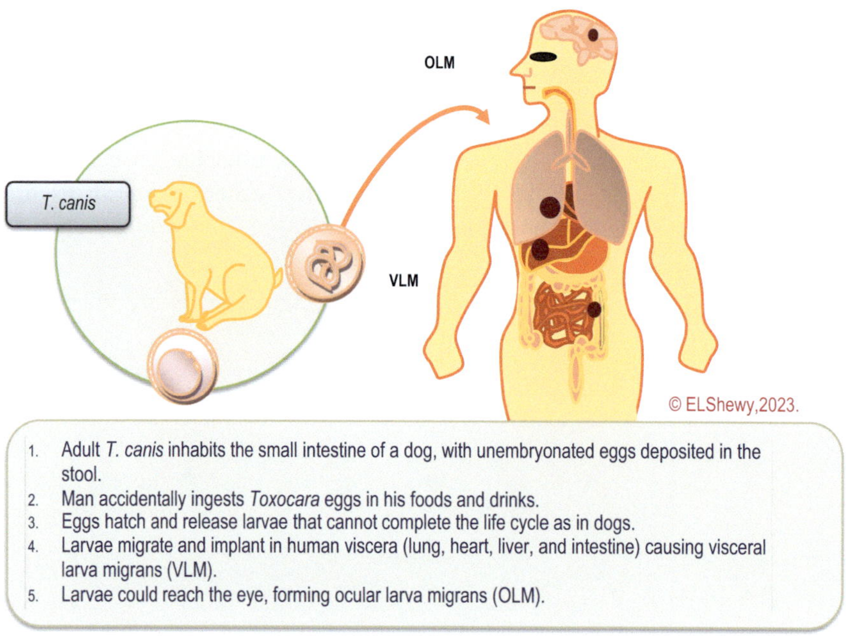

Fig. 11.3 *Toxocara canis*: life cycle (visceral larva migrans)

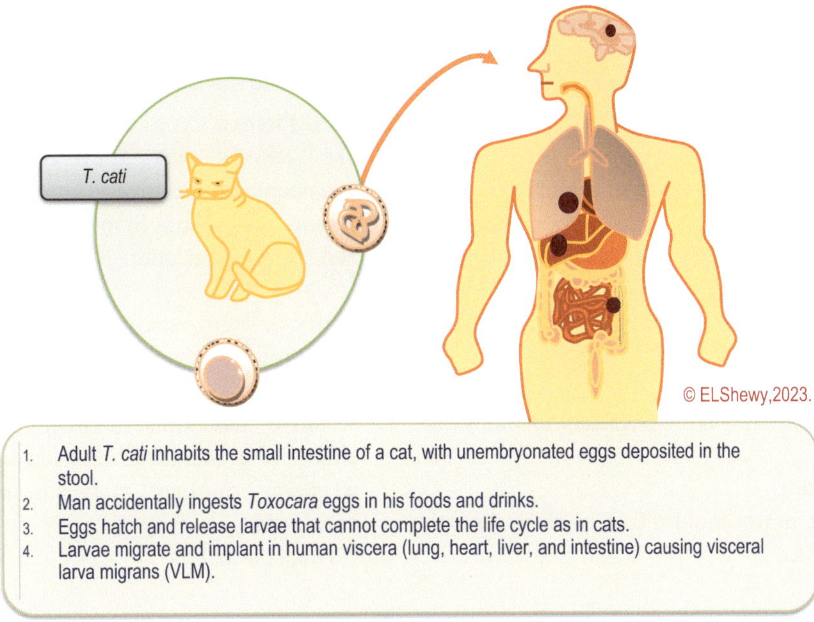

Fig. 11.4 *Toxocara cati*: life cycle (visceral larva migrans)

House Dust Mites (HDM)

Dermatophagoides spp.

The term "house dust mites" has been applied to a large number of mites found in association with dust in the environment. House dust mites are arthropods that are not capable of biting or stinging humans. Their significance is due to the powerful allergens in their cast skins, fecal material, and secretions.

Geographical Distribution
House dust mites are cosmopolitan in distribution. ***Dermatophagoides farinae*** represents more than 60% of mites, followed by ***D. pteronyssinus***. Generally, they flourish in warm humid environments.

Parasite Morphology

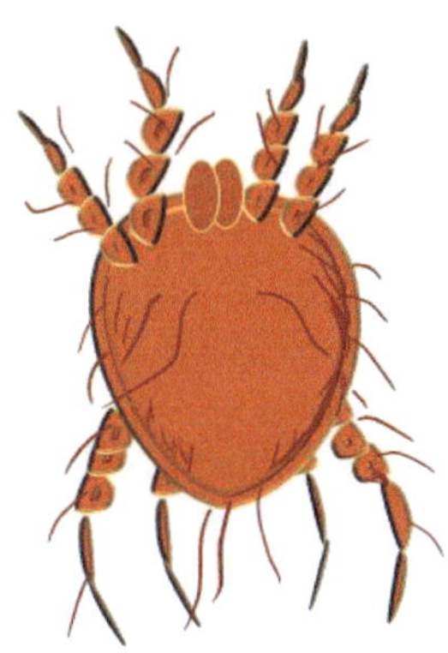

Dermatophagoides
(House dust mite)

- Bothv male and female adult mites are globular in shape.
- They are creamy white and have a striated cuticle.
- The female measures approximately 0.4 mm in length and 0.3 mm in width, and the male is smaller.
- They have four pairs of hairy legs, no eyes, no antennae.
- The mouth is in the anterior end (it resembles the head).
- They have long setae (bristle-like hairs) that extend from the outer margins of the body.

Transmission
House dust mites could affect humans through inhalation of:

- The whole mite.
- Mite secretions and fecal matter.

Habitat
House dust mites infect many mammals, particularly humans. They could be found in mattresses, carpets, sofas, and chairs. They can also be found in bird nests and occasionally in beehives.

Life Cycle
- The life cycle of HDM starts with the deposition of eggs in the environment, which develop into larvae, nymphs, and adults.
- Depending on humidity and temperature, dust mites need from 15 to 30 days to complete their life cycle.
- Mated females live for about 2 months.
- House dust mites feed on the humans' shed skin.

Pathogenesis
- The allergenic proteins responsible for causing allergic symptoms are contained within the mites (alive or dead), in their shed skins and their feces.
- These allergic components stimulate Th2 cytokine production, which induces the production of IgE antibodies, leading to different allergic manifestations.

Clinical Picture
Unlike "seasonal" allergies caused by molds and pollen, people who are allergic to dust mites have symptoms year-round. Symptoms of the allergy caused by HDM include:

1. Stuffy or runny nose, sneezing, coughing, or watery eyes.
2. Inhalation of dust mite allergens by hypersensitive individuals can result in acute attacks of bronchial asthma, accompanied by wheezing, shortness of breath, and perhaps even death.
3. Contact dermatitis.

Treatment
There are two basic approaches to managing dust mite allergy:

1. **Treatment of the patient:**
   The use of antiallergic drugs, e.g., corticosteroids, antileukotriens, antihistaminic, and the use of specific immunotherapy, which includes:
   (a) Injecting gradually increasing concentrations of mite extracts over time desensitizes the affected individual.
   (b) Allergy shots: The use of allergen extracts as subcutaneous immunotherapy (SCIT) and sublingual immunotherapy (SLIT).
   (c) Probiotics increase the number of Treg cells by enhancing IgA production from B cells.
2. **Controlling the patients' environment** to minimize exposure to the mites. (See prevention and control).

Diagnostic Procedures
- To diagnose a dust mite allergy, the following two tests could be used:
  1. Skin Prick Test (SPT).
  2. Specific IgE blood Test.

- To separate mites from dust collected from vacuuming mattresses, carpets, sofas, and chairs, the following could be used:
  1. A small amount of dust is laid on the surface of the water and examined under 20× magnifications.
  2. A sieving and flotation technique: 0.05 g samples of dust in 30 ml of a saturated NaCl solution with 5 drops of detergent are mixed. The suspension is rinsed through a 45-micron mesh sieve, stained with crystal violet, and examined under the microscope.
  3. Dust mites are attracted to white sheets of paper placed where the dust mites are suspected. Paper is examined over a lamp. The mites cast a shadow.

Prevention and Control

The following are steps that should be taken to minimize exposure to HDM:

1. Use an air conditioner or dehumidifier to keep the relative humidity in the house between 30 and 50%.
2. Use a vacuum cleaner equipped with a HEPA (High Efficiency Particulate Arrestor) filtration system.
3. Use only washable toys, and wash them often. Keep toys off beds.
4. Consider treating carpets with an acaricide (mite-killing products containing benzyl benzoate, e.g., Acarosan).
5. Replace carpeting with wood, tile, linoleum, or vinyl flooring, if possible.

Dirofilaria issmmitis (D. immitis)

(Heart Dog Worm)

D. immitis is a filarial nematode worm of dogs and other canines. It infects the heart and lungs of the dogs. Occasionally, it could infect humans through mosquitoes causing a pulmonary artery disease.

Geographical Distribution

There are two species commonly found all over the world:

1. *D. immitis* (pulmonary dirofilariasis).
2. *D. repens*.

Parasite Morphology

1. Adult females of *D. immitis* are larviparous, about 25–30 cm long, by 0.3 mm wide.
2. The adult male is about 12–15 cm long by 0.3 mm wide and with a spirally coiled tail.
3. Adults can live for 5–10 years.

Transmission

1. Transmission from dog to dog and from dog to human through the bite of a mosquito (*Aedes, Culex,* and *Anopheles*).
2. Human is not a source of transmission.

Habitat

3. Transmission from dog to dog and from dog to human through the bite of a mosquito (*Aedes, Culex,* and *Anopheles*).
4. Human is not a source of transmission.

Life Cycle

- Mosquito (*Aedes, Culex,* and *Anopheles*) bites dogs and takes microfilariae with their blood.
- When biting a new host, mosquito introduces filarial larvae into the skin.
- In dogs, larvae undergo two molts to become adults.
- Adults reside in the pulmonary arteries of the dog and, occasionally, in the right ventricle of its heart, where the female produces microfilariae.
- Microfilariae are released in peripheral blood to be taken by mosquitoes.
- In humans, larvae of *D. immitis* migrate as in the dogs, ending up in the lungs, where they often lodge in the narrow vessels.
- **Adult worms in humans cannot mature sexually, so humans have no role in the transmission of *D. immitis*.**

Pathogenesis

- Death of *D. immitis* worms in the pulmonary artery and its branches stimulates the production of granulomas with monocellular infiltrate.
- Granulomas appear as coin lesions (small, round abnormalities) on chest X-rays.

Clinical Picture

Most individuals with dirofilariasis have no symptoms. However, the following could be detected:

1. Fever.
2. Cough and hemoptysis.
3. Chest pain.
4. Dyspnea and may be pleural effusion.

Diagnostic Procedures

1. Imaging: Coin lesions and nodules are the characteristic pictures of dirofilariasis. The picture could be misdiagnosed as lung cancer.
2. Lung biopsy and histological examination are conclusive.

Treatment

Surgical removal of the lesions is the curative procedure for human dirofilariasis. Drug treatment is not effective.

Prevention and Control

Preventive measures against mosquitoes: use of mosquito nets, repellants, insecticides, biological control measures against mosquitoes, and treatment of dogs.

Tropical Pulmonary Eosinophilia (TPE)

Tropical pulmonary eosinophilia (TPE) is a rare syndrome characterized by pulmonary interstitial infiltrates and marked peripheral eosinophilia. It is detected during infection with *Wuchereria bancrofti* and *Brugia* species.

Geographical Distribution

TPE is more widely recognized in filariasis-endemic regions, such as India, Africa, Asia, and South America.

Pathogenesis

Hypersensitive immunological reaction against microfilariae of *W. bancrofti* and *Brugia malayi* is the cause of TPE with the following sequences:

- Adult worms release their microfilariae, which are trapped in the pulmonary circulation.
- Microfilariae release their antigenic constituents, triggering a local inflammatory process.
- This inflammatory process is associated with interstitial, histocytes, and acute eosinophilic infiltrations.
- This is followed by interstitial fibrosis with a picture of alveolitis.
- Released antigens in the blood stimulate a marked eosinophilia.

Clinical Picture

A typical TPE patient is an **Indian young adult male** with the following:

1. Asthmatic manifestations (nocturnal cough, wheezy chest, hemoptysis, chest pain, and dyspnea).
2. Symptoms at night are probably related to the nocturnal periodicity of microfilariae.
3. Sputum is scanty, viscous, and mucoid, with clumps of eosinophils.
4. Enlarged spleen and lymphadenopathy.
5. Previous history of exposure to lymphatic filariasis.
6. Good response to diethylcarbamazine (DEC).

N.B. Chronic TPE should be differentiated from tuberculosis and larvae of roundworms. (Loffler, VLM).

Diagnostic Procedures

- Blood picture.
- Serum IgE level.
- Antifilarial antibodies.
- Blood film for microfilariae.

Laboratory investigations include the following data:

1. High eosinophilia.
2. An elevated serum IgE level.
3. High antifilarial antibodies.
4. Negative blood for microfilariae.

Treatment

Diethylcarbamazine (DEC) (Hetrazan):

- 6 mg/kg/body weight/day in three divided doses for 21 days.
- **S.E.** headache, fever, pruritus, and gastrointestinal upset.
- Eosinophil count often falls dramatically within 7–10 days of starting treatment.

Parasitic Pneumonia

(Loffler's Syndrome)

Loffler's syndrome is defined as a disorder characterized by transient pulmonary infiltrates accompanied by peripheral blood eosinophilia in asymptomatic or mildly ill patients. William Loffler was the first to describe this syndrome in Swiss patients.

Etiology

- Parasitic Loffler syndrome is mainly caused by infection with three parasites:
  Ascaris, Hookworms (*A. duodenale* and *N. americanus*), and *Strongyloides*.
- Larvae of these parasites pass through the lungs, entering via the bloodstream, penetrating alveoli, and then ascending the airway to transit down the esophagus into the small bowel.

Pathogenesis

Pathological changes are caused by larvae that pass through the lungs as part of the parasite's initial developmental cycle in humans. This results in:

1. An exudate into the alveoli with eosinophilic infiltration of both the alveoli and the interstitial tissue.
2. Inflammatory involvement of the pleura and the interlobar fissures.

N.B. Almost all pathological changes are transient, so occasionally, Loffler syndrome is termed "**transient pulmonary eosinophilia**" or "**simple pulmonary eosinophilia (SPE).**"

Clinical Picture

- Symptoms of Loffler's syndrome are usually mild or absent.
- Symptoms appear 2 to 3 weeks after ingestion of *Ascaris* eggs or penetration of the infective larvae of *N. americanus, A. duodenale,* or *S. stercoralis.*
- Symptoms tend to spontaneously resolve after 3–4 weeks.
- The most common symptoms are fever, malaise, dry unproductive cough, wheezing, and dyspnea.
- Less commonly, the patient may present with myalgia, anorexia, and urticaria.

Diagnostic Procedures

1. Social and travel history to identify risk factors of exposure to parasites.
2. Microscopic sputum examination to reveal the characteristic larvae of the offending parasites.
3. Microscopic stool examination to detect the possible parasitic infection. Parasites can be found in the stool 6–12 weeks after the initial parasitic infection. However, pulmonary symptoms usually resolve by the time parasitic forms are found in the stool.
4. A complete blood picture shows mild blood eosinophilia, usually 5–20%; however, it may be increased to 40%.
5. Plain X-ray and CT scans reveal transient opacities, which typically resolve within a month.
6. Occasionally, crackles or wheezes may be heard in the lungs.

Treatment

1. Loffler's syndrome is self-limited, and most treatment is not required.
2. Symptomatic medication may be taken to relieve symptoms.

3. Anti-helminthic drugs for the underlying parasitic cause.
4. Repeat chest X-ray and CBC to check prognosis.

Review Questions

1. **The following biological specimens could contain eggs of *Paragonimus* except**
   (a) stool
   (b) blood
   (c) sputum
   (d) bronchial lavage
2. **House dust mite is characterized by**
   (a) long live
   (b) having biting mouth parts
   (c) having wings
   (d) secretion of allergic products
3. **Which of the following worms could be detected in the heart?**
   (a) *Paragonimus Westermani.*
   (b) *Dirofilaria immitis.*
   (c) *Brugia timori.*
   (d) *Wuchereria bancrofti.*
4. **The following parasites could cause cystic lesions except**
   (a) *Paragonimus Westermani*
   (b) *Entamoeba histolytica*
   (c) *Taenia solium*
   (d) *Wuchereria bancrofti*
5. **The pulmonary artery could be affected during infection with**
   (a) *Paragonimus westermani*
   (b) *Dirofilaria immitis*
   (c) *Toxocara canis*
   (d) *Wuchereria bancrofti*
6. **Allergy caused by house dust mites is**
   (a) detected only in children
   (b) represented as dermatitis only
   (c) caused only by mite secretion
   (d) not seasonal
7. **Visceral larva migrans (VLM) is caused by**
   (a) *Paragonimus westermani*
   (b) *Dirofilaria immitis*
   (c) *Toxocara canis*
   (d) *Wuchereria bancrofti*
8. **Parasitic pneumonia is mainly noticed during infection with**
   (a) *Schistosoma mansoni*
   (b) *Schistosoma haematobium*
   (c) *Strongyloides stercoralis*
   (d) *Trichostrongylus colubriformis*
9. **Tropical pulmonary eosinophilia is characterized by the following laboratory findings except**
   (a) high eosinophilia
   (b) microfilaremia
   (c) elevated serum IgE level
   (d) high antifilarial antibodies
10. **All of the following worms have heart-to-lung circulation except**
   (a) *Trichostrongylus colubriformis*
   (b) *Ancylostoma duodenale*
   (c) *Strongyloides stercoralis*
   (d) *Ascaris lumbricoides*

12 Blood and Lymphatic Parasites

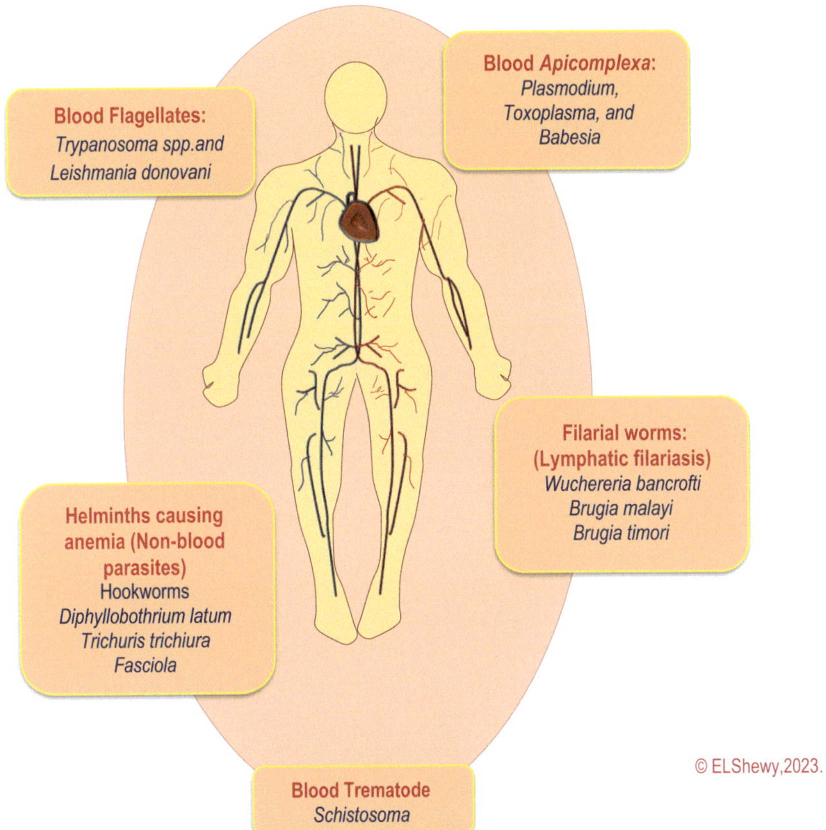

Blood and lymphatic parasites

K. ElShewy, *Medical Parasitology*, https://doi.org/10.1007/978-3-031-64736-9_12

Test your learning and check your understanding of this book's contents: use the "Springer Nature Flashcards" app to access questions using ► https://sn.pub/oyFLIx. To use the app, please follow the instructions in Chap. 1.

Introduction

Blood and lymphatic parasites are those that inhabit the host circulatory system and could be transmitted to another host through blood (blood transfusion, sharing needles, organ transplant, and tissue graft). Most but not all of these parasites are transmitted through vectors. Blood-borne parasites are found worldwide, causing thousands of deaths annually, especially in developing countries. Blood parasites could cause the following:

1. **Hepatosplenomegaly:** e.g., *Trypanosoma, Leishmania donovani, Toxoplasma, Plasmodium, Babesia*, and filarial worms
2. **Leukopenia:** e.g., *Trypanosoma, Leishmania donovani*, and *Plasmodium*
3. **Brain edema, coma and meningoencephalitis***:* e.g., *Plasmodium* and *Trypanosoma*
4. **Bone marrow depression:** e.g., *Trypanosoma* and *Leishmania donovani*
5. **Lymphangitis and lymphadenitis (lymphatic filariasis):** e.g., *Wuchereria bancrofti, Brugia malayi*, and *Brugia timori*
6. **Anemia:** e.g., *Plasmodium, Babesia, Trypanosoma*, and *Leishmania donovani*
7. **Multisystem effects:** e.g., *Trypanosoma, Leishmania donovani, Toxoplasma*, and *Plasmodium*
8. **General symptoms and signs of infection** (fever, malaise, headache, muscle ache, loss of appetite)**:** e.g., most blood-borne parasites during their acute stage of infection

Parasites affecting blood and lymphatics could be one of the following:

1. **Parasites spend most or all of their life cycle in the bloodstream:** e.g., *Plasmodium* and *Babesia*
2. **Parasites inhabiting other organs and that migrate through the blood:** e.g., *Trypanosoma* (Chap. 10), *Leishmania donovani, Toxoplasma* (Chap. 14)
3. **Parasites inhabiting lymphatic vessels and organs:** e.g., *Wuchereria bancrofti, Brugia malayi*, and *Brugia timori*
4. **Non-blood-borne parasites causing alteration of the blood picture of the host:** hookworms *(A. duodenal* and *N. americanus)* (Chap. 2), *Schistosoma, Trichuris trichiura* (Chap. 7), *Fasciola* (Chap. 8)

In this chapter, we will discuss the main blood parasites; other parasites are mentioned in other chapters.

Plasmodia (*P. vivax, P. ovale, P. malariae, and P. falciparum*)

(Malaria)

Plasmodium is an *Apicomplexan* blood-borne protozoan parasite. It is an obligatory intracellular organism inhabiting human blood and tissue cells, causing malaria. Malaria is one of the most serious parasitic diseases in the world. There are four common human species of *Plasmodium* causing different patterns of malaria, *i.e.*, tertian, quartan, subtertian, and malignant malaria, in addition to many animal species that could infect humans.

Geographical Distribution

In 2021, more than 247 million people were infected with malaria, with around 619,000 deaths per year as estimated by the World Health Organization (WHO). Distribution of *Plasmodia* varies according to:

1. Prevalence of vectors (*Anopheles* mosquitoes)
2. Altitudes
3. Climate changes
4. Economics
5. Health facilities

Recently, the geographical distribution of *Plasmodia* has been reported as follows:

1. *P. vivax* is detected in most parts of the world.
2. *P. ovale* is prevalent in Africa, especially in sub-Saharan areas.
3. *P. malariae* is a **less** frequent **rarest** species.
4. *P. falciparum* is **the most dangerous and most prevalent** species and is restricted mainly to the tropics and subtropics although there are sporadic cases in the Mediterranean.
5. *P. knowlesi* is an animal species that affects humans in Indonesia, Malaysia, and Thailand (the fifth human malaria). Also, *P. ovale wallikeri*, *curtisi*, and *cynomolgi* are reported to infrequently infect humans (the sixth human malaria).

Parasite Morphology

Plasmodia could be found in the blood in one or more of the following morphological stages: **ring**, **trophozoite**, **schizont**, and **gametocyte**. Meanwhile, **sporozoite** is the infective stage of *Plasmodium* detected in the mouthparts of the vector *Anopheles* mosquito (Fig. 12.1).

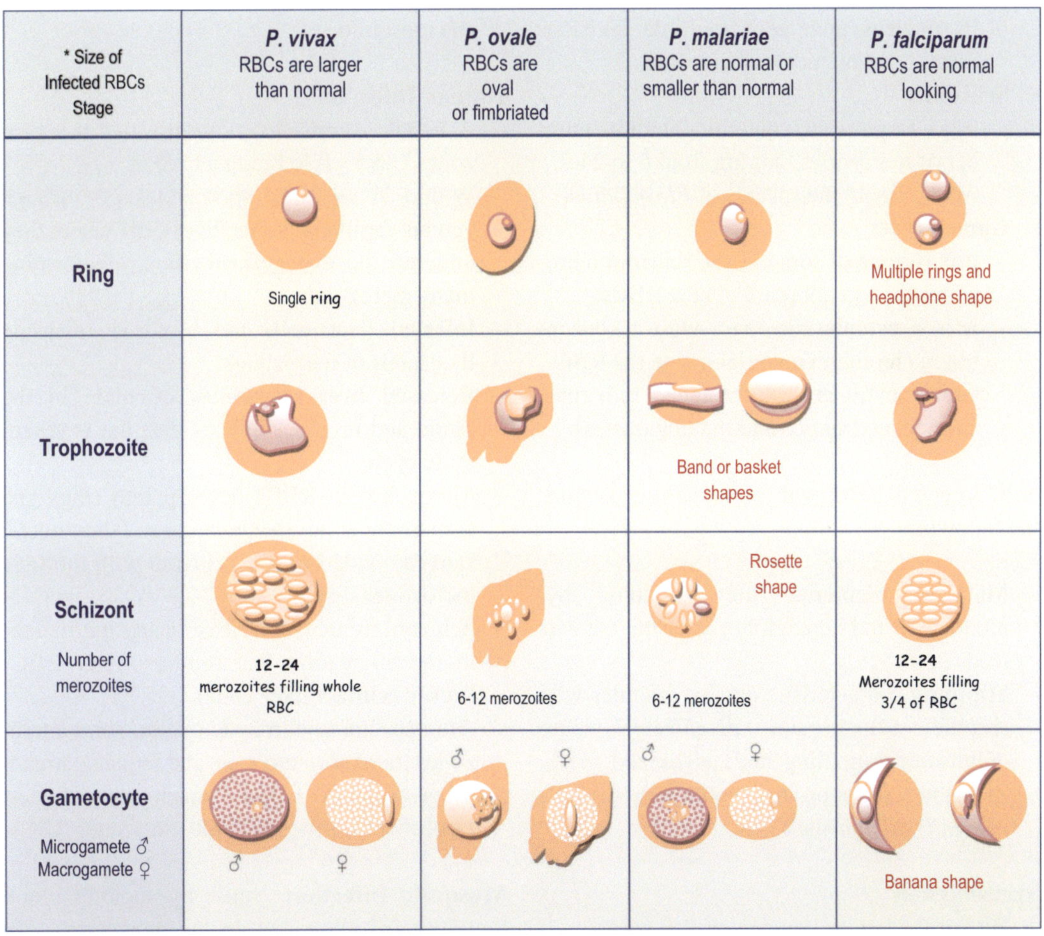

* *P. vivax* infect young RBCs, Infected RBCs with *P. ovale* are ovale in shape with fimbriae, Infected RBCs with *P. malariae* are normal or smaller than normal while RBCs infected with *P. falciparum* are normal looking in size.

Fig. 12.1 Malaria parasites: morphology

- **Sporozoite:**
  - Spindle shape, about 2–5 µm in size.
  - It contains an apex with an apical complex, helping in the penetration of RBCs.
  - It is injected by a mosquito, and once in a liver cell, it is transformed into merozoite.
- **Ring stage:**
  - Merozoite in RBC becomes vacuolated, with a cytoplasmic ring and a chromatin dot.
  - It feeds on the hemoglobin of the infected RBC.
- **Trophozoite:**
  - It is the ring stage after its enlargement and maturation. It is ameboid in shape.
  - Its nucleus undergoes multiple divisions forming many merozoites inside.
- **Schizont:**
  - It is a trophozoite-containing definite number of merozoites ranging from 6 to 24. According to the species of *Plasmodium.*
- **Gametocyte:**
  - It is formed of some of the schizont merozoites that do not reinfect new RBCs.
  - It is a cytoplasmic mass with a concentrated chromatin material inside the RBC.
  - Gametocytes are differentiated into male and female (micro- and macrogametes).

N.B. Infected RBC with *plasmodium* contains the following:

- **Malarial pigments**: the remaining byproducts of the hemoglobin after the parasite feeds on it.
- **Stippling** characteristic dots or granules with specific colors and diagnostic values (Schüffner's stippling in *P. vivax* and *ovale*; Ziemann's stippling in *P. malariae*; Maurer's dots in *P. falciparum*).

Transmission

1. Bite of the female *Anopheles* mosquito
2. Blood transfusion (transfusion malaria)
3. Tattooing
4. Transplacental transmission
5. Sharing contaminated needles
6. Organ transplant
7. Tissue grafting

Habitat

In humans, the parasite resides in red blood corpuscles (RBCs) and in the liver cells.

Life Cycle

The *Plasmodium* life cycle includes three multiplication cycles as follows (Fig. 12.2):

1. Two asexual reproduction (**schizogony**) cycles: take place in the human's liver and RBCs (exoerythrocytic and erythrocytic), leading to many merozoites' formation.
2. One sexual reproduction (**sporogony or gametogony**) cycle: in the mosquito, leading to sporozoite formation in the salivary gland of the mosquito.

Human Infection

- A female *Anopheles* mosquito injects sporozoites (I.S.) while biting a human.
- Within ½ an hour, sporozoites go through venous capillaries to the liver cells where they undergo the **exoerythrocytic cycle** forming many merozoites.
- Infected liver cells are ruptured releasing thousands of merozoites.
- Released liver merozoites circulate in the blood and invade RBCs to start the **erythrocytic cycle.**
- Merozoites in RBCs develop into rings and then trophozoites that enlarge, and their nuclei start to divide forming schizont with multiple merozoites inside.
- Schizont ruptures to release many merozoites in the circulation that reinfect other RBCs (**never reinfect liver cells**).
- After several erythrocytic cycles, some merozoites transform into male and female gametocytes (micro- and macro-gametocytes, respectively) instead of infecting other RBCs.

Mosquito Infection: Starts by sucking blood that contains all stages. In the mosquito's gut, all stages are digested except micro- and macrogametes. They unite forming a zygote and then an ookinete. Ookinete develops into an oocyst containing many sporozoites in the mosquito's salivary gland to infect a new host on biting.

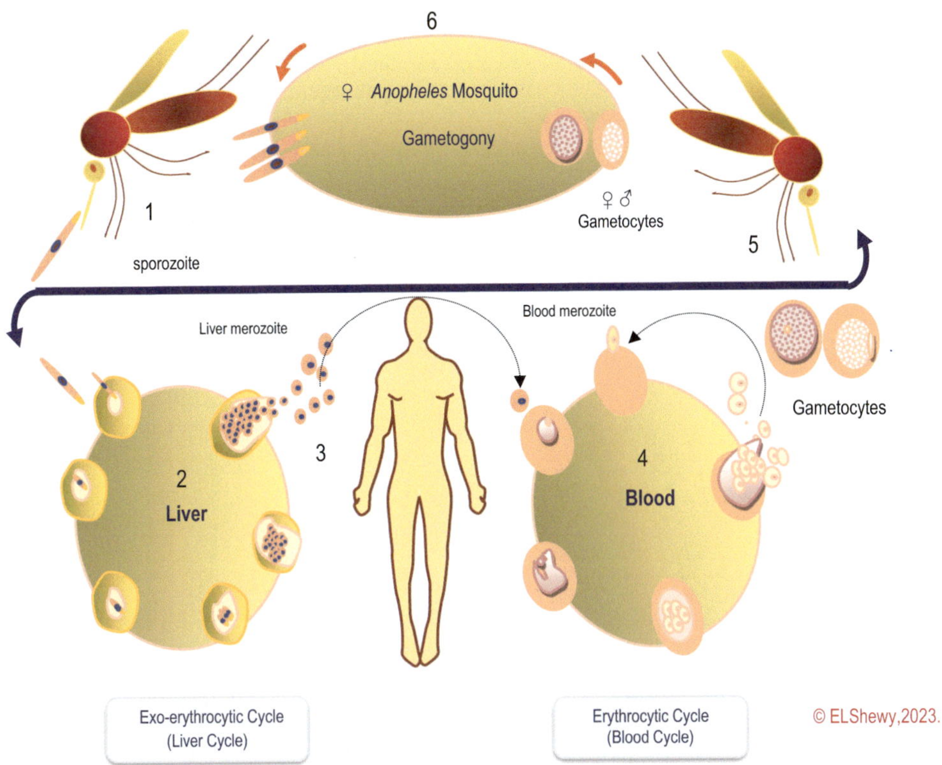

1. Sporozoite injected by ♀ *Anopheles* mosquito into humans.
2. Asexual reproduction in the liver cells (schizogony), forming many liver merozoites.
3. Liver merozoites are released from the liver and infect RBCs.
4. Asexual reproduction in RBCs (schizogony) with formation of many blood merozoites, which reinfect new RBCs (but never reinfect the liver again).
5. ♀+♂ gametes are formed in the blood to be taken by mosquitoes.
6. Sexual reproduction in mosquito (gametogony/Sporogony) to form many sporozoites in the mouth parts of the mosquito.

Fig. 12.2 *Plasmodium spp.:* life cycle

Final hosts:	Female *Anopheles* mosquito
Infective stage:	Sporozoites for humans and for mosquitoes the gametocytes in the blood of an infected person
Diagnostic stage:	Different *Plasmodium* stages in the blood or tissues of the patient
Intermediate hosts:	Humans

N.B.

- All liver merozoites are released to infect RBCs; however, some merozoites of *P. vivax* and *P. ovale* enter a latency period of resting in the liver cells **(hypnozoites),** to be released later causing reactivation **(relapse)** of the infection after complete cure.
  – Blood merozoites **never** reinfect liver cells and only reinfect RBCs, so transfusion malaria does not have liver stages.

Pathogenesis

1. Sporozoites of *Plasmodium* enter human skin (**traversal**) and migrate to invade hepatocytes and Van Kupffer cells, using their apical apparatus and circumsporozoite protein (CSP). This is aided by specific receptors on these cells.

2. After multiplication in hepatocytes, liver merozoites egress to go into circulation and invade RBCs (**egression**) (Fig. 12.3).
3. Infected RBC undergoes changes in its hemoglobin, ferrous store, and surface. It finally ruptures and releases corpuscular stroma, antigens, and hemoglobin content.
4. The pathogenic effects of the *Plasmodium* are mainly due to the following:
   a. Direct parasitic effects:

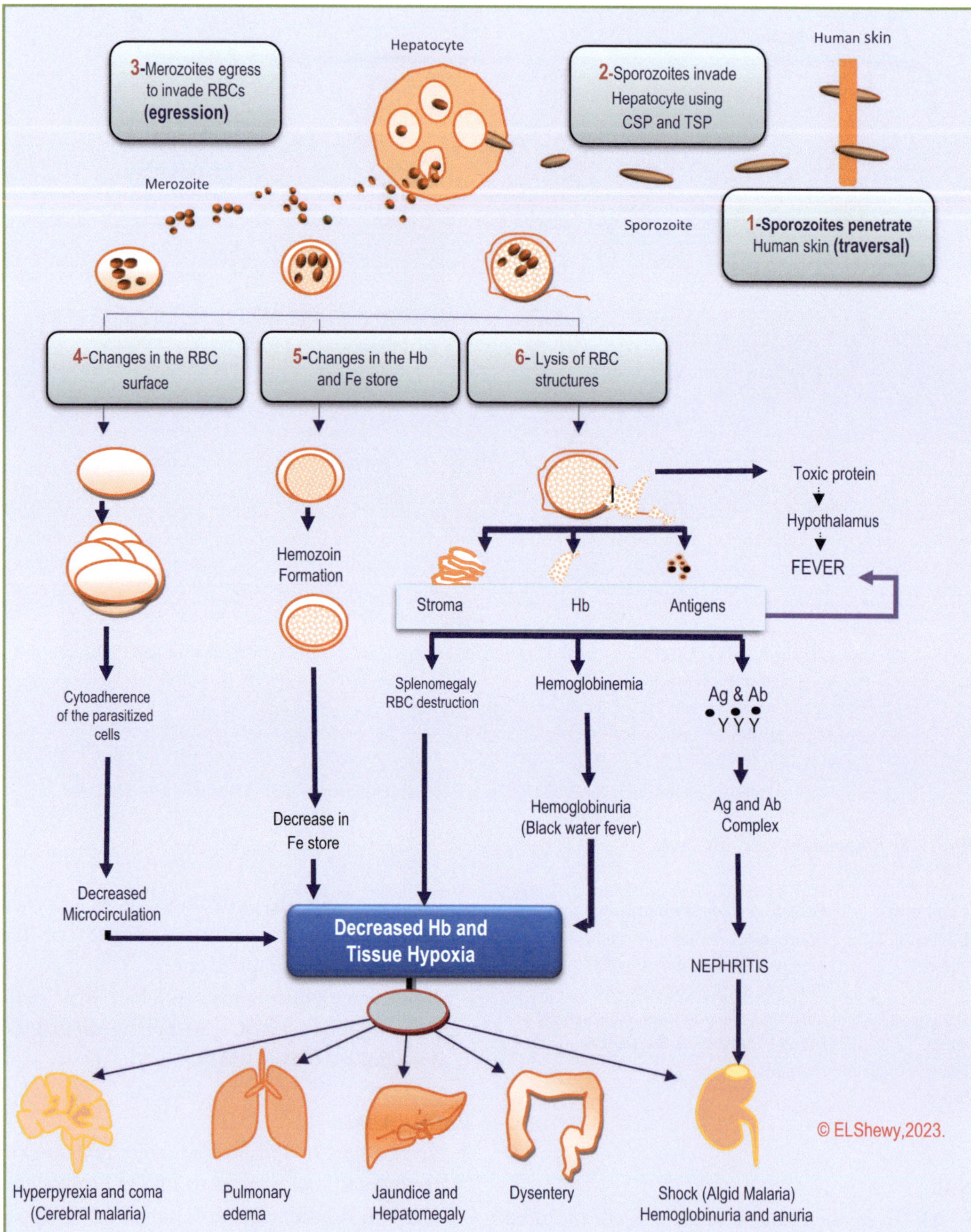

Fig. 12.3 Pathogenesis of malaria

- Hemolysis of the infected and uninfected RBC
- Liberation of the parasite metabolites (secretory and excretory products)
- Malarial pigments
- Cytoadherence of the infected RBCs

b. Indirect effects through the host immune mediators:
- Cytokines
- Tumor necrosis factors (TNFs)
- Interleukins (ILs)

5. All the abovementioned factors share in the formation of severe **anemia** and **hypoxia** with ischemic manifestations in different body organs.
6. The characteristic clinical manifestations of malaria (paroxysm) are mainly due to the multiplication of the parasites in patient RBCs and their subsequent rupture.

Clinical Picture

The clinical picture of malaria includes both uncomplicated and complicated manifestations:

A. **Uncomplicated malaria:** The cardinal symptoms and signs of malaria are fever, chill, paroxysms, splenomegaly, and anemia.
   1. The **incubation period** of malaria infection ranges from a few days (*P. falciparum*) to many weeks (*P. malariae*).
   2. **Prodrome** in the form of headache, photophobia, muscle aches, and vomiting.
   3. **Malarial paroxysm** including:
      - **Cold phase** (cold sensation, shivering, and pale cyanotic skin for 1 h)
      - **Hot phase** (fever, flushed skin, delirious, shaking, and rigor for 1–4 h)
      - **Sweat phase** (decrease in temperature, weakness, exhaustion for 1–4 h)
   4. **Repetition** of the paroxysm according to the erythrocytic cycle: on the third day in *P. vivax* and *ovale* (tertian malaria), on the fourth day in *P. malariae* (quartan malaria), and every 36 h in *P. falciparum* (subtertian or malignant malaria).
   5. **Relapse** of the malarial attack occurs during infection with *P. vivax* and *P. ovale*.
   6. **Recrudescence** occurs with *P. malariae* infection.
      - **Relapse** means repetition of the malarial infection after complete initial clearing of the erythrocytic infection. It implies reinvasion of the RBCs by hypnozoites from the liver cycle. This occurs only in *P. vivax* and *P. ovale*.
      - **Recrudescence** means a recurrence of symptoms in a patient having a low level of parasitemia not sufficient to cause infection. It occurs mainly with *P. falciparum* and *malariae*, and also after infection with drug-resistant strains of *P. vivax* and *ovale*.

B. **Complicated Malaria:**

The following complications are reported during severe uncontrolled malarial attacks, particularly with *P. falciparum* (Fig. 12.3):

1. **Blackwater fever:** due to massive RBC destruction with the appearance of hemoglobin in urine. It is commonly seen during paroxysm of the *P. falciparum*.
2. **Cerebral malaria**: sudden onset of severe headache, drowsiness, confusion, coma, and cerebral edema (steroid is contraindicated).
3. **Anemia:** hemolytic anemia, especially in *P. falciparum*.
4. **Renal diseases**: chronic *P. malariae* infection in children causes deposition of the immune complex on the glandular walls leading to nephritis. *P. falciparum* causes renal ischemia, renal necrosis, and acute renal failure.
5. **Hypoglycemia:** as the result of the antimalarial drugs that stimulate the insulin to be released or due to deficient gluconeogenesis (liver impairment).
6. **Dysenteric malaria**: chronic intestinal ischemia leading to abdominal pain and dysentery.
7. **Algid malaria**: ischemia of the adrenals due to a *falciparum* malarial attack with rapid hypotension, rapid weak pulse, pale cold skin, and deliriousness.

8. **Pulmonary edema**: due to disseminated intravascular coagulation, pulmonary ischemia, and oligemic anemia.
9. **Tropical splenomegaly syndrome (TSS)**: chronic splenomegaly, increased IgM, decreased T-cells, abdominal distension, reticulocytosis, and thrombocytopenia. Usually, patients with TSS receive antimalarial drugs for life.

P. vivax and *ovale* are relatively benign. However, infection with *P. falciparum* is considered malignant (fatal), causing localized capillary obstruction, severe tissue hypoxia, and infarction. These effects during *P. falciparum* malaria are due to:

1. Multiple RBC infections (two or more than two rings infecting one RBC)
2. Accollé formation with the tendency of the infected RBCs to adhere together
3. Increased intracapillary clot formation
4. Short incubation period with prolongation of the primary attack
5. Multiple and frequent paroxysms (every 36 h)
6. High parasitemia (*P. falciparum* could infect any type of RBCs)

Immune Response

Both natural innate and acquired immunity share in immunity against malaria:

1. Natural immunity against malarial infection is noticed in the following cases:
   - Duffy blood group gives resistance against *plasmodium* in West Africa.
   - Patients with sickle cell anemia (HbS) are resistant to *P. vivax*.
   - Patients with G-6-phosphate dehydrogenase deficiency are resistant to *P. falciparum*.
2. Infants during the first year of life gain a relative resistance against malaria as the result of the maternal antibodies.
3. Acquired immunity against malarial infection could be gained. However, acquired immunity is species specific and not absolute.

Diagnostic Procedures

In addition to the clinical picture characterizing infection with malaria, the following procedures are commonly used for confirming diagnosis:

- **Direct:**
  1. Microscopic examination of **thin and thick blood films** stained with Giemsa stain. Blood film is made by using a finger brick or venous blood.
  2. Concentration of the venous blood samples using microhematocrit centrifugation, buffy coat method, and Knott's concentration and membrane filtration method.
  3. **Provocation test:** to increase the detection level, a small dose of adrenaline is injected before blood sampling.
  4. **Image analyzer:** depending on artificial intelligence and database to read microscopic slides.

N.B.

- Blood film is the gold standard for the detection of malaria infection.
- Every species of malaria has its characteristic blood stage(s), e.g., band-shape trophozoite and rosette-shape schizont in *P. malariae*; multiple rings and banana-shape gametocyte in *falciparum, etc.*
- Blood stages (schizonts and gametocytes) of *P. falciparum* are rarely seen in the peripheral blood; instead, they are detected in the tissues (Table 12.1).

- **Indirect Immunological:**

  Detection of specific malarial antigens (**biomarkers**) in the patient's blood is done through various tests. These tests are called rapid diagnostic tests (RDTs) including:

  a. **ParaSight-F test (HRP2):** A dipstick test used for capturing malarial antigenic enzyme: histidine-rich protein 2 (HRP) in a blood sample. It takes only 10 minutes. It could give a positive result after the cure.
  b. **OptiMAL test:** A dipstick used for capturing malarial antigenic enzyme (LDH) in a

Table 12.1 Clinical and parasitological differences between different *Plasmodia* species

	P. vivax	*P. ovale*	*P. malariae*	*P. falciparum*
Distribution	All over the tropics, subtropics, and temperate regions	West Africa, especially in sub-Saharan areas	Less frequent in the tropics, subtropics (less common form of malaria)	Limited to the tropics and subtropics (The most common form of malaria)
Disease	Benign tertian malaria	Tertian malaria	Quartan malaria	Subtertian malignant malaria
Prepatent period[a]	8–12 days	8–12 days	15–18 days	5–9 days
Incubation period[b]	8–27 days	8–27 days	16 days–8 weeks	6–25 days
Hypnozoites	+ve	+ve	−ve	−ve
Relapse	+ve	+ve	−ve	−ve
Recrudescence	−ve	−ve	+ve	−ve
Sporogony	9–12 days	14–15 days	15–21 days	11–14 days
Exoerythrocytic (liver) cycle	8 days	9 days	15 days	5–7 days
Erythrocytic cycle	44–48 h	44–48 h	72 h	36 h
Host RBCs	Young RBCs—infected RBC is enlarged	Young RBCs—infected RBC is oval and fimbriated	Older RBC—infected RBC is not enlarged	Any RBC can be infected
Stippling	Schüffner's granules	James's dots	Ziemann's dots	Maurer's clefts
Malarial pigments	Fine golden–brown granules	Coarse brown granules	Brown or black coarse granules	Black to dark brown
Ring stage	Single ring	Single ring	Single ring	Multiple rings
Trophozoite	Ameboid	Ameboid	Typical band/basket form	Small, delicate, appliqué/accollé forms, double-chromatin dots
Schizont	12–24 merozoites	6–12 merozoites, with irregular arrangement	6–12 merozoites daisy-head (rosette)	12–24 merozoites only in tissues
Gametocyte	Occupies most of the RBCs	Occupies most of the RBCs, but smaller size	Uncommon round, occupies the entire RBC	♀Crescent shape, central mass of chromatin and pigment ♂ Sausage shape, diffuse chromatin, and pigment

[a]Prepatent period: the time between infection and appearance of the parasite in the blood
[b]Incubation period: the time from infection to the appearance of clinical manifestation
Schizonts and gametocytes of *P. falciparum* are not found in blood mainly detected in tissues

blood sample. It is only positive during the active disease (needs living parasites).

c. **ICI Malaria Pf**: A dipstick used for capturing malarial antigens of *P. falciparum* in the blood sample.
d. **BinaxNow**: Detects both aldolase found in all species and HRP of *P. falciparum.*

- **Molecular Diagnosis:**
  - PCR is commonly used for the amplification of the malarial DNA in blood or tissue samples, e.g., loop-mediated isothermal amplification (LAMP).
  - It is mostly used for the confirmation of diagnosis and in research work, not for field work. It is used for the detection of drug-resistant malaria.
  - It helps in the identification of new species and target genes for treatment, vaccine development, and control.
- **Other Investigations:**

1. Ultrasonography
2. Plain X-rays
3. Renal and liver function tests

Treatment

- **Supportive measures:** bed rest, fluid intake, antipyretics, and cold fomentation.
- **Antimalarial chemotherapeutics:** according to their effects, antimalaria drugs are grouped as follows:
  1. **Suppressive drugs**(4-aminoquinolines):
     - Drugs that suppress parasite multiplication when they enter the body of an **infected** person.
     - Examples: quinine, quinidine, chloroquine, hydroxychloroquine, amodiaquine, pyrimethamine, mefloquine, tetracycline, and norfloxacin.
  2. **Drugs causing clinical cure** (4-aminoquinolines):
     - Drugs capable of elimination of the erythrocytic blood stages of the parasites **(blood schizonticides).**
     - Examples: the same as the suppressive drugs but given in higher doses.
  3. **Drugs causing radical cure** (8-aminoquinolines):
     - These drugs eliminate both the blood and tissue stages of the parasites **(tissue schizonticides).** Examples: primaquine and tafenoquine.
  4. **Gametocidal drugs** (transmission-blocking drugs):
     - Drugs that kill gametocytes, rendering patients not infective to the mosquitoes.
     - Examples: chloroquine, amodiaquine, and primaquine (an effective gametocidal drug for all malaria species).
  5. **Prophylactic drugs:**
     - Drugs suppress parasites when they enter the body of an **uninfected** person.
     - They are taken before groining, during and after leaving the malarious areas.
     - Examples: daraprim, proguanil, lariam, fansidar, and oxytetracycline.
     - **Daraprim** destroys the early stages of the parasites as they enter the liver.
     - **Fansidar** and lariam are given for prophylaxis in individuals with CRPF.
  6. **Drugs for chloroquine-resistant *Plasmodium falciparum*** (CRPF):
     - Drugs that overcome *P. falciparum* resistance to antimalarial drugs.
     - Examples: lariam, tetracycline, doxycycline, norfloxacin, and ciprofloxacin.
     - Pyrimethamine-sulfadoxine-efloquine (**Fansimef**) is a single-dose treatment (two tablets once).
  7. **Artemisinin** (artemether, artesunate, artemether): Artemisinin is derived from the "sweet wormwood" plant *Artemisia annua*; it is the first choice for the treatment of uncomplicated and complicated malaria.
  8. **Artemisinin compound therapy (ACT):** Artemisinin is combined with other drugs to avoid the drug resistance recently detected. It is combined with lumefantrine, mefloquine, amodiaquine, sulfadoxine/pyrimethamine, piperaquine, and chlorproguanil/dapsone.
  9. **Triple-artemisinin compound therapy (TACT)**: Artemether-lumefantrine-amodiaquine has recently been used to overcome the resistance of ACT.
  10. **The following are recent drugs used in malaria:**
      a. Atovaquone-proguanil (**Malarone**)
      b. Artemether-lumefantrine (**Coartem**)
      c. **Clindamycin** (used in combination with quinine)
      d. Tafenoquine **(Arakoda and Krintafel)** (alternatives to primaquine)

Prevention and Control

A. **Chemoprophylaxis:**
   - Chemotherapeutic drugs should be administered before going to the malarious areas and after leaving for at least 4 weeks. Proguanil (Paludrine) is taken as 200 mg once daily. Other drugs like mefloquine and doxycycline are also used.

B. **Mosquito control:**
   - The different ways include breeding of mosquito's predators, release of a vast number of sterile male mosquitoes, and

use of mosquito nets, mosquito screens, repellents, and insecticides.

C. **Malarial vaccines:**
   - There are many forms of antimalaria vaccines: pre-erythrocytic vaccines, i.e., anti-sporozoites (anti-infection); anti-gametocyte vaccines (transmission blocking); and anti-merozoite vaccines (anti-morbidity).
   - RTS, S and RTS, S/AS01 (commercial name: **Mosquirix**), and **R21** are the most recently developed recombinant vaccines. They are the first of the current generation of malaria vaccines to show more than 50% effectiveness in reducing malaria.

Babesia (*B. microti, B. divergens, B. duncani*)

Babesia species are blood *Apicomplexan* parasites. They are the causative agents of tick-borne disease of domestic animals and humans called babesiosis. Babesiosis closely mimics malaria but with a self-limited, nonfatal pattern except in splenectomized patients.

Geographical Distribution

There are many species of *Babesia,* but only a few infect humans: *B. divergens, B. microti*, and *B. duncani.* They have a worldwide distribution at the sheep-raising countries of New England, in Europe, and in many areas of Africa and Asia.

Parasite Morphology

- *Babesia* is an obligate intracorpuscular parasite.
- *Babesia* exists only as a ring in the infected RBCs, which should be carefully differentiated from that of malaria parasites.
- Ring stages of *Babesia* are morphologically similar with the following differences:

- ***B. microti*** **ring stage:**
  - It resembles the ring stage of *P. falciparum.*
  - It has a variable sized vacuole with a large chromatin dot.
  - Infected RBC is not enlarged with no pigments, nor stippling.
- ***B. divergens*** **ring stage:**
  - It is larger than *B. microti.*
  - It takes an oval teardrop (rabbit ear) shape or a tetrad form (Maltese cross).
  - There are no pigments, nor stippling.

Transmission

Babesia spp. are transmitted by several ixodid ticks including *Ixodes scapularis* (a vector of Lyme disease) for *B. microti* and *Ixodes boophilus* for *B. divergens.*

Other methods for transmission of *Babesia* are:

1. Blood transfusion
2. Tattooing
3. Transplacental transmission
4. Sharing contaminated needles

Habitat

Babesia spp. inhabit RBCs of humans and domestic animals.

Life Cycle

The life cycle of *Babesia* comprises sexual reproduction in the ticks and asexual reproduction in the mammalian host (human, rodents, and cattle) (Fig. 12.4):

- It starts by injecting the infective stage (sporozoite or ovule) by tick bite.
- In humans, *Babesia* sporozoite invades RBCs and undergoes asexual reproduction.
- Ticks acquire the parasite by feeding on *Babesia*-infected mammals (rodents and cattle) and transmit the parasite to their offspring by transovarial passage.
- Within the ticks, the sexual cycle (sporogony) takes place and the infective sporozoites (ovules) are formed.

Final hosts:	Ticks
Infective stage:	For humans, it is the sporozoite (ovule) inoculated by the ticks, and for the tick, ring stages ingested from the infected host
Diagnostic stage:	Ring stages in the blood
Intermediate hosts:	Cattle, rodents, dogs, and humans

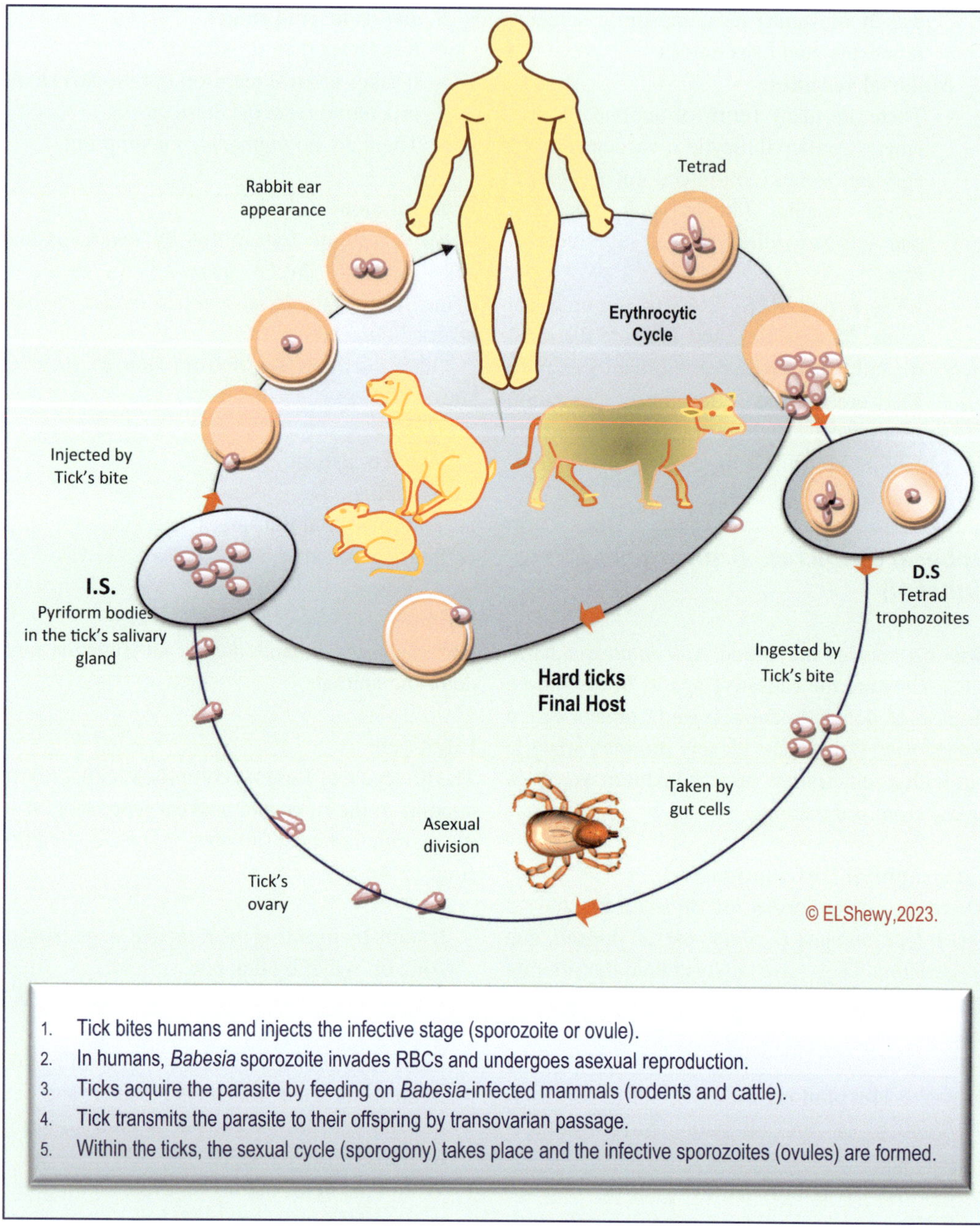

Fig. 12.4 *Babesia*: life cycle

Pathogenesis

1. *B. divergens* is more dangerous than *B. microti*.
2. Infection with *Babesia* results in hemolysis of the infected RBCs.
3. Hepatomegaly with increased serum bilirubin and transaminase enzymes.
4. Splenomegaly as the result of marked RBC destruction.
5. Disseminated intravascular coagulopathy (DIC) has been detected in many cases of babesiosis.

Clinical Picture

1. High fever, malaise, muscle and bone ache, sweating, chest pain, shortness of breath (air hunger), anemia, jaundice, and renal troubles.
2. Later, hepatosplenomegaly with marked deterioration of the liver functions.
3. No periodicity (paroxysms), as in malaria.
4. In splenectomized patients, the disuse is usually fatal.

Diagnostic Procedures

- **Direct:**
  1. Microscopic examination of thin and thick blood films stained with Giemsa stain. Peripheral blood samples could be used for blood smears. Bone marrow, liver, and spleen aspiration could be used for microscopy.
  2. Animal inoculation.
- **Indirect Immunological:**
  Detection of specific serum antibodies using ELISA and IHA.
- **Other Investigations:**
  1. Ultrasonography.
  2. Renal and liver function tests.

Treatment

1. **Quinine-clindamycin combination:**
   - Clindamycin is given as 300–750 mg every 6 h by intravenous injection (I.V.).
   - Quinine is given in a dose of 600 mg/6 h for 10 days orally.
   - This combination decreases parasitemia and causes relief of the symptoms.
2. **Azithromycin:**
   In severe cases, give I.V. plus oral quinine for 7–10 days.
3. **Atovaquone:**

Prevention and Control

1. Avoidance of tick bites in endemic areas by wearing protective clothes.
2. Tick's repellents and insecticides.
3. Blood examination immediately following suspected exposure.

Leishmania

- *Leishmania* is an obligate intracellular blood and tissue flagellate. It is transmitted mainly by different species of sandfly vectors.

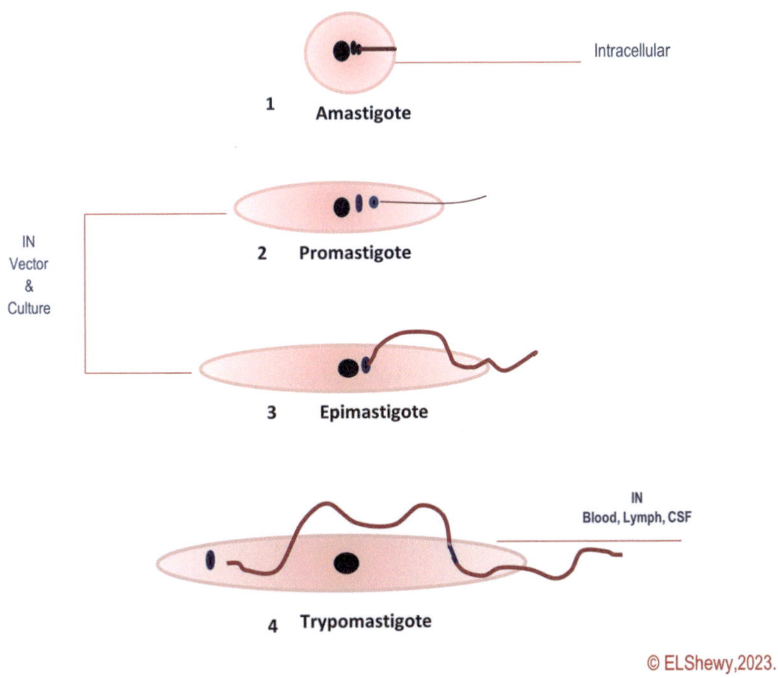

Fig. 12.5 Morphological forms of blood flagellates

- *Leishmania* acquires two forms, from the four morphological forms of the blood flagellates: amastigote intracellularly and promastigote in vector and culture (Fig. 12.5).
- *Leishmania* includes different species, each of which forms together a specific complex with the same life cycle. Each complex is responsible for specific disease as follows:
  a. Cutaneous leishmaniasis caused by *L. tropica* and *L. mexicana* complexes.
  b. Mucocutaneous leishmaniasis caused by *L. braziliensis* complex (Chap. 16).
  c. Visceral leishmaniasis (kala-azar) caused by *L. donovani* complex.
- *L. donovani* complex is the only blood-borne *Leishmania spp.*, with a viscerotropic tendency to infect lymphoid organs and cells.

Leishmania donovani (*L. donovani*)

Visceral Leishmaniasis (Kala-Azar)

L. donovani is a blood flagellate causing visceral leishmaniasis (**kala-azar**). It inhabits the reticuloendothelial cells and organs of humans and dogs (**zoonotic**). Visceral leishmaniasis is endemic in certain localities, transmitted by sandfly vectors and through blood.

Geographical Distribution

Leishmania donovani complex comprises three species:

1. ***L. donovani*** in India, Pakistan, and parts of Africa
2. ***L. infantum*** in the Mediterranean, Europe, and North Africa
3. ***L. chagasi*** in Central and South America

L. donovani is endemic in India, and occasionally, it is detected as an epidemic in East African countries (Ethiopia, Kenya, South Sudan, and Sudan) with high morbidity and mortality in these communities.

Parasite Morphology

L. donovani acquires the amastigote and promastigote forms of the blood flagellates (Fig. 12.5).

- Amastigote:
  - Oval to round body, 3–5 μm.
  - It has one eccentric nucleus.
  - The kinetoplast consists of a dot-like pattern.
  - Blepharoplast with small axonemes and parabasal body.
  - No flagellum.
- Promastigote:
  - Long slender body, 9–15 μm.
  - It has one central nucleus.
  - Kinetoplast at the anterior end with no undulating membrane.
  - Single anterior free flagellum.

Transmission

L. donovani could be transmitted through:

1. Sandfly: there are two forms of transmission of *L. donovani* through sandfly:
   a. **Urban:** from human to human through sandfly
   b. **Rural:** from animals (dogs) to humans through sandfly
2. Blood transfusion
3. Tissue grafting
4. Organ transplant

Habitat

- *Leishmania* **amastigote** forms are detected intracellularly in the following sites:
  a. Reticuloendothelial cells (macrophages, monocytes, and other phagocytic cells)
  b. Lymphoid tissues and organs (liver, spleen, lymph nodes, and bone marrow)
- The **promastigote** forms are detected only in the culture and in the sandfly vector.

Life Cycle

1. A female sandfly becomes infected by drawing a blood meal containing amastigotes from an infected host.
2. Inside sandfly, amastigotes transform to promastigotes and multiply.
3. Sandfly feeds again on another host and injects promastigotes to be engulfed by reticuloendothelial cells (RECs).
4. In REC, promastigotes metamorphose into amastigotes and multiply until infected cells rupture to reinfect other adjacent cells (Fig. 12.6).

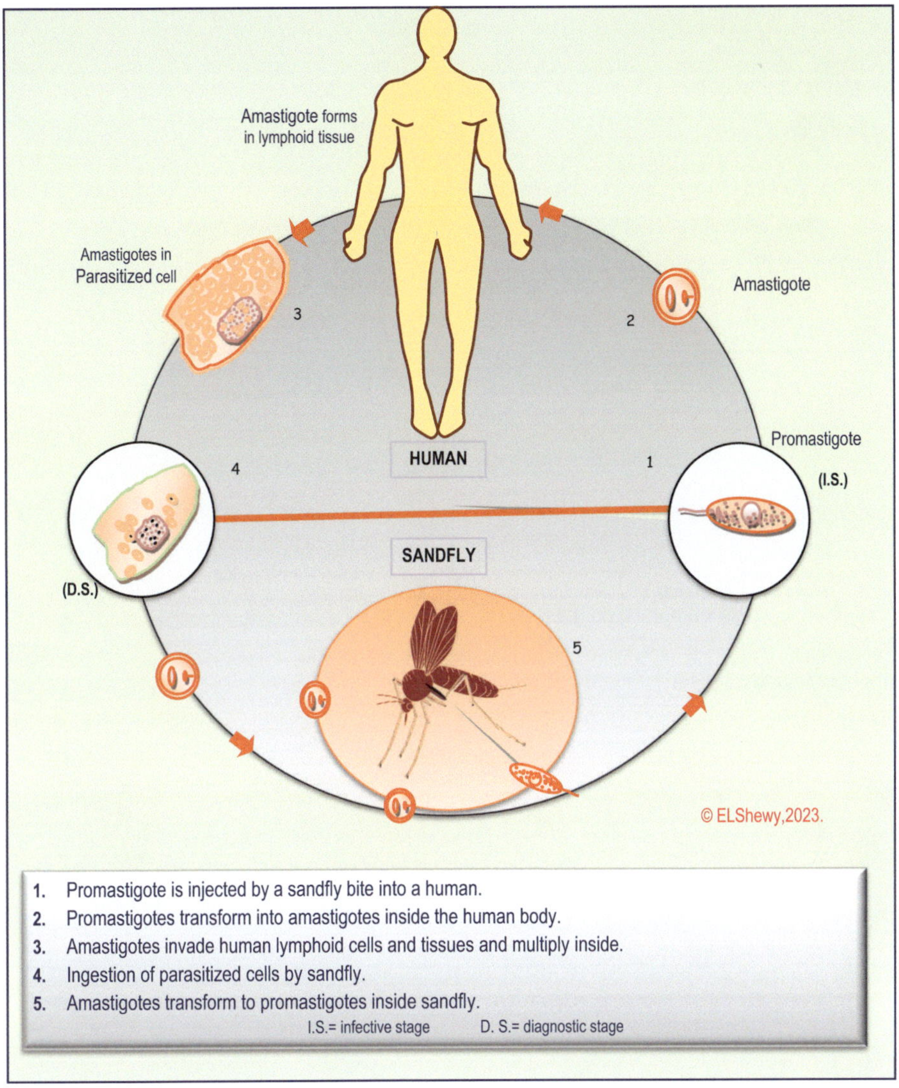

Fig. 12.6 *Leishmania spp.*: life cycle

Final hosts:	Humans and dogs
Infective stage:	For humans, promastigote is inoculated by the sandfly, and for sandflies, amastigote is taken from the infected host
Diagnostic stage:	Amastigote
Intermediate host:	Sandfly
Reservoir hosts:	Dogs are the main reservoir hosts for *L. donovani*

Pathogenesis

The following pathological events are demonstrated during infection with *L. donovani* (Fig. 12.7):

1. Sandfly injects promastigotes onto human skin which are taken by skin phagocytic cells and transform intracellularly into amastigotes.
2. Intracellular amastigotes proliferate rapidly.

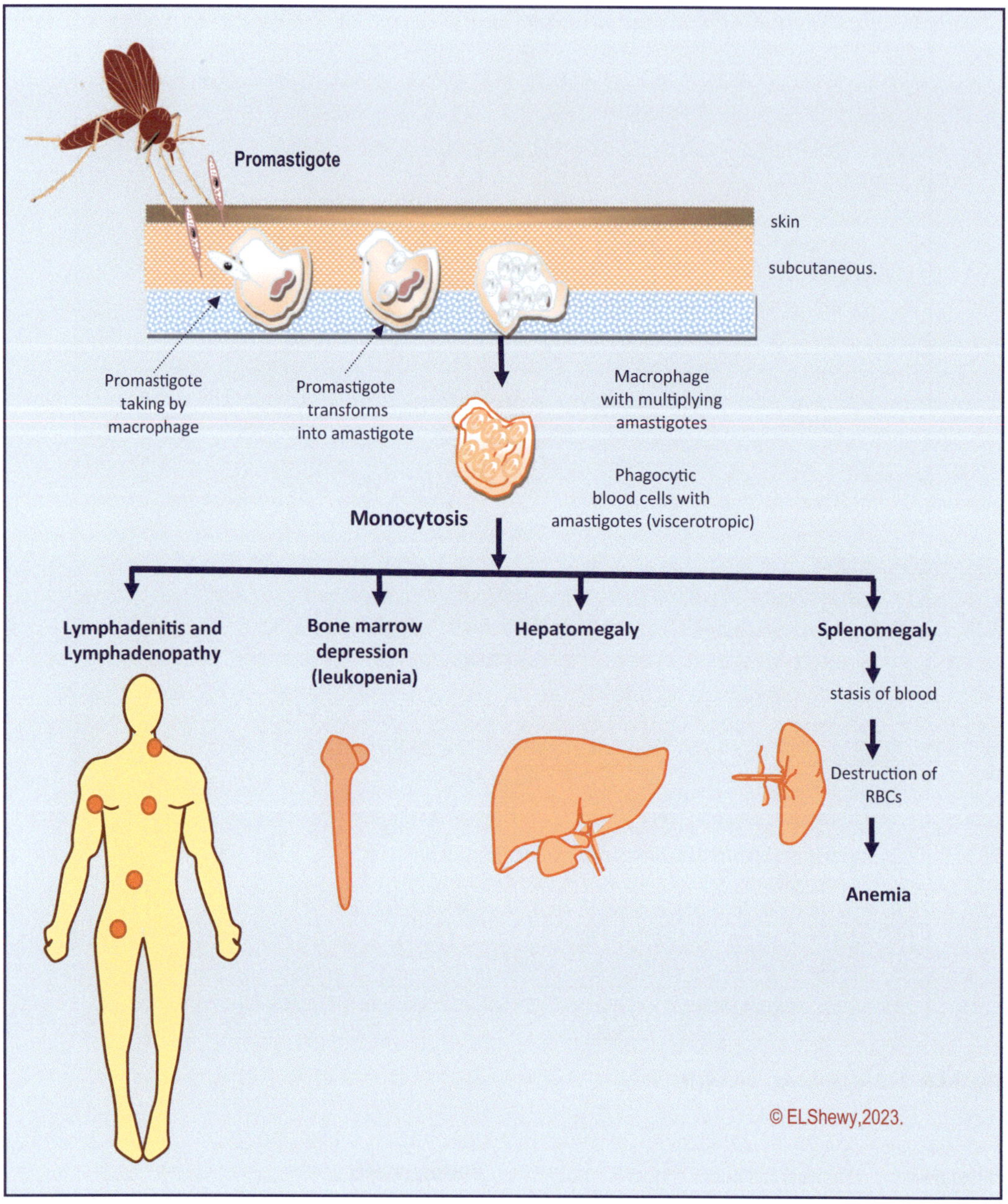

Fig. 12.7 Pathogenesis of visceral leishmaniasis

3. The infected phagocytes (mainly monocytes) rupture releasing many amastigotes through the bloodstream to infect the lymphoid tissues and cells all over the body, especially the liver, spleen, and bone marrow.
4. Rapid proliferation of the amastigotes in infected cells leads to hypertrophy of the tissues containing them.
5. Hypertrophy of the liver and spleen causes **hepatosplenomegaly**.
6. Splenomegaly leads to stasis of the blood in its sinusoids with **hypersplenism** and increased RBC destruction causing **anemia**.
7. Bone marrow hyperplasia leads to hemopoietic disturbances, e.g., **leukopenia** and

monocytosis, with depressed immunity during active infection.
8. Recovery is followed by a **long-lasting immunity** in most cases of visceral leishmaniasis.

Clinical Picture

Visceral leishmaniasis, also known as black skin fever **(kala-azar)** in India, is characterized by the following:

1. The incubation period is around 2 weeks to 2 years after the sandfly bite.
2. Fever (double- or triple-peak fever), for 2–4 weeks.
3. Lymphadenopathy.
4. Marked anemia and weight loss.
5. Hepatosplenomegaly with ascites.
6. Diarrhea with dysenteric manifestation.
7. Darkening of the forehead skin (black skin fever)

Diagnostic Procedures

In endemic areas, diagnosis of visceral leishmaniasis depends on the clinical picture together with the following laboratory diagnostic procedures:

- **Direct:**
  1. **Microscopic** demonstration of the parasites in a blood film or in tissue samples taken from either the spleen, liver, sternum, or buffy coat layer of a venous blood sample.
  2. **Staining** of blood film or tissue samples with Leishman or Giemsa stain for demonstration of the amastigotes (leishmanoid bodies).
  3. **Culture** of the specimen for demonstration of the promastigotes in Nicole, Navy, and Neil (NNN) medium.
  4. **Animal inoculation** of the tissue samples intraperitoneally in the hamster. The infection is noticed later on the animal.
- **Indirect Immunological:**
  1. **Montenegro skin test** (Leishmanin test)**:** It is a hypersensitivity skin test including intradermal injection of a killed suspension of promastigotes. Elevated skin with induration is a positive reaction. It is negative in active kala-azar (depressed cell-mediated immunity) but becomes positive after 2 months of successful treatment (test for the effect of treatment or test of cure).
  2. **ELISA, IFA**.
  3. Direct agglutination test (**DAT**) is a rapid dipstick test for detection of leishmaniasis in survey and field study.

Treatment

In addition to the general lines for the treatment of infection (antipyretics, foments, ample fluids, bed rest, and suitable feeding), the following drugs are commonly used:

1. **Sodium stibogluconate** (Pentostam)**:**
   - 10 mg/kg body weight daily for 10 days, by I.V. or I.M. injection
   - Up to three courses every 10-day interval.
2. **Pentamidine:**
   - 5 mg/kg/once a day/10 days, by I.M. injection.
3. **Allopurinol:**
   - 20 mg/kg body weight three times for one day.
   - It could be used in AIDS patients with visceral leishmaniasis.

The most recent drugs for visceral leishmaniasis include both amphotericin B (currently recommended as the drug of choice for kala-azar) and miltefosine:

1. **AmBisome:**
   - Total dose 7.5–21 mg/kg over 6 days.
2. **Amphocil:**
   - Total dose 7.5 mg/kg over 6 days.
   - A low dose (0.5–1 mg/kg) is given on the first day, increasing to 1–2 mg/kg on the second day, followed by 1.5–3 mg/kg on the third and subsequent days.
3. **Miltefosine:**
   - It is the first oral treatment for this disease. It could be taken alone.
   - The cure rate of miltefosine is 95%.

- In addition, amphotericin B single dose with 7 days of oral miltefosine or miltefosine plus paromomycin for 10 days is very effective.

Prevention and Control

1. Vaccination by using anti-*Leishmania* vaccine (**LEISH–F3 + GLA-SE**), which is a highly purified, recombinant vaccine. It is the first ever vaccine to prevent visceral leishmaniasis.
2. Controlling sandfly is important through the following:
   - The use of insecticides in endemic areas is important for travelers.
   - House and space spraying.
   - Pyrethroid-impregnated bed nets.
   - Destruction of rodent reservoirs.
   - Pyrethroid-impregnated curtains.
3. Screening and treatment of the infected cases.
4. Combating the dogs and other animal reservoirs.

- **Parasitology**: *L. donovani* a blood flagellate, which exists in amastigote form in RECs and lymphoid tissues and in the promastigote form in sandfly vector and in NNN culture. It is transmitted by blood transfusion and congenital and organ transplant.
- **Clinical:** Visceral leishmaniasis (kala-azar), manifested by fever, hepatosplenomegaly, lymphadenopathy, dysentery, ascites, bone marrow depression, and leukopenia. Infection is followed with a long-lasting immunity.

Filarial Worms

Filarial worms are tissue threadlike nematodes that could infect lymphoid, skin, eye, and subcutaneous tissues. They are characterized by being heterophroditic, and their females are larviparous (laying larvae called microfilariae).

The following are the general criteria for filarial worms:

Adult Filarial Worms:

- They are long threadlike nematodes, with lengths ranging from a few millimeters to many centimeters.
- They need arthropod vectors for their transmission.
- Adult female produces microfilariae that circulate in the peripheral blood, lymph, and lymphatics or accumulate in the skin and subcutaneous tissues.

Microfilariae:

- Their lengths range from 150 to 350 μm, and their widths range from 4 to 10 μm.
- They may have or not have sheath (sheathed and unsheathed microfilariae).
- They could have a specific periodicity inside the human host.
- Tail morphology is a characteristic.
- They have a characteristic nuclei arrangement.

The most common human filarial worms include:

1. *Wuchereria bancrofti (W. bancrofti)*, infecting lymphatic vessels and organs.
2. *Brugia malayi (B. malayi)* and *Brugia timori (B. timori)* infect the lymphatic system (vessels and organs).
3. *Loa loa (L. loa)* infects the eye and skin.
4. *Onchocerca volvulus (O. volvulus)* infecting the skin, subcutaneous tissues, and eye.
5. *Mansonella spp*. (less common filarial worms infecting blood and skin).

In this section, we will discuss *W. bancrofti*, *B. malayi*, and *B. timori* as parasites infecting blood and the lymphatic system causing lymphatic **filariasis**.

Other filarial worms will be discussed in the section of special senses as they could infect eye and skin tissues.

A. *Wuchereria bancrofti* (*W. bancrofti*)

(Lymphatic filariasis)

W. bancrofti is a filarial tissue nematode, transmitted by mosquitoes. It causes a chronic long-standing disease entity termed bancroftian filariasis (lymphatic filariasis or elephantiasis).

Geographical Distribution

W. bancrofti has an extensive distribution throughout the tropical and subtropical regions of Central Africa, Asia, the Pacific Islands, and Central and South America.

Parasite Morphology

W. bancrofti harbors the general morphological criteria of the filarial worms in addition to the following (Table 12.2):

- Adult worm (♀):
  - Long threadlike worm, with a size of 100 × 0.25 mm.
  - It has two uteri.
  - Its posterior end has a pointed tip.
- Adult worm (♂):
  - Shorter threadlike worm with a size of 40 × 0.1 mm.
  - The posterior end has a curved tail.
  - It has two unequal posterior spicules.
- Microfilaria:
  - A small threadlike **sheathed** body with a size of 230–320 × 10 μm.
  - The body demonstrates graceful curves in stained preparations with small discrete nuclei.
  - Tail is free from nuclei.
  - It demonstrates serpentine motility in a wet mount.
  - It has a **nocturnal periodicity**.

Transmission

W. bancrofti is transmitted mainly through mosquito (*Culex, Aedes*, and *Anopheles*) bite, inoculating infective microfilariae into the human body; other methods of infection are:

1. Blood transfusion
2. Tattooing
3. Organ transplants
4. Sharing contaminated needles

Habitat

Adult *W. bancrofti* could inhabit any lymph nodes, particularly the axillary, epitrochlear, inguinal, and pelvic lymph nodes, together with the lymphatic distal to them, while their microfilariae could be detected in:

- **Peripheral blood** at night (nocturnal periodicity) and in **pulmonary capillaries** in the daytime
- The lymph and hydrocele fluid

Life Cycle

In mosquitoes, sheathed microfilariae are taken by the mosquito during a blood meal from an infected person.

Table 12.2 Human filarial worms

	Disease	Microfilaria	Vector	Habitat of adult worm	Periodicity
Wuchereria bancrofti	Lymphatic filariasis	**Sheathed** (loose) **In blood**	*Culex, Aedes, Anopheles*	Lymphatics	**Nocturnal**
Brugia malayi	Lymphatic filariasis	**Sheathed** (loose) **In blood**	*Mansonia, Culex, Aedes*	Lymphatics	**Nocturnal**
Loa loa	Blindness Calabar swelling	**Sheathed** (tight) **In blood**	*Chrysops*	Skin and subcutaneous	**Diurnal**
Mansonella ozzardi	Occasional serosal and peritoneal infection	**Unsheathed** In blood	*Midge*	Peritoneal cavity	**No periodicity**
Mansonella perstans	Occasional serosal and peritoneal infection	**Unsheathed** In blood	*Midge*	Peritoneal cavity	**No periodicity**
Onchocerca volvulus	River blindness Onchodermatitis (sowda) Onchocercoma	**Unsheathed** In skin and subcutaneous	*Simulium*	Skin and subcutaneous	**No periodicity**

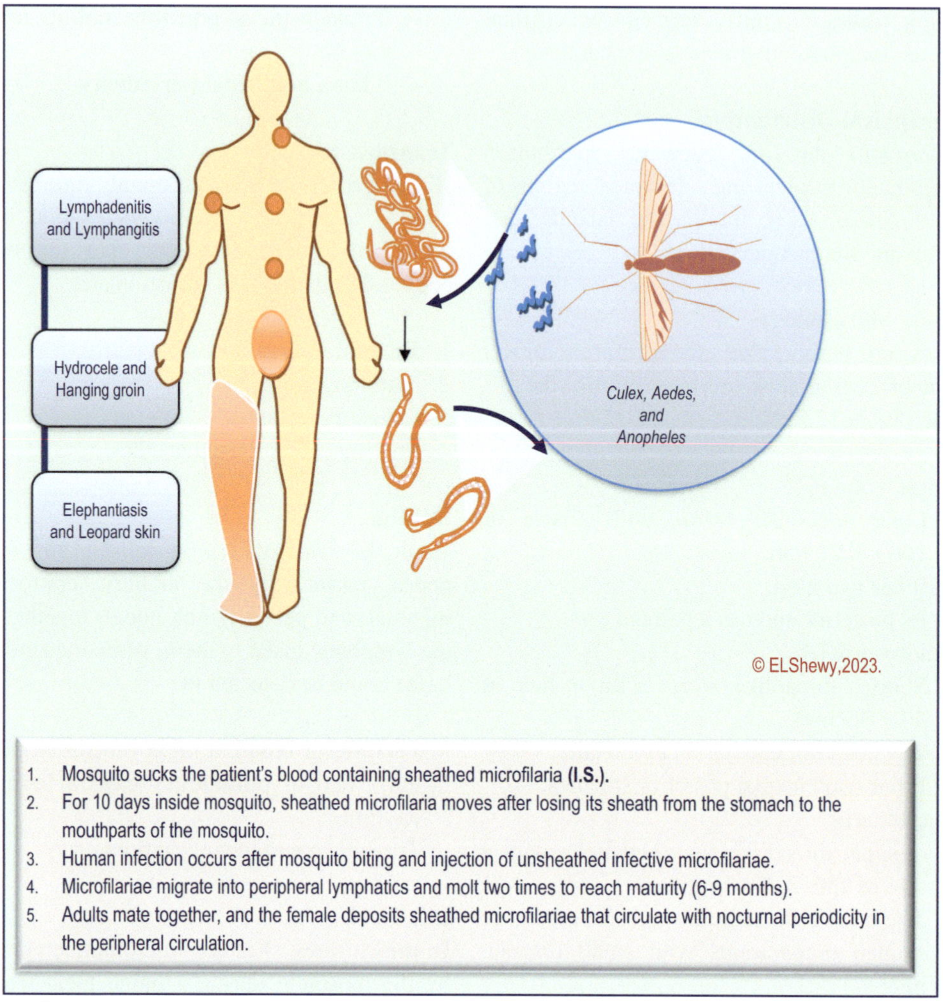

Fig. 12.8 *Wuchereria bancrofti*: life cycle

- Microfilariae lose their sheaths and penetrate the stomach wall of the mosquito to go to its thoracic musculature.
- After 10 days, they mature to be infective microfilariae and travel to the mouthparts of the mosquito.
- When a mosquito bites another host, it injects infective microfilariae into his body.

In humans, infective microfilariae pass to the peripheral lymphatics where they undergo two molts and then pass to lymph nodes and vessels to mature (6–9 months) to be male and female. After mating, the pregnant female lays sheathed microfilariae periodically into the blood (Fig. 12.8).

Final hosts:	Humans
Infective stage:	Infective microfilariae
Diagnostic stage:	Sheathed microfilariae in the blood or lymph
Intermediate hosts:	Mosquitoes *(Culex, Aedes*, and *Anopheles)*

N.B.

- Microfilariae appear in the peripheral blood after 1 year from infection. They have a nocturnal periodicity, i.e., appear during night (10 p.m.–2 a.m.) and disappear during daytime.
- The exact explanation of this nocturnal periodicity is still unknown, but it may be due to mosquito biting habit (nocturnal biter) or it

may be related to human activity; during the day, the microfilariae hide in the pulmonary structure and at the lower limbs. At night and due to laying position, they exit from these sites to the peripheral blood. Another assumption is related to the increased O_2 concentration during the daytimes at the pulmonary tissue due to human activity.

- Adult *W. bancrofti* can live up to 7 years in the human body.

Pathogenesis

The following pathological changes are demonstrated during bancroftian filariasis:

1. Adult worms stimulate marked inflammatory reactions in lymph nodes together with the lymphatics distal to them.
2. Bacterial infection and/or worm metabolites are the stimuli for such inflammatory reactions, leading to **lymphangitis** and **lymphadenitis.**
3. Repeated inflammations cause vessel dilatation with lymph stagnation (lymphatic varices) with subsequent **lymphedema.**
4. Lymphedema first is intermittent and then becomes chronic and persistent, with subsequent thickening and hyperplasia of the connective tissues under the skin.
5. By time, woody fibrous indurations with thickening and verrucous changes give the picture of elephant skin. This could develop in any limb, the scrotum, the breast, or the vulva.
6. Infection of scrotal lymphatics leads to **hydrocele.**
7. Lymphatic varices may be formed in any affected vessel that could be ruptured. (Ruptured varices in the urinary bladder result in the passage of chyle in the urine giving it a white color **(chyluria)**.)
8. The adult worms die and calcify; sometimes, an abscess is formed around them.
9. Filarial antigens could cause transient pulmonary eosinophilia. A condition is known as tropical pulmonary eosinophilia (**TPE**).

Clinical Picture

Bancroftian filariasis could be acquired for many years, and the symptoms may not appear. The following are signs and symptoms of bancroftian filariasis:

1. At the start of infection, fever, malaise, headache with lymphangitis, and lymphadenitis in the limbs, breast, or scrotum are commonly seen manifestations.
2. Limb lymphangitis is centrifugal, *i.e.*, starts in the region of the LN and extends distally.
3. Occasionally, lymphadenitis could be associated with an abscess containing remnants of the worms.
4. Scrotal lymphangitis, with thickening of the spermatic cord, hydrocele, and chyluria (white urine) could be detected.
5. Elephantiasis (dermal hypertrophy and verrucous changes (leopard skin)), although uncommon, could be seen in the repeated chronic attacks affecting limbs, breast, scrotum, vulva, and other affected organs.

N.B.

- The two most serious sequelae of filarial infection are **elephantiasis** and **hydrocele** that only affect individuals exposed to repeated infections over many years.
- Tropical pulmonary eosinophilia (TPE) is a syndrome detected in patients having filarial infection without filaremia. It is characterized by persistent hypereosinophilia, nocturnal cough, wheezing, low-grade fever, adenopathy, raised IgE, and high titer of antifilarial antibodies in blood and in bronchoalveolar lavage. There is a good response to DEC.

Diagnostic Procedures

Diagnosis of bancroftian filariasis depends on its clinical picture together with the following investigations:

- **Direct:**
  Microscopic demonstration of sheathed microfilariae in peripheral blood or lymphatic aspirate; the following should be considered:
  - **Thin and thick blood films** stained with Giemsa. Blood samples should be taken between 10 p.m. and 2 a.m. (nocturnal periodicity).

- **Provocative test:** Oral administration of 100 mg diethylcarbamazine (DEC), 1–2 h before sampling stimulates microfilariae to appear in the daytime.
- **Concentration** methods are helpful in the detection of light infection, e.g., centrifugation of fluid fixed in 2% formalin (**Knott's technique**) or filtration of citrated blood or fluid specimen through a 5-micron pore-size polycarbonate microfilter (nucleopore).

• **Indirect Immunological:**
1. Hypersensitivity skin test: injection of an extract of dog microfilariae (*Dirofilaria immitis*); positive reaction denotes infection with *W. bancrofti*.
2. IHA, IFA, and ELISA are used for detection of the specific antifilarial antibodies.
3. **A rapid antigen detection test:** It is now recognized as the method of choice for the detection of *W. bancrofti* infection. They have the following advantages:
   a. Antigen tests can be performed with blood collected during the day or night.
   b. Very easy to perform compared to serological tests. Recommended for mapping and survey of *W. bancrofti*.

   The following are common antigen detection tests:
   1. **Filariasis test strip (FTS)**
   2. **Immunochromatographic test (ICT)**

• **Other Investigations:**
1. Molecular detection of specific DNA in humans and mosquitoes.
2. X-ray examination and lymphangiography.
3. Doppler ultrasonography.
4. Blood picture demonstrates a significant high eosinophilia.

Treatment
1. **Diethylcarbamazine** (DEC) (Hetrazan):
   - Drug of choice to be given as 2 mg/kg/b.w., three times/day for 2–4 weeks.
   - **S.E.**: allergic reactions, malaise, vertigo, headache, nausea.
   - In mass treatment, it is given as an oral single dose of 6 mg/kg/b.w.
2. **Ivermectin** (IVR):
   - 100 μg/kg/b.w., as single dose.
3. **Combination IVR and DEC:**
   - Proved to be highly effective on adult worms and microfilariae. The combination is a very powerful tool for the control of lymphatic filariasis.
4. **Antihistaminic and acetylsalicylic acid:**
   - These are adjuvant drugs administered during attacks of lymphangitis.

N.B. Surgical intervention is used for skin changes (elephantiasis) together with corticoids mostly for cosmetic reasons.

Prevention and Control
1. Mass treatment with DEC.
2. Tracing and repeated treatment for the infected persons.
3. Personal protection from mosquito bites during travel to endemic areas.
4. Vector control (mosquito nets, repellents, insecticides, and biological control).

- **Parasitology:** *W. bancrofti* is a filarial nematode. It inhabits the lymph nodes and lymphatics; it is transmitted through mosquito bite injecting microfilariae. Microfilariae have nocturnal periodicity.
- **Clinical:** Lymphangitis, lymphadenitis, elephantiasis, hydrocele, and chyluria are common manifestations.

B. *Brugia malayi* (*B. malayi*)

B. malayi is one of the filarial nematodes with sheathed microfilariae and with nocturnal periodicity. It is responsible for Malayan filariasis (lymphatic filariasis). It generally looks like bancroftian filariasis but with different geographical and vector patterns.

Geographical Distribution
B. malayi has an extensive distribution in the regions of Malawi, Malaysia, Sri Lanka, Indonesia, and India, as well as in some parts of China, Korea, and Japan.

Parasite Morphology

B. malayi is morphologically similar to *W. bancrofti*; however, its microfilaria is characterized by the following:

- A sheathed body with a size of 200–250 × 5 µm.
- The tail is full of nuclei with two terminal nuclei separated from the others in the tail.
- Cephalic space (length to width) 2:1.
- It has a nocturnal periodicity.

Transmission

B. malayi is transmitted by inoculation of the infective microfilariae through mosquito biting, mainly ***Mansonia*** species, in addition to *Aedes* and *Anopheles*.

Life Cycle

B. malayi has a similar life cycle as *W. bancrofti* but with *Mansonia* species as a vector in addition to *Aedes* and *Anopheles* (Fig. 12.9).

Diagnostic Procedures

1. Thin and thick blood film.
2. The *Brugia* Rapid point-of-care cassette test (BRT) is used for the diagnosis of lymphatic filariasis. It detects IgG4 antibody against *Brugia spp.* in human blood samples.

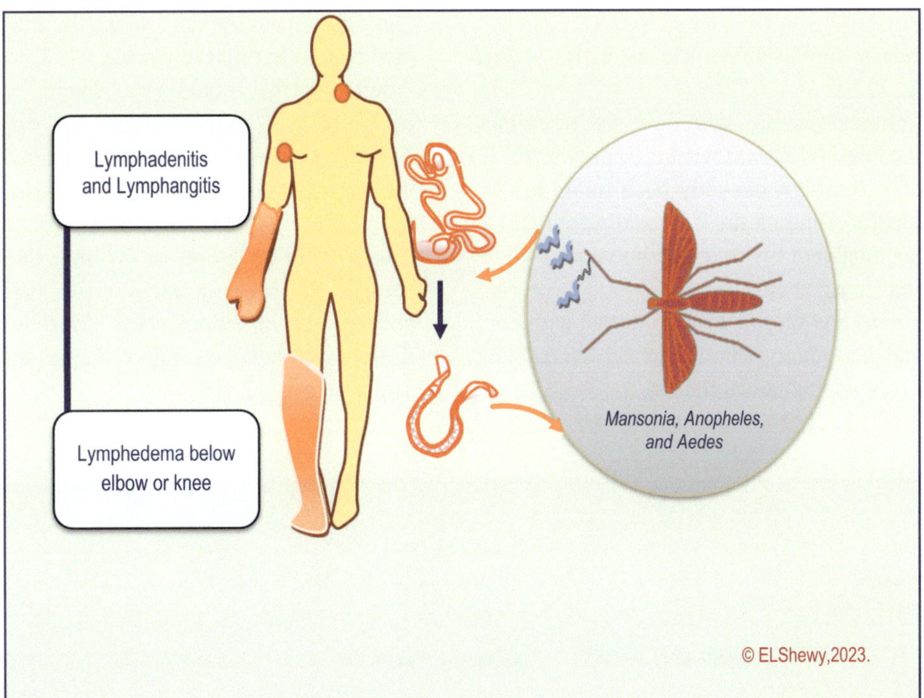

1. Mosquito sucks the patient's blood containing sheathed microfilaria (I.S.).
2. For 10 days inside a mosquito, sheathed microfilaria moves after losing its sheath from the stomach to the mouthparts of the mosquito.
3. Human infection occurs after mosquito biting and injection of unsheathed infective microfilaria.
4. Microfilariae migrate into peripheral lymphatics and molt two times to reach maturity (6-9 months).
5. Adult female deposits sheathed microfilariae that circulate with nocturnal periodicity in the peripheral circulation.

Fig. 12.9 *Brugia malayi* and *Brugia timori:* life cycle

Pathogenesis and Clinical Picture

B. malayi demonstrates the same basic pathologic changes seen during bancroftian filariasis with the following differences:

1. Elephantiasis is commonly developed in the lower limb below the knee or less commonly below the elbow in the upper limb.
2. Effects on scrotal lymphatics and hydrocele are less frequent.
3. Lymphatic varices and chyluria are rare.

C. *Brugia timori* (*B. timori*)

B. timori is the third filarial worm causing lymphatic filariasis.

1. It has a similar life cycle as in *B. malayi* (Fig. 12.9).
2. Its clinical picture more closely resembles that caused by *B. malayi* and *W. bancrofti*.
3. So far, *B. timori* has only been found in certain parts of Indonesia. It is locally confined to areas inhabited by its mosquito vector, which breeds in rice fields.
4. *B. timori* has sheathed microfilariae with nocturnal periodicity, similar to *B. malayi* with the following characteristics:
   - They are longer than that of *B. malayi* and *W. bancrofti*.
   - They have nuclei that extend to the tip of the tail.
   - It has a nocturnal periodicity.
   - The sheath of *B. timori* does not stain pink with Giemsa stain as with *B. malayi* and *W. bancrofti*.

D. *Mansonella ozzardi, perstans,* and *streptocerca*

- These are a group of filarial nematodes with less pathogenic effects than other filarial worms and with smaller sizes (Table 12.2).
- They have different geographical and morphological patterns and should be differentiated from other filarial worms.
- Adult worms could be found in the peritoneum.
- Their microfilariae are unsheathed and detected in blood or in tissues like skin.
- Mansonellosis is a term used for disease caused by this group of worms; they may cause no diseases or, occasionally, peritoneal and serosal infections. N.B. Morphological differences between microfilariae are presented in Fig. 12.10.

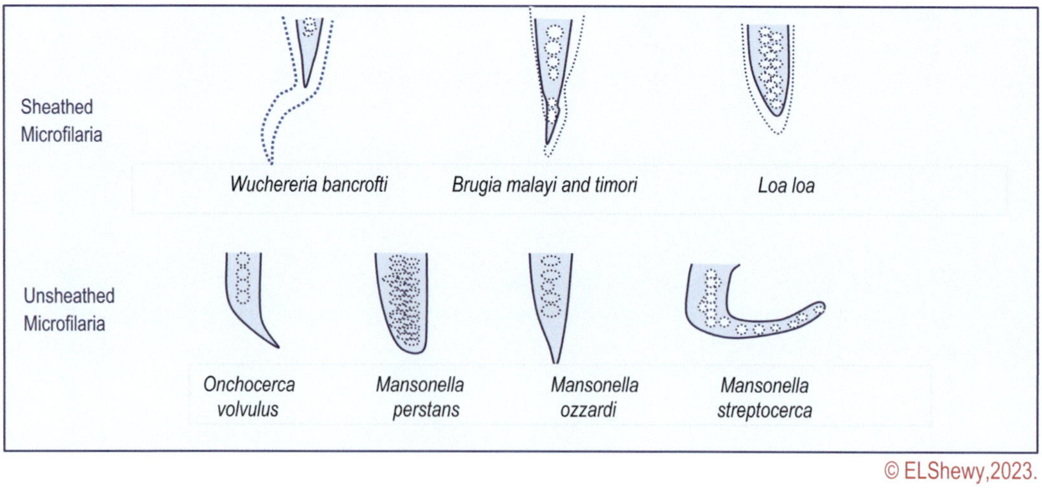

Fig. 12.10 Sheathed and unsheathed microfilariae

Review Questions

1. ***Leishmania* species tend to proliferate within**
   a. Red blood cells
   b. Phagocytic cells
   c. Platelets
   d. Muscle cells
2. **OptiMAL test is used for detection of**
   a. *Leishmania donovani*
   b. *Plasmodium malariae*
   c. *Entamoeba histolytica*
   d. *Toxoplasma*
3. **Kala-azar is characterized by the following except**
   a. Fever
   b. Black forehead skin
   c. Splenomegaly
   d. Short-lived immunity
4. **The following forms of *Leishmania* could be detected within NNN medium:**
   a. Epimastigotes
   b. Trypomastigotes
   c. Amastigotes
   d. Promastigotes
5. **Malarial paroxysm caused by *P. malariae* occurs every**
   a. 12 hours
   b. 24 hours
   c. 72 hours
   d. 48 hours
6. **Banana-shape gametocytes in a blood film identify infection with *Plasmodium***
   a. *vivax*
   b. *ovale*
   c. *malariae*
   d. *falciparum*
7. **Multiple infections with small ring stages in a normal-looking RBC characterize infection with *Plasmodium***
   a. *vivax*
   b. *ovale*
   c. *malariae*
   d. *falciparum*
8. **Algid malaria means**
   a. Tertian malaria with severe anemia
   b. Malignant malaria with severe hypotension and shock
   c. Tertian malaria with relapse
   d. Quartan malaria with recrudescence
9. ***Mansonia* is the vector of**
   a. *Wuchereria bancrofti*
   b. *Plasmodium*
   c. *Leishmania*
   d. *Brugia malayi*
10. **Which one of the following *Plasmodia* species is responsible for quartan malaria? *Plasmodium***
    a. *vivax*
    b. *ovale*
    c. *malariae*
    d. *falciparum*
11. **Leukopenia could be associated with the following parasitic disease:**
    a. Giardiasis
    b. Leishmaniasis
    c. Amebiasis
    d. Balantidiasis
12. **Dogs are reservoir hosts of**
    a. *Wuchereria bancrofti*
    b. *Plasmodium malariae*
    c. *Brugia timori*
    d. *Leishmania donovani*

Parasites Infecting the Musculoskeletal System

13

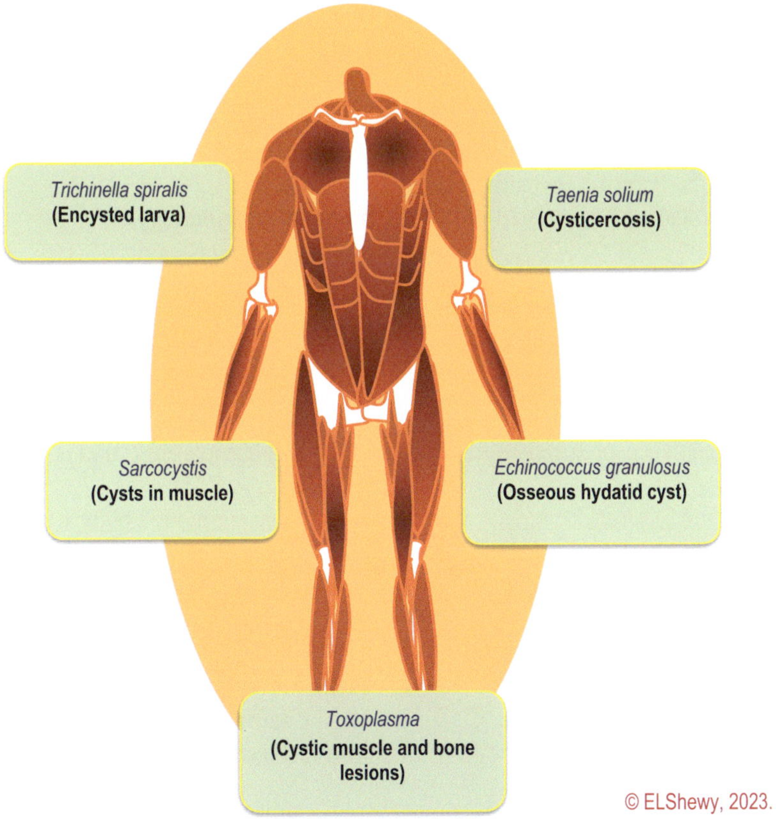

Parasites infecting the musculoskeletal system

K. ElShewy, *Medical Parasitology*, https://doi.org/10.1007/978-3-031-64736-9_13

Test your learning and check your understanding of this book's contents: use the "Springer Nature Flashcards" app to access questions using ▶ https://sn.pub/oyFLIx. To use the app, please follow the instructions in Chap. 1.

Introductory

- The following parasites could affect the musculoskeletal system:
  1. *Trichinella spiralis* (encysted larvae).
  2. *Sarcocystis.*
  3. *Taenia solium* (cysticercosis) **(Chap. 3).**
  4. *Toxoplasma* (cystic lesions) **(Chap. 14).**
  5. *Echinococcus granulosus* **(Chap. 8).**
- Almost all parasitic infections of the muscles result in cystic muscle lesions associated with **parasitic myositis.** Parasitic myositis is often diffuse or multifocal and associated with the following:
  1. Muscle pain.
  2. Tenderness on movement or palpation.
  3. Swellings and/or weakness.
  4. Painless calcified muscle swellings.
- **Parasitic myositis is diagnosed by**:
  1. History of residence or travel to endemic area.
  2. Clinical picture.
  3. High eosinophilia.
  4. Imaging (X-ray, Sonar, CT, and MIR).
  5. Parasitological, serologic, and molecular methods.
  6. Histopathological examination of tissue biopsies.

Trichinella spiralis and *Sarcocystis* are discussed in this chapter in detail; all other musculoskeletal parasites have been mentioned in other chapters.

Trichinella spiralis (T. spiralis)

(Pork Worm)

T. spiralis is a tissue nematode worm infecting both humans and animals. Humans are infected by eating undercooked ham containing encysted larvae. Fever, diarrhea, myositis, and eosinophilia are cardinal manifestations of trichinosis.

Geographical Distribution

T. spiralis has a worldwide distribution.

Parasite Morphology

- Adult worm (♀) (Fig. 13.1):
  - It has a rounded body with a tapered anterior end, 4 mm × 60 µm.
  - Its esophagus (cellular type esophagus) extends ½ the body.
  - Its uterus fills most of the body.
  - It is larviparous (larva is 100 µm).
- Adult worm (♂):
  - It has a rounded body with a pointed anterior end; it is smaller than the female.
  - Its posterior end has two fleshy papillae.

Transmission

T. spiralis infection is transmitted through eating raw or undercooked ham containing encysted *T. spiralis* larvae.

Fig. 13.1 *Trichinella spiralis:* morphology

Habitat

1. *T. spiralis* adult worms inhabit the small intestinal mucosa of man and pigs.
2. Larvae of *T. spiralis* migrate to and settle in the active striated muscles, e.g., skeletal muscle of the limbs, diaphragm, tongue, masseter, intercostal, ocular, and laryngeal muscles.

Life Cycle

In humans, infection with *T. spiralis* is acquired through eating undercooked ham containing encysted larvae (Fig. 13.2).

- In the small intestine of humans, larvae are liberated from the digested muscle to undergo four molts to become adults (after 30–40 h).
- Mating takes place in the lumen of the intestine, and the female lays larvae (1500 during its life), while the male dies after mating (intestinal phase).
- Larvae penetrate the intestinal mucosa to pass through the circulation to the active skeletal muscles where they encyst (tissue phase).

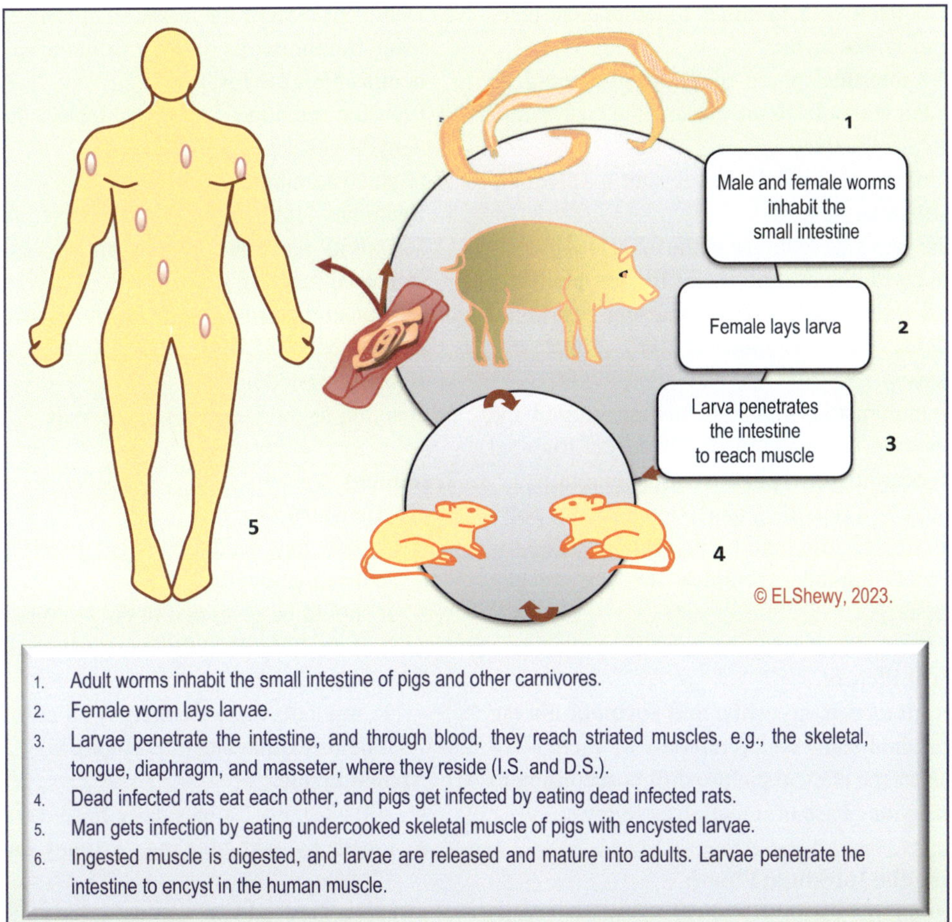

Fig. 13.2 *Trichinella spiralis*: life cycle

Final hosts:	Pigs, cats, dogs, rats, and humans.
Infective stage:	Encysted larvae in the striated muscle.
Diagnostic stage:	Larvae inside the striated muscle.
Reservoir hosts:	Pigs, cats, dogs, and rats.

N.B.

- Man is considered a dead-end host in the life cycle of *T. spiralis*.
- Carnivorous animals, rats, and other flesh-eating animals are considered reservoir hosts for *T. spiralis*.
- Encysted larvae can live for many years viable; after that, calcification can occur.
- Hosts of *T. spiralis* act as both final and intermediate hosts.

Pathogenesis

1. The presence of *T. spiralis* in the intestine is associated with minimal inflammatory reactions with diarrhea.
2. The intestinal phase of *T. spiralis* could be ended without a tissue phase if the host immunity is effective.
3. If the immunity fails, larvae start to infect the skeletal muscle.
4. The presence of larvae in the muscle is associated with the destruction of the muscle fibers and splinter hemorrhage (causing periorbital edema and vasculitis if they affect eye tissues).
5. Granulomatous tissue formation around the encysted larvae could lead to severe trichinal myocarditis, encephalitis, or pneumonitis.
6. There is a possible connection between polyarteritis nodosa and the deposition of the trichinal immune complex in the related vessels.

Clinical Picture

Fever, diarrhea, myositis, and eosinophilia are the cardinal signs and symptoms of trichinosis. However, the clinical picture differs according to the parasitic phase of infection as follows:

During the Intestinal Phase:

1. Symptoms of gastroenteritis: nausea, vomiting, and diarrhea with or without abdominal pain may last for days.
2. The outbreak of gastroenteritis from 2 to 7 days after eating uncooked pork should suggest *Trichinella* infection.

During Muscle Phase:

1. Fever and eosinophilia increase to their maximum after 3–4 weeks from infection, then decline later. Leukocytosis.
2. Mild myositis (myalgia) appears early, then it turns severe on the 12–20th days.
3. Periorbital edema was demonstrated 1 week after the infection.
4. Sometimes, symptoms of meningoencephalitis and/or pneumonitis could be manifested.

Diagnostic Procedures

1. Demonstration of the larvae in muscle biopsy using trichinoscope (usually from the gastrocnemius after the first week)
2. Elevated creatine kinase, aldolase, lactase dehydrogenase, and transaminase.
3. High eosinophilia.
4. Bachman's intradermal test (a delayed hypersensitivity reaction against *T. spiralis* extract) (not used now).
5. Demonstration of the specific anti-*Trichinella* antibodies by ELISA and IHA (after 2 weeks from suggested infection).
6. Imaging using X-ray, CAT, and MIR.

Treatment

1. **Corticosteroids:**
   - 20–60 mg prednisone is given to be reduced gradually.
   - It should be restricted to only severe cases as it could prolong the intestinal phase.
2. **Thiabendazole:**
   - 25 mg/kg/b.w., twice daily for 7 days.
   - It helps to alleviate the symptoms.
3. **Mebendazole.**
   - 200–400 mg, 3 times daily for 3 days, followed by 400–500 mg, 3 times daily × 10 days.
   - It is effective in preventing tissue and muscle phases.
4. **Albendazole.**
   - 400 mg, twice a day by mouth for 8–14 days.

Prevention and Control

1. Proper, safe cooking and storage of meat.
2. Pork and pork products should be properly inspected for *T. spiralis* infection.
3. Combating rats.

- **Parasitology**: *T. spiralis* is a zoonotic nematode worm transmitted by eating undercooked infected pork with larvae. It is diagnosed mainly by muscle biopsy and serology.
- **Clinical:** Fever, diarrhea, muscle pain, and eosinophilia are cardinal manifestations of trichinosis after a pork meal.

Sarcocystis spp., (S. suihominis, S. bovis, S. lindemani)

Sarcocystis is an *Apicomplexan* parasite infecting both humans and animals. Infection with *Sarcocystis* is mostly asymptomatic. However, some species of *Sarcocystis* cause severe enteritis (enteric infection), while others cause myositis. *Sarcocystis* causes a life-threatening disease among immunocompromised individuals.

Geographical Distribution

There are sporadic cases of sarcocystosis worldwide in areas raising livestock. Medically important species of *Sarcocystis* are:

1. *Sarcocystis bovihominis in* cattle.
2. *Sarcocystis suihominis* in pigs.
3. *Sarcocystis lindemanni* in human.

Parasite Morphology

There are two main morphological forms of *Sarcocystis* parasite (Fig. 13.3):

1. Oocyst (sporocyst), passed mature and infective with the stool of the final host.
2. Sarcocyst that is found in the muscle of the intermediate host.

- Mature Oocyst (sporocyst):
  - It is oval or elliptical. It has a very thin wall.
  - 25–33 µm long and 10–20 µm width.
  - Mature oocyst contains two sporocysts, each containing four sporozoites.
  - Sporocysts are usually released from oocysts and seen singly.
  - Oocyst is infective when passed with stool.
- Sarcocyst:
  - It may be oval or elongate in shape.
  - 100–320 µm long, it may reach up to 5 cm (looks like long tubules).
  - Thin-walled, surrounded with little inflammatory cells.
  - It contains many bradyzoites.

N.B.

- In most cases, mature oocyst passes in the stool without a cyst wall and with ruptured sporocysts, so it is referred to as sporocyst.

Habitat

Sarcocystis bovihominis and *suihominis* are the two species that inhabit the small intestine of their host, while *Sarcocystis lindemanni* inhabits human striated muscle, forming muscle cysts (sarcocysts).

Transmission

There are two routes for transmission of *Sarcocystis*:

1. Consumption of meat containing sarcocysts leads to enteric infection.
2. Consumption of food and drinks contaminated with the mature oocysts of *Sarcocystis* resulting in muscle infection.

Life Cycle

In the life cycle of *Sarcocystis*, humans can act as an intermediate and final host:

If a man ingests undercooked meat (cattle or pigs) containing the sarcocysts, sarcocysts are digested and undergo sexual reproduction (gametogony) to form mature oocysts containing two sporocysts each with four sporozoites ready to infect.

On the other hand, if humans or other mammals ingest mature oocysts (sporocysts) in their foods, the sporocysts are digested, and sporozoites are released. Sporozoites penetrate intesti-

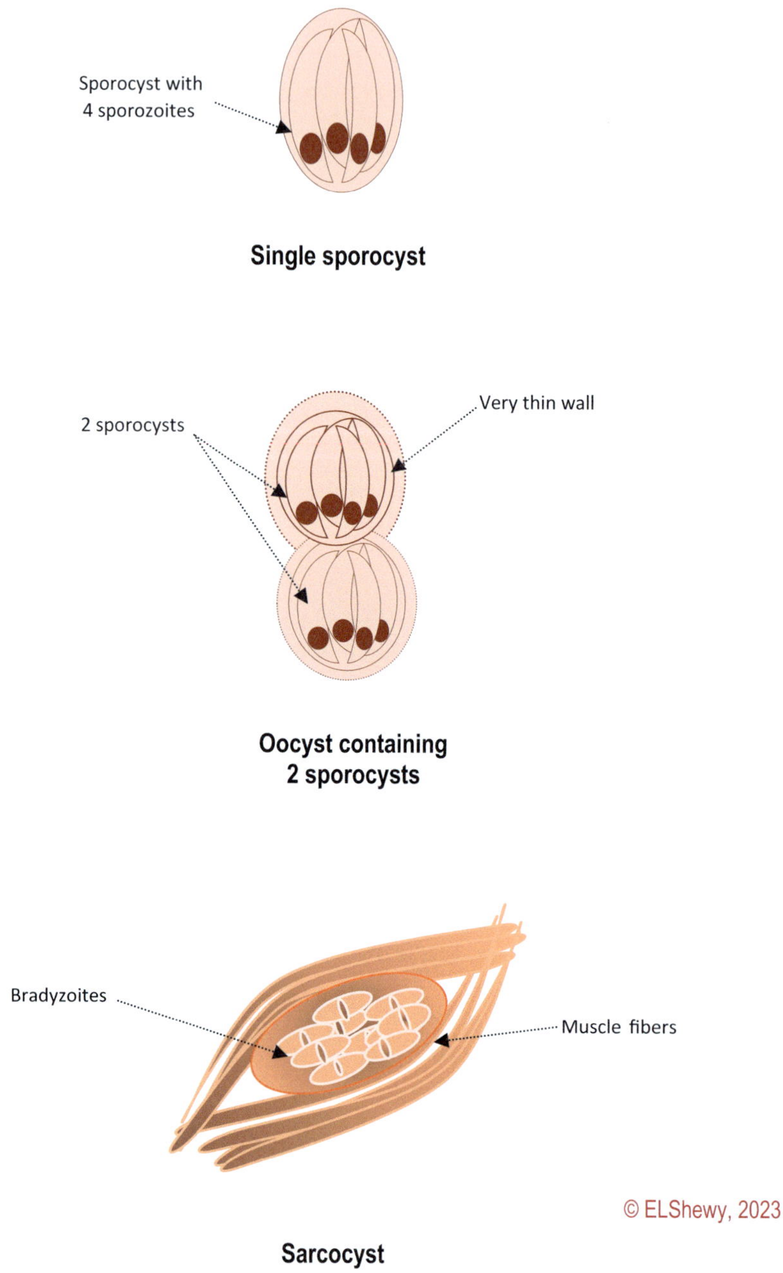

Fig. 13.3 *Sarcocystis* spp.: morphology

nal epithelial cells and undergo an asexual cycle (schizogony) with the formation of merozoites. Merozoites penetrate the muscle cells of their host to encyst and multiply by (endodyogony) forming sarcocysts in the host muscle (Fig. 13.4).

Final hosts:	Humans, cattle, and pigs.
Infective stage:	Mature oocysts (muscle infection) or sarcocysts in the undercooked pork (intestinal infection).
Diagnostic stage:	Oocysts in the stool OR sarcocysts in the muscle.
Intermediate hosts:	Humans and pigs for *Suihominis* and cattle for *S. Bovihominis.*
Reservoir hosts:	Humans, cattle, and pigs.

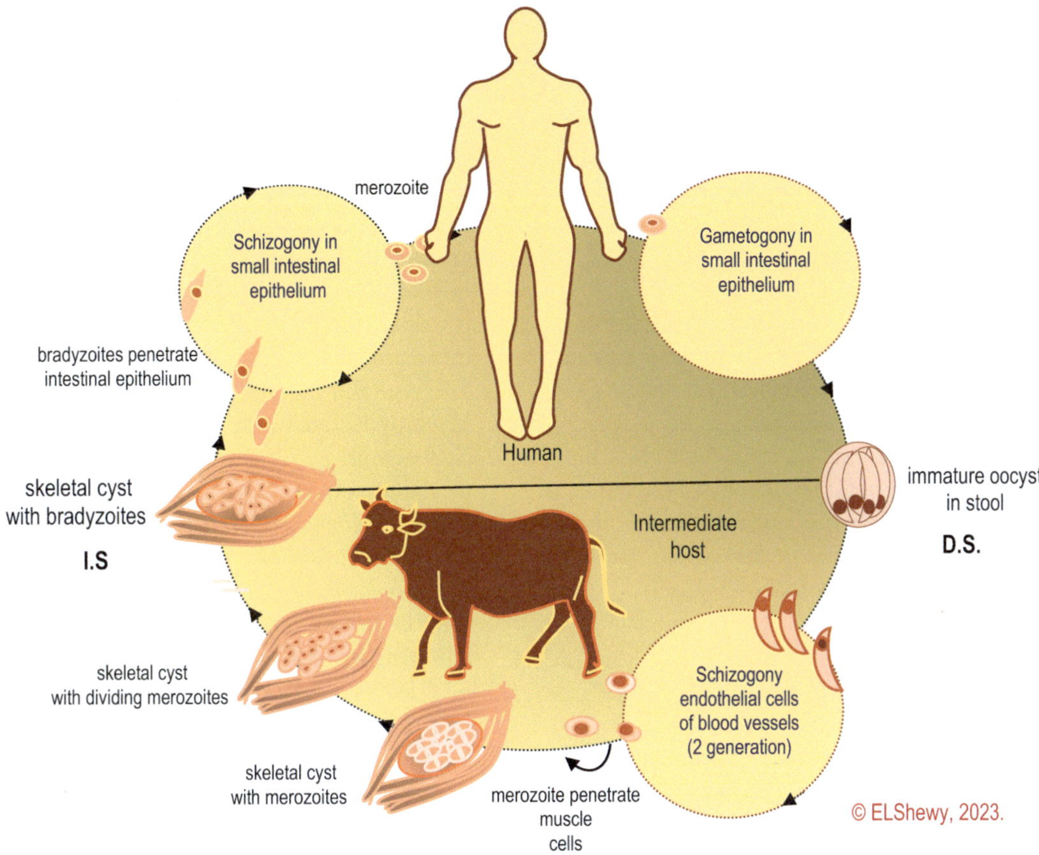

1. Ingestion of meat containing sarcocysts causes enteric sarcocystosis.
2. Ingestion of foods or drinks contaminated with oocysts or sporocysts leads to muscle infection.
3. Humans can act as an intermediate and final host for *Sarcocystis.*

I.S.= infective stage D.S.= diagnostic stage

Fig. 13.4 The life cycle of *Sarcocystis*

Pathogenesis

1. The presence of the *Sarcocystis* in the intestine of the host results in intestinal malabsorption, electrolyte imbalance, and gastroenteritis.
2. The presence of the sarcocysts in the host muscle stimulates cellular inflammatory reactions and inflammatory by-products resulting in myositis and eosinophilia.

Clinical Picture

1. **Enteric infection:**

   Usually, sarcocystosis is asymptomatic; however, in the case of immunocompromised patients, the parasite causes a severe form of gastroenteritis with fever, nausea, vomiting, respiratory distress, diarrhea, and abdominal pain.

2. **Muscle Infection**
   The infection is mostly asymptomatic in the immunocompetent individuals. However, in the case of immunocompromised patients, the parasite causes a severe form of myositis, muscle weakness, fever, and marked eosinophilia.

Diagnostic Procedures

1. Enteric infection:
   Demonstration of the characteristic sporulated oocysts either by direct stool examination or after concentration and staining with modified Ziehl-Neelsen stain.
2. Muscle infection:
   The finding of the sarcocysts (tissue cyst) using muscle biopsy and microscopic examination. Bradyzoites of *Sarcocystis* are larger than those of *Toxoplasma* (5–7 μm).

Treatment

1. Asymptomatic sarcocystosis needs no treatment.
2. There is no effective treatment for muscle sarcocystosis.
3. In case of severe infection or immunocompromised, these drugs could be used.
   - **Trimethoprim and sulfamethoxazole.**
   - **Sulfadiazine-pyrimethamine.**

Prevention and Control

1. Proper meat cooking and abstaining from eating ham.
2. Good personal hygiene and standards of sanitary conditions.
3. Safe and clean water supply.

Review Questions

1. **Fever, diarrhea, myositis, and eosinophilia are cardinal signs of**
   a. filariasis
   b. paragonimiasis
   c. trichinosis
   d. oxyuriasis
2. **Myositis is caused by the following parasites except**
   a. *Taenia solium*
   b. *Paragonimus westermani*
   c. *Trichinella spiralis*
   d. *Toxoplasma*
3. **Man is considered a dead-end host in the following infections except**
   a. *Trichinella spiralis*
   b. *Toxoplasma*
   c. *Taenia saginata*
   d. *Echinococcus granulosus*
4. **An Osseous cyst is caused during infection with**
   a. *Trichinella spiralis*
   b. *Echinococcus granulosus*
   c. *Toxoplasma*
   d. *Paragonimus westermani*

14 Parasites of the Reproductive System

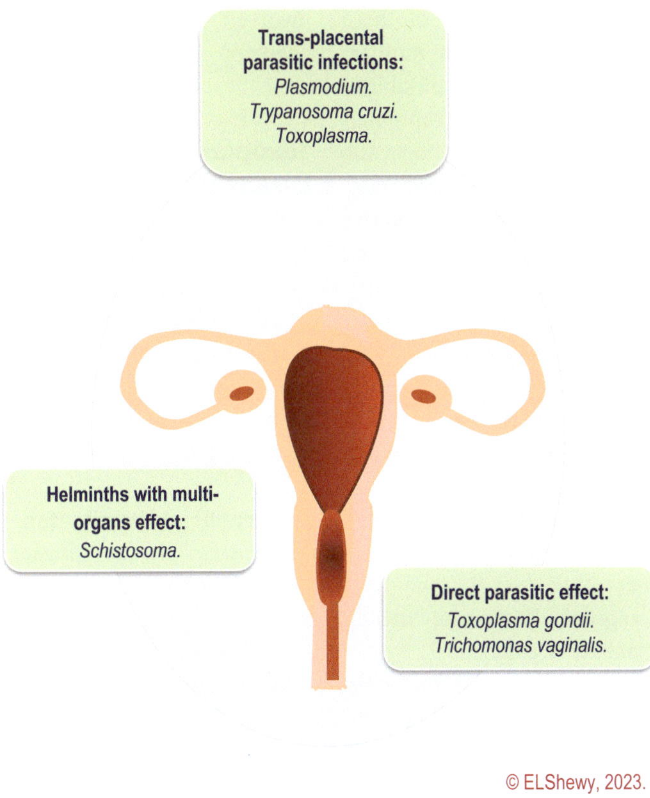

Parasites of the reproductive system

K. ElShewy, *Medical Parasitology*, https://doi.org/10.1007/978-3-031-64736-9_14

Test your learning and check your understanding of this book's contents: use the "Springer Nature Flashcards" app to access questions using ► https://sn.pub/oyFLIx. To use the app, please follow the instructions in Chap. 1.

Introductory
Parasites infecting the reproductive system are increasingly recognized as serious health problems with an impact on women and men. These infections can have severe consequences, including infertility and miscarriage. Some of these parasites could also be transmitted from mothers to children during pregnancy and childbirth. These infections affect more than just reproductive health; the resultant morbidity affects the economic productivity of the whole community.

Parasites infecting the female reproductive system could cause the following manifestations:

1. **Infertility:** caused by *Toxoplasma gondii, Trichomonas, and Trypanosoma cruzi.*
2. **Abortion:** e.g., *Toxoplasma gondii.*
3. **Congenital malformations (microcephaly, blindness,):** e.g., *Toxoplasma.*
4. **Endometritis and salpingitis**: e.g., *Toxoplasma gondii, Schistosoma,* and rarely *Entamoeba histolytica.*
5. **Vaginitis and vaginal discharge:** mainly caused by *Trichomonas vaginalis.*
6. **Cervical neoplasia:** e.g., *Trichomonas vaginalis.*

Manifestations caused by parasites infecting the male reproductive system:

1. **Prostatitis and penile discharge:** caused by *Trichomonas vaginalis* in males.
2. **Hydrocele and varicoceles:** e.g., **lymphatic** filarial worms (*W. bancrofti*).
3. **Epididymitis, spermatic and testicular inflammation:** e.g., *Schistosoma.*

Parasites infecting the reproductive system could be one of the following:

1. **Parasites directly infect the reproductive organs:** e.g., *Toxoplasma gondii, Trichomonas vaginalis,* and *Trypanosoma* .
2. **Parasites transmitted through the placenta:** e.g., *Toxoplasma gondii, Plasmodium* (Chap. 12), and *Trypanosoma cruzi.*
3. **Parasites with multi-organs effects:** e.g., *Trypanosoma, W. bancrofti* (Chap. 12).

In this chapter, we focus on parasites that directly affect the reproductive system, and other involved parasites are mentioned in other chapters.

Toxoplasma gondii (*T. gondii*) (Cat Disease)

T. gondii is an obligatory intracellular *Apicomplexan* parasite. It is primarily a parasite of cats and other felines. It could infect every nucleated cell in the human body, causing toxoplasmosis. Toxoplasmosis is generally asymptomatic except among pregnant women and immunocompromised patients.

Geographical Distribution
Toxoplasma is a parasite of cosmopolitan distribution.

- It infects a wide range of mammalian hosts.
- The seropositive prevalence of toxoplasmosis ranges from 30 to 90%.
- Among immunocompromised, toxoplasmosis is very common either as de novo or as a reactivated old infection.

Parasite Morphology
T. gondii has two phases of the life cycle: an intestinal sexual cycle in the definitive host (cat) and an asexual tissue cycle in intermediate hosts. Each cycle has particular morphological patterns as follows (Figs. 14.1 and 14.2):

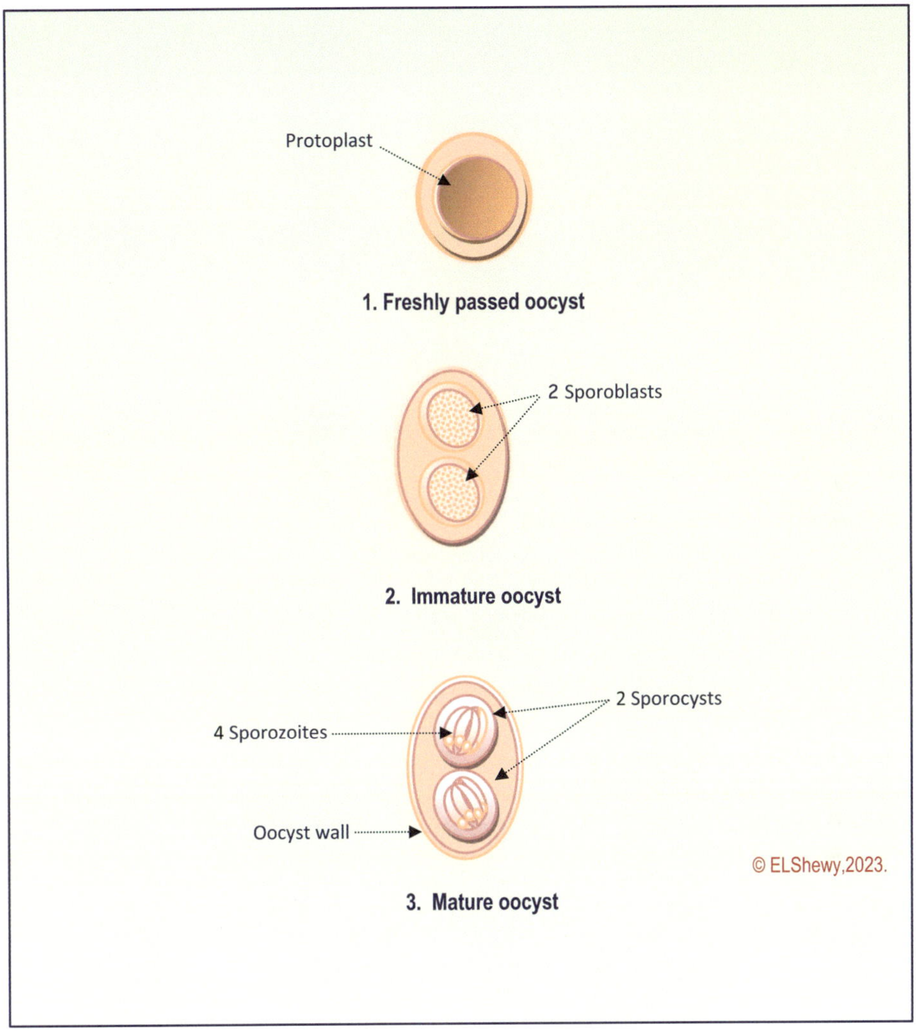

Fig. 14.1 Oocysts of *Toxoplasma gondii*: morphology

A. **Intestinal phase (oocyst)**
   Oocyst passes in the cat feces as immature, unsporulated, and not infective.
   - Immature oocyst:
     – It is an elongated, oval with a thin wall, 10–12 μm in size.
     – It has one protoplasmic mass.
     – It divides into two sporoblasts to be sporocysts after 5 days from its passage.
     – It is **not infective** at this stage.
   - Mature oocyst:
     – It is an elongated, oval with a thick wall, 10–12 μm in size.
     – It has two sporocysts, each with four sporozoites.
     – It remains **infective** in moist, shady soil for many months or even years.

B. **Tissue phase (tachyzoites, bradyzoites, pseudo and true-cysts)**
   Usually found in rodents, birds, humans, and other mammals.
   - Tachyzoite:
     – It is a rapidly multiplying active stage of the parasite.
     – Crescent in shape, 3–5 μm, single nucleus.
     – It develops in every cell except RBCs.

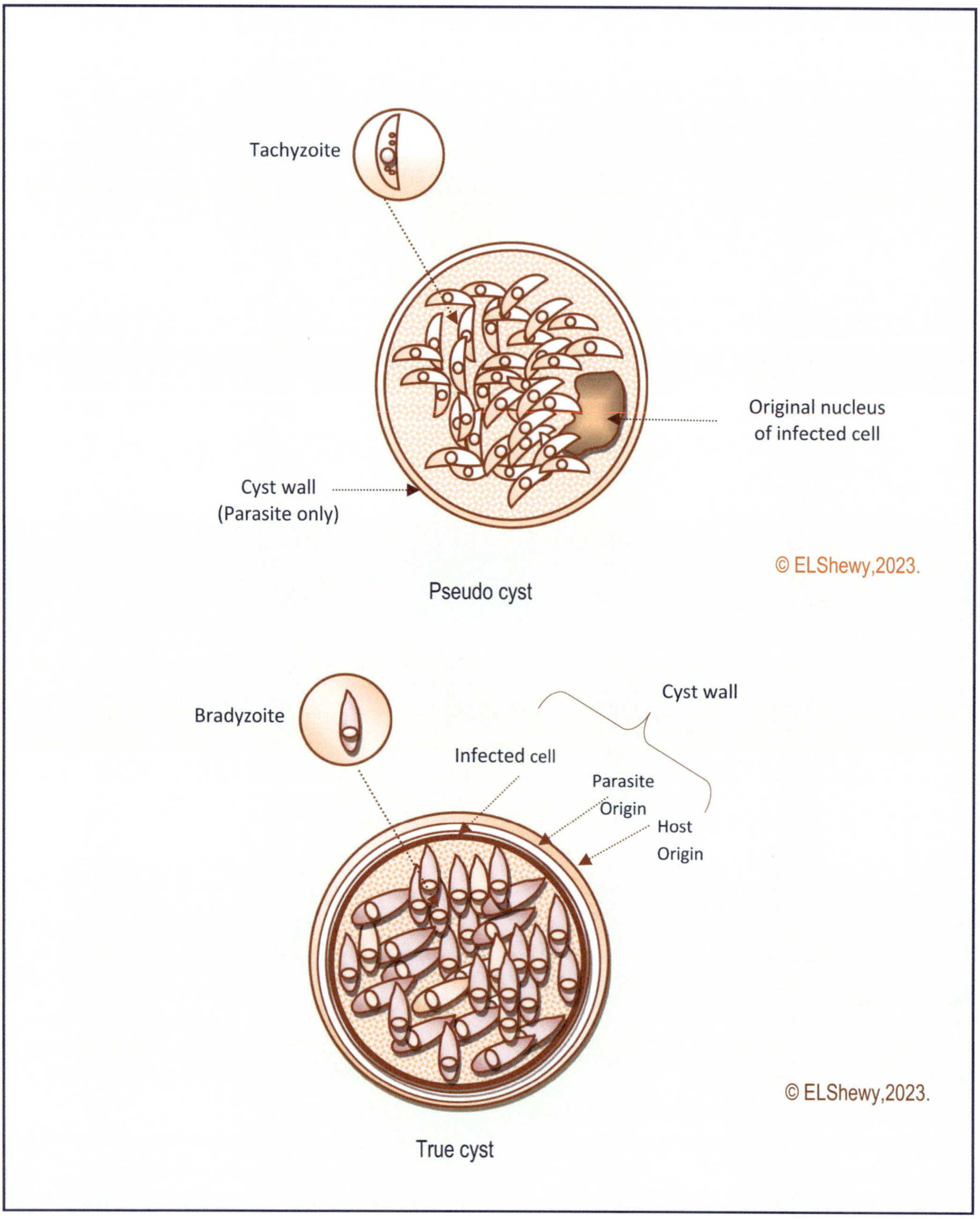

Fig. 14.2 Cysts of *Toxoplasma gondii*: morphology

  - Proliferation and multiplication of this form in the tissue results in a pseudo-cyst.
- Bradyzoite:
  - Slowly multiplying and resting stage of the parasite.
  - Long slender, 2–7 µm. Single nucleus.
  - It multiplies, forming a true cyst.
  - Pseudo cyst:
  - Rounded cyst, 12–100 µm, is formed inside the infected host cell.
  - It contains multiplying tachyzoites in the infected host cell.
  - The nucleus of the original host cell could be seen in this cyst.
  - It has a wall formed only from the host tissues.
  - It is responsible for the acute illness.
- True Cyst:
  - Rounded, 12–100 µm, it is formed inside the infected host cell.
  - Contains multiplying bradyzoites.
  - The nucleus of the original host cell is no longer seen in this cyst.
  - It has a wall formed from both the host tissues and the parasite.
  - It is responsible for the chronic resting lesion of toxoplasmosis.

Habitat

- There is no or little host specificity for *T. gondii;* therefore, it could infect man, animals, birds, reptiles, and others.
- The intestinal stages of the parasite inhabit the intestinal epithelium of cats and other felines.
- The tissue stages could be seen in every nucleated cell in the human **(except in the RBCs)** and other mammals.

Transmission

T. gondii is transmitted through:

1. Eating undercooked meat containing *Toxoplasma* cysts.
2. Ingestion of food and beverages contaminated with the mature oocysts.
3. Transplacental.
4. Blood transfusion.
5. Organ transplantation.
6. Tissue grafting.

Life Cycle

Sexual reproduction, in the form of gametogony, occurs in cats only. Meanwhile, asexual reproduction in the form of endodyogony occurs in the intermediate hosts, including humans (Fig. 14.3):

Human:

- Ingestion of *Toxoplasma* oocysts, or eating undercooked infected meat, is followed by their digestion in the stomach with the release of their sporozoites (I.S).
- Sporozoites penetrate the intestinal wall and through the bloodstream; they could invade every nucleated cell in the human body.
- Inside the invaded cells, parasites are transformed into tachyzoites (rapidly dividing) that multiply rapidly by endodyogony forming a pseudo cyst **(acute phase).**
- By time, tachyzoites are transformed into bradyzoites (slowly dividing), forming a true cyst **(chronic phase).**

Cat:

- Cat gets infection by eating infected meat or foods (with cysts and mature oocysts).
- Both stages are digested, releasing bradyzoites and sporozoites, respectively.
- Bradyzoites or sporozoites invade the intestinal epithelial cells of the cat, starting the asexual reproductive cycle (schizogony) and then a sexual gametogony cycle.
- Finally, immature oocysts are formed.

In the external environment:

- Cats discharge immature oocysts that need from 1 to 5 days to be mature and infective. A mature oocyst contains two sporocysts, each have four sporozoites.

Final hosts:	Cats and other felines.
Infective stage:	*Toxoplasma* oocysts in foods or drinks or bradyzoites encysted in meats.
Diagnostic stage:	Tissue stages.
Intermediate hosts:	Humans, rodents, birds, lambs, cattle, and sheep.
Reservoir hosts:	Cats, rats, and mice.

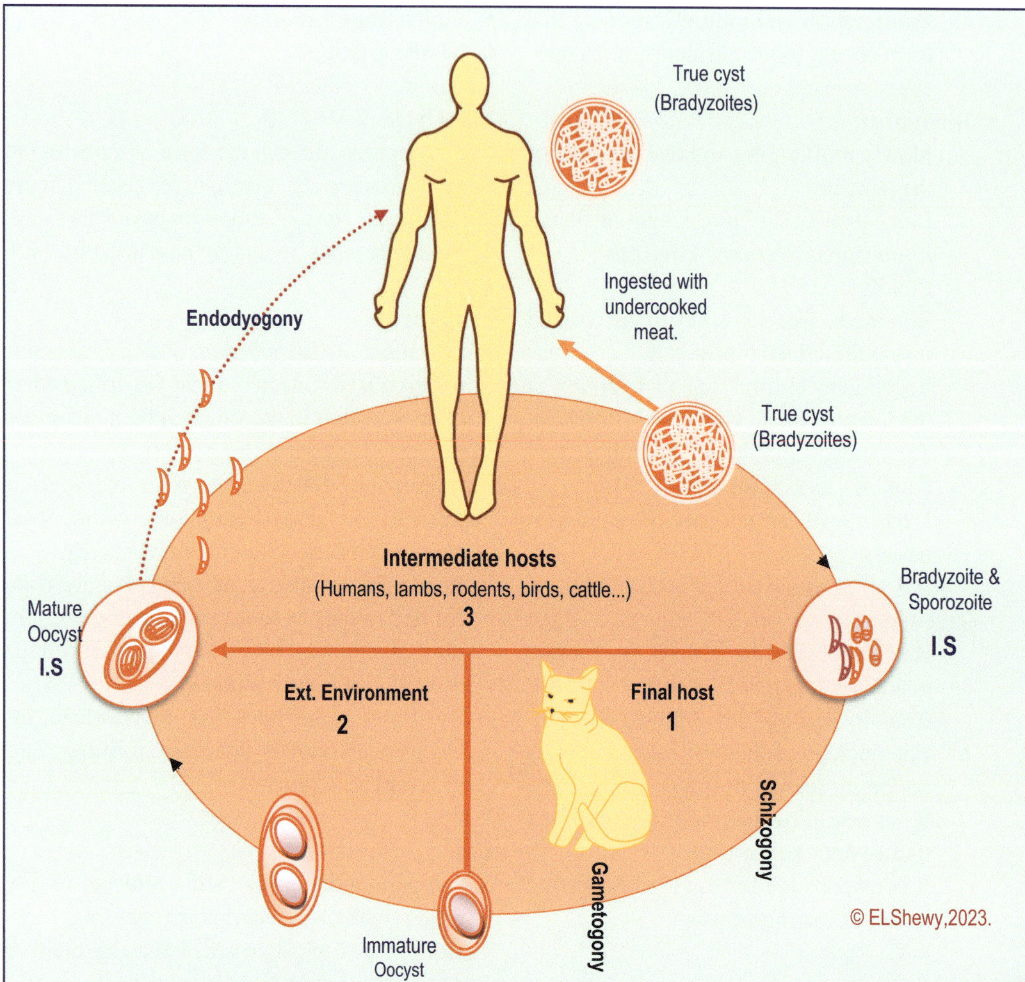

1. The cat is infected by ingestion of bradyzoites, mature oocysts with sporozoites (in cats, schizogony and gametogony occur, forming immature oocysts excreted in the cat stool).
2. Immature oocysts excreted with the cat's stool take 5 days to mature in the external environment.
3. Humans, lambs, cattle, and rodents ingest mature oocysts containing sporozoites, which start infection (in the intermediate host, sporozoites penetrate everycell except RBCs to multiply by endodyogony to form pseudo cysts and then true chronic cysts).
4. Humans and other intermediate hosts could be infected by eating undercooked meat containing cysts with bradyzoites.

Fig. 14.3 *Toxoplasma gondii*: life cycle

Pathogenesis

1. Once *Toxoplasma* sporozoites penetrate the intestinal wall, they spread among different organs and tissues via the bloodstream.
2. Spread to organs and tissues occurs through wandering reticuloendothelial cells (RECs), including monocytes, macrophages, and neutrophils, where sporozoites transform into tachyzoites, which multiply rapidly.
3. Multiplication of the parasites in lymphoid organs leads to their hypertrophy, e.g., **hepatomegaly, splenomegaly, and lymphadenopathy.**

4. Focal areas of necrosis and inflammatory cell infiltration appear in areas where tachyzoites start to multiply (in muscle, endometrium, retina, and bone marrow).
5. As the result of the host immune reaction, a cyst wall is formed from host tissues, parasites, and remnants of the infected cells, thus forming a chronic true cyst.
6. The **brain, eye, bone, and muscles** are commonly infected with *T. gondii* tissue cysts.
7. Toxoplasmosis could infect the retina, forming acute inflammatory spots and leading to macular lesions with a permanent decrease in visual acuity.
8. Chronic cysts may be associated with scattered necrotic areas, which later calcify, especially in the eye and brain.
9. Infection in the brain has been proven to interfere with neurotransmitters, leading to **behavioral, personality, and mentality changes.**

Clinical Picture

Clinically, toxoplasmosis varies greatly among immunocompetent normal individuals and immunocompromised also during pregnancy as follows:

1. **Early intrauterine infection:**
   During the first trimester, the embryo could suffer of:
   - Retinochoroiditis, encephalomyelitis, and hydro or microcephaly.
   - Spontaneous abortion is inevitable at this stage.
2. **Late intrauterine infection:**
   During the last trimester, the newborn could be normal; later, he could suffer from:
   - Retinochoroiditis, cerebral infection, with later cerebral calcification.
   - Stillbirth is very common in these cases.
3. **Newborn infection:**
   Infection among newborns is represented by:
   - Fever, pneumonitis, hepatosplenomegaly, blindness, mental retardation, convulsion, and neonatal jaundice.
   - Sometimes, the baby is born normal, then later, he may develop retinochoroiditis and mental retardation.
4. **Immunopotent adults and children:**
   - The infection started with chill, fever, headache, myalgia, lymphadenitis, and extreme fatigue. This clinical picture mimics infectious mononucleosis.
5. **Infection in the immunocompromised individuals:**
   Immunocompromised individuals, particularly those with AIDS, suffer from either:
   - Reactivation of an old toxoplasmosis or getting a new fulminating infection.
   - These individuals suffer headaches, ataxia, convulsions, hemiparesis, orchitis, and increased IgG in their CSF, with normal serum levels.

Diagnostic Procedures

- **Direct:**
  Specimens for detection of toxoplasmosis could be taken from:
  Blood, buffy coat layer, sputum, bone marrow, CSF, lymph node exudates, tonsils, striated muscle biopsy, amniotic fluid, and placenta.
  1. Demonstration of the organism in tissue sections or smear of body fluids by:
     - Optical microscopic examination of stained smears with hematoxylin-eosin, aqueous methylene blue, Periodic acid Schiff (PAS), or peroxidase.
     - Electron microscopy.
     - Immunofluorescence microscopy.
  2. Recovery of the parasites by:
     - **Animal inoculation:** test samples could be inoculated intraperitoneally (IP) in rats, mice, and hamsters. After 14 days, an intraperitoneal (IP) wash is performed to recover the organisms.
     - **Culture:** in Hela cells, human or tissue cultures, an extract from the culture is microscopically tested for the presence of the parasites.
- **Indirect Immunological:**
  1. **Sabin Feldman test** (dye test):
     This test depends on the inhibition of staining of *Toxoplasma* organisms with methylene blue if reacted with serum taken from *Toxoplasma*-infected patients.

Table 14.1 Immunoglobulins in toxoplasmosis

IgM	IgG	Interpretation
+ve	–ve	Recent toxoplasmosis.
Normal	+ve	Old infection.
High +ve	+ve	Recent on top of old infection.
+ve (In fetal blood)	–ve (In fetal blood)	Intrauterine infection.
–ve (In fetal blood)	+ve (In fetal blood)	Maternal and not fetal infection.
Low +ve (in blood)	High +ve (in CSF) and normal in serum	Reactivation of latent infection.

2. **Toxoplasmin test** (skin hypersensitivity test):
   It includes the injection of an extract of *Toxoplasma* in the forearm of suspected patients. A positive reaction is associated with raised indurated skin after 2 days.
3. **Polymerase Chain Reaction (PCR):**
   It is applied for the detection of the time and intensity of infection, especially during intrauterine infection.
4. **Detection of specific Immunoglobulins (Table 14.1):**
   This is considered the **main diagnostic** method of toxoplasmosis. IHA, ELISA, and IFA could be used for the detection of specific antibodies against toxoplasmosis.

Treatment

The following are commonly used drugs for the treatment of toxoplasmosis:

1. **Pyrimethamine and Trisulfapyrimidine:**
   - 50 mg Pyrimethamine and 2–6 g Trisulfapyrimidine, daily for 1 month.
   - This combination should be administered with folinic acid to combat bone marrow depression. Contraindicated in pregnancy.
2. **Clindamycin:**
   - It should be given parentally in patients with AIDS.
3. **Spiramycin** (Rovamycin):
   - It is used during pregnancy.
   - A 2–3 g/3 times daily for a month is recommended even in the 1st trimester.
4. **Folic acid, Calcium, and Leucovorin.**

Prevention and Control

1. Raw and undercooked meat should not be consumed.
2. Proper meat cooking before consumption.
3. Contact with cat feces in a litter pan or soil should be done with caution, particularly for females in the childbearing period.
4. Avoidance of contamination of food or beverages with cat feces or soil contaminated with these feces. Veterinary care of cats.
5. Proper screening of blood donors.

- **Parasitology**: Humans are the intermediate host, and cats are the final host of the *Toxoplasma* parasite. Mature oocysts in contaminated foods or undercooked meat transmit the infection. *Toxoplasma* sporozoites could infect any cells except RBCs. Diagnosis of toxoplasmosis is mainly through elevated specific IgM with rising titer.
- **Clinical:** Toxoplasmosis is asymptomatic in immunocompetent individuals. Disseminated infection in immunocompromised. Abortion in intrauterine infection, congenital malformation, mental retardation, and stillbirth in the newborn.

Trichomonas vaginalis (*T. vaginalis*)

T. vaginalis is the only human urogenital flagellate. It is responsible for vaginal and urinary trichomoniasis (vaginitis, urethritis, and acute and chronic vulvovaginitis). It is a sexually transmitted disease, as it could infect both males and females.

Geographical Distribution

T. vaginalis has a worldwide distribution.

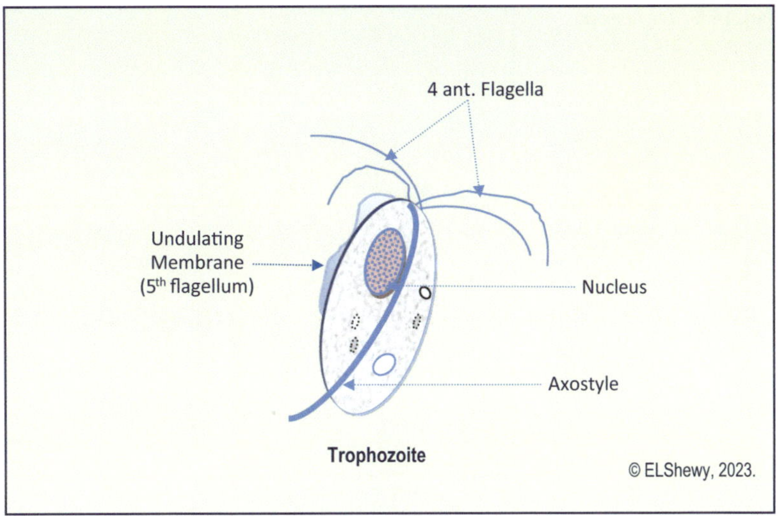

Fig. 14.4 *Trichomonas vaginalis:* morphology

Parasite Morphology

The parasite has only a trophozoite stage with no cyst (Fig. 14.4):

- Trophozoite:
  - It is an ovoid in shape, 10–20 μm length × 7–10 μm width.
  - It has four free anterior flagella and a fifth flagellum lying posteriorly, forming an undulating membrane extending to ½ the trophozoite body.
  - An axostyle extends along the whole length and projects from the posterior end.
  - It has a jerky motility.

Transmission

1. Sexual contact.
2. During delivery, the neonates may catch the infection from their infected mothers.

Habitat

T. vaginalis inhabits the urogenital tract of both males and females (vagina, urethra, and prostate).

Life Cycle

The trophozoite of *T. vaginalis* (I.S.) resides on the mucous membrane of the vagina, feeds on bacteria and white blood cells, and multiplies by binary fission. During sexual contact, trophozoites are transmitted to the other partner. In males, the trophozoites reside in the prostatic gland and at the urethral epithelia (Fig. 14.5).

Final host:	Humans.
Infective stage:	Trophozoite.
Diagnostic stage:	Trophozoite.
Intermediate host:	None.

Pathogenesis

The following sequences of pathological changes occur during infection:

1. Trophozoites come in contact with the mucosal epithelia and adhere to them.
2. Cell adherence of *T. vaginalis* to their infected cells occurs through the production of four types of surface adhesive proteins.
3. The adhesion to infected cells results in contact-dependent cytopathic effects with subsequent killing of these cells without being phagocytosed by the trophozoites.
4. Trophozoites then secrete cell-detaching factors that cause infected cells to detach and pass as a vaginal discharge.

Clinical Picture

Infection with *T. vaginalis* is mostly asymptomatic; however, the following could be associated with *T. vaginalis* infection:

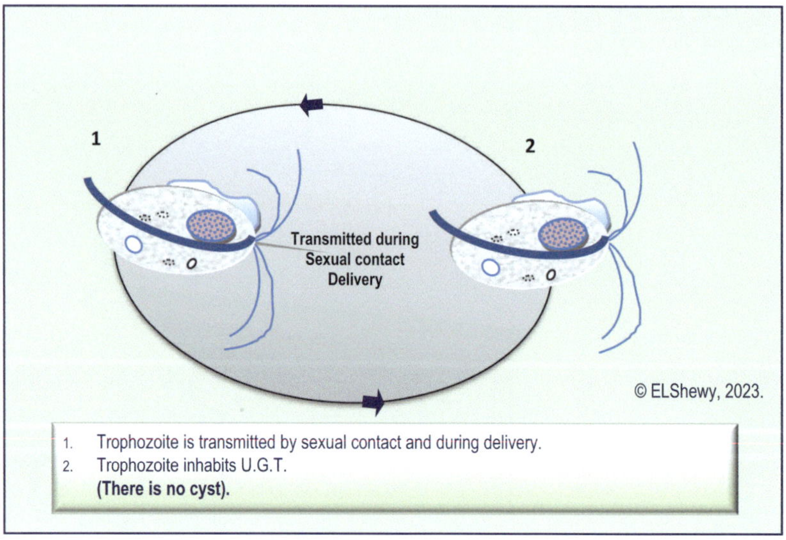

Fig. 14.5 *Trichomonas vaginalis:* life cycle

1. **Vaginal discharge** is the main manifestation of trichomoniasis in females. The discharge is profuse and associated with pruritus and itching.
2. **Dysuria, itching, burning**, redness, and soreness of the genitalia that could spread to the cervix.
3. Trichomoniasis increases the risk of infection with HIV and is associated with pregnancy complications such as preterm birth and infertility.
4. Males experience itching, irritation, discharge from the penis, burning after urination or ejaculation, prostatitis, tender prostate, and frequency of micturition.

Diagnostic Procedures

- **Direct:**

  A swab is taken by a cotton stick from the vaginal or urethral discharge. This swab could be used for:
  1. Microscopic examination of a wet preparation to demonstrate *T. vaginalis* trophozoites.
  2. Staining with trichrome stain.
  3. Culture in **Diamond's** medium.
- **Indirect Immunological:**
  1. Detection of the specific antibodies using IHA and gel diffusion test (GD).
  2. Identification of the trophozoites in the samples using IFA and monoclonal Abs.
  3. **OSOM *Trichomonas* Rapid Test:** could be used. It is an immunochromatographic dipstick test that detects antigens in the urine of male and female patients. Results are read within 10 min. Freezing and transportation of specimens do not appreciably alter the test results.
- **Molecular: Nucleic acid Amplification Testing (NAAT)** proved its high sensitivity. Samples for NAAT could be taken from urine, vaginal, or endocervical specimens for women and with urine and urethral swabs for men.

Treatment

1. **Metronidazole** (Flagyl):
   - 250 mg, three times a day for 5 days or 4 tablets of 400 mg as a single oral dose.
   - Not to be used in the 1st and 2nd trimester of pregnancy.
   - **S.E.**: nausea, diarrhea, and headache.
2. **Clotrimazole** (Nizoral):
   - 100 mg intravaginally at bedtime for 7 days.
3. **Tinidazole** (Fasigyn):
   - 2 g as a single oral dose for men.

N.B.

- Topical antiseptic lotions and powder could be used with/or if the medical treatment is contraindicated.

– The patient and his sexual partner should be treated.

Prevention and Control

1. Prompt treatment of the cases even if asymptomatic.
2. Personal hygiene with safe sex.
3. Proper antenatal care with antiseptic measures before delivery.

- **Parasitology**: *T. vaginalis* is an oval-shaped trophozoite with a single nucleus, four anterior and one posterior flagellum forming ½ body undulating membrane. No cyst is transmitted through sexual contact and during delivery.
- **Clinical:** in females, vaginitis and vulvovaginitis with yellowish offensive discharge and pruritus. Urethritis and prostatitis with frequency in micturition in males.

Congenital Parasitic Infections (Transplacental Parasitic Infections)

The most common parasitic infections that could be transmitted through the placenta are *Toxoplasma*, *Trypanosoma cruzi*, and *Plasmodium*. Helminths like filarial worms are very rare in being transmitted congenitally.

A. ***Toxoplasma:***
   (See above)
B. ***Trypanosoma cruzi*:**
   T. cruzi could be responsible for either infertility, prematurity, and/or congenital trypanosomiasis. *T. cruzi* could cause infertility or prematurity through the following possible mechanisms:
   1. Inhibits cell division of the embryo.
   2. Impairs normal implantation and development of the placenta.
   3. Decreases the gestation rate.
   4. Destruction of hormone-producing glands.
   5. Invasion of the placental tissues.
   6. Production of many harmful inflammatory cytokines.

   Congenital transmission of *T. cruzi* could be manifested as:

   1. Asymptomatic.
   2. Stillbirth.
   3. Acute fatal disease.
   4. Hepatosplenomegaly and jaundice.
   5. Myocarditis.
   6. Meningoencephalitis.
   7. Mega disease.

   Diagnosis of congenital trypanosomiasis through the following:

   1. History of trypanosomiasis during the last trimester of pregnancy.
   2. Seropositive mother.
   3. Seropositive babies during the first year of life.
   4. Detection of the parasites in the infant blood.

C. ***Plasmodium*:**
   Although infection with malaria is uncommon during the first year because of maternal immunity, however many cases of congenital malaria have been reported. Both *P. vivax* and *P. falciparum* could be transmitted congenitally. Infants with congenital malaria are misdiagnosed initially because of the lack of specific symptoms. However, the following characteristics are suggestive of congenital infection:
   1. A history of maternal exposure to malaria could be obvious or obscured.
   2. The onset of symptoms usually occurs 2–4 weeks after birth, but in rare cases, it may be as late as 15 months.
   3. Fever is uniformly present.
   4. Initially, the birth weight and feeding pattern may be normal; later, the infant may become irritable, lethargic, and anorexic as the disease progresses.
   5. Seizures may occur secondary to fever or as a consequence of cerebral malaria.
   6. Hepatomegaly or splenomegaly may not be present during the first few days of symptoms but will develop rapidly, particularly in *P. vivax* and *P. falciparum* malaria.

Review Questions

1. **The following are true about Trichomonas vaginalis except, it**
   a. inhabits the urogenital tact
   b. has a cyst form
   c. could be stained with trichrome stain
   d. is a flagellated parasite
2. **House flies could transmit the following protozoan parasites except**
   a. *Entamoeba coli*
   b. *Giardia lamblia*
   c. *Trichomonas vaginalis*
   d. *Entamoeba histolytica*
3. ***Toxoplasma* oocysts in the cat' stool become mature and infective after**
   a. 2 days
   b. 4 days
   c. 5 days
   d. 7 days
4. **The following are congenitally transmitted parasites except**
   a. *Toxoplasma*
   b. *Trypanosoma cruzi*
   c. *Trichomonas vaginalis*
   d. *Plasmodium*
5. **The commonest laboratory investigation for the detection of toxoplasmosis is**
   a. stool examination
   b. antibodies detection
   c. tissue culture
   d. animal inoculation
6. **High IgM anti-*Toxoplasma* antibodies with normal level of IgG antibodies in a patient's serum means**
   a. an old infection
   b. a recent infection on top of an old infection
   c. false infection
   d. a recent toxoplasmosis
7. **Sabin Feldman test is used for the detection of**
   a. ascariasis
   b. toxoplasmosis
   c. enterobiasis
   d. amebic dysentery

15 Parasites Infecting the Skin

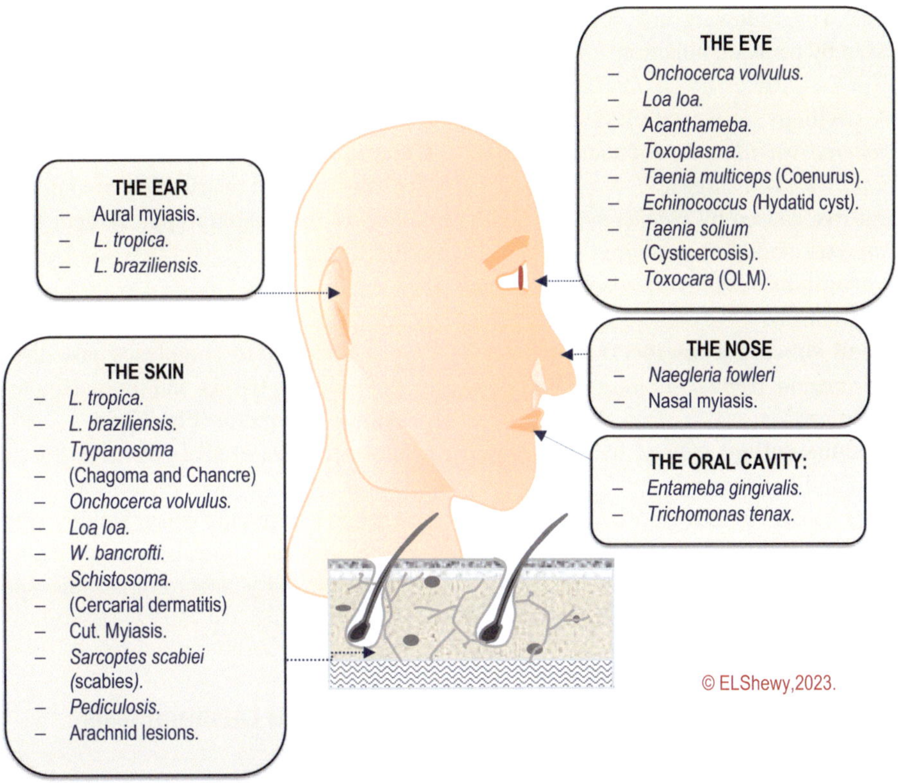

Parasites infecting organs of special senses

K. ElShewy, *Medical Parasitology*, https://doi.org/10.1007/978-3-031-64736-9_15

Test your learning and check your understanding of this book's contents: use the "Springer Nature Flashcards" app to access questions using ► https://sn.pub/oyFLIx. To use the app, please follow the instructions in Chap. 1.

Introductory
Skin and subcutaneous tissues are very common sites for a variety of arthropods, protozoa, and helminths. Arthropods are the most commonly encountered parasites in the skin and subcutaneous tissues. The following are skin manifestations caused by parasitic infection:

1. **Skin swellings and nodules:** caused by *Onchocerca volvulus*, *Loa,* cutaneous myiasis, *Sarcoptes,* and chigger.
2. **Dermatitis**: caused by *Onchocerca volvulus*, *Schistosoma,* and Medina worm.
3. **Skin eruptions:** caused by insect bites and arachnid stings.
4. **Skin and subcutaneous ulceration**: caused by *Leishmania tropica, L. braziliensis,* and *Trypanosoma.*
5. **Skin pigmentation:** caused by *Onchocerca volvulus* and *L. donovani.*
6. **Itching:** caused by *Sarcoptes scabiei*, fleas, ticks, mites, and lice.

Parasites affecting skin and/or mucocutaneous tissues could be one of the following:

1. **Parasites infect only the skin and subcutaneous tissues:** e.g., *Sarcoptes scabiei,* cutaneous myiasis-causing fly larvae, *Tunga penetrans, Demodex,* cutaneous larva migrans, *L. tropica, L. braziliensis,* and *Dracunculus medinensis.*
2. **Parasites infect the skin and other special organs:** e.g., *Onchocerca volvulus* and *Loa loa.*
3. **Parasites infect the skin as a part of its life cycle:** e.g., *Trypanosoma* (Chap. 10) and *Schistosoma.*

In this chapter, we will discuss the most common parasites infecting the skin and subcutaneous tissues; other parasites are mentioned in detail in other chapters.

Leishmania

(*L. tropica, L. mexicana,* and *L. braziliensis*) Cutaneous and Mucocutaneous Leishmaniasis

Leishmania infecting the skin and mucocutaneous tissues belongs to two complexes: *L. tropica and L. mexicana* complexes causing cutaneous leishmaniasis and *L. braziliensis* complex causing mucocutaneous leishmaniasis.

Geographical Distribution
The geographical distribution of skin and mucocutaneous *Leishmania spp.* is demonstrated in Table 15.1.

- Morphologically, cutaneous and mucocutaneous *Leishmania* complexes take the amastigote form in tissues and promastigote form in culture and vector **(Fig. 12.5).**
- The life cycle of all *Leishmania* species is the same; however, the vector species could be changed within the genus Sandfly **(Fig. 12.6).**
- Pathogenesis, clinical picture, diagnosis, and treatment vary according to the species of *Leishmania.*

A. Cutaneous Leishmaniasis

L. tropica and *L. mexicana* complexes are responsible for cutaneous leishmaniasis characterized by chronic unhealed skin lesions. Both species are endemic in countries where the sandfly vector exists.

Pathogenesis
The following pathological events are demonstrated during infection with *L. tropica* (Fig. 15.1):

Table 15.1 Geographical distribution of cutaneous and mucocutaneous *Leishmania spp.*

Leishmania complex	Species	Clinical disease	Geographical distribution
L. tropica complex	*L. tropica*	Old World Cutaneous Leishmaniasis	Mediterranean, Middle East, Armenia, and India.
	L. ethiopica		Ethiopia, Kenya, and Yemen.
	L. major		Northern Africa, the Sahara, Iran, Syria, Egypt, Jordan, and Palestine
L. mexicana complex	*L. mexicana*	New World Cutaneous Leishmaniasis	Guatemala.
	L. amazonensis		Amazon river and Brazil.
	L. venezulensis		Venezuela.
L. braziliensis complex	*L. braziliensis*	Mucocutaneous Leishmaniasis	Mexico to Argentina.
	L. panamensis		Panama and Columbia.
	L. peruviana		Peruvia

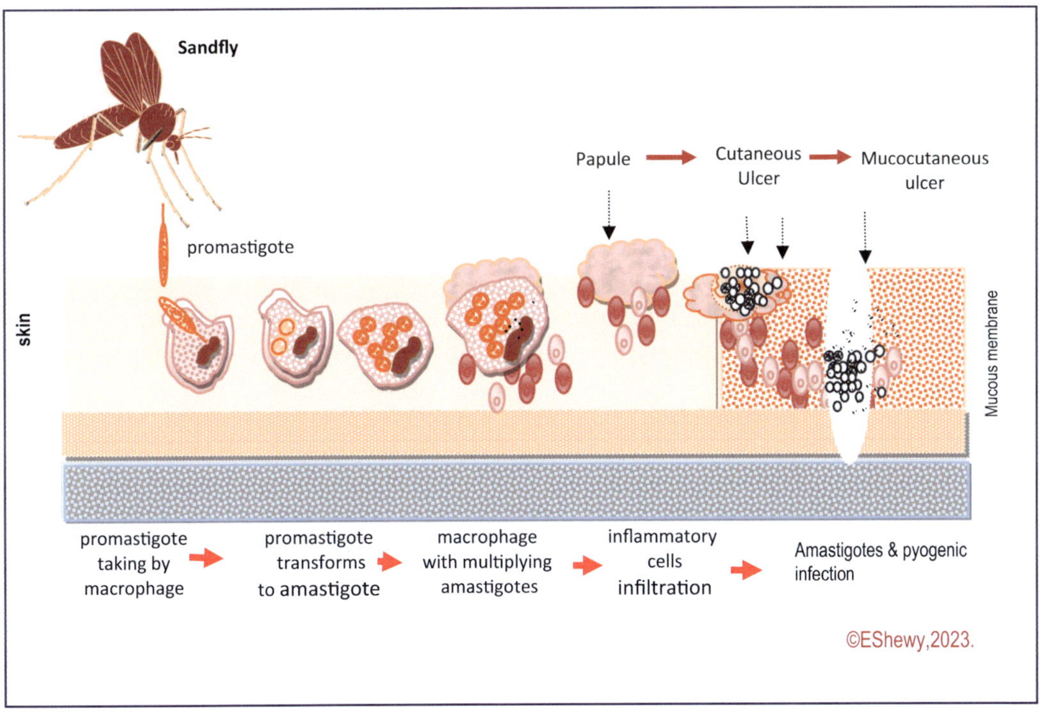

Fig. 15.1 Pathogenesis of cutaneous and mucocutaneous leishmaniasis

1. Promastigote (infective stage) is inoculated in the host skin during sandfly bite.
2. Promastigotes are taken by skin phagocytic cells and transform into amastigotes. Amastigotes proliferate rapidly within the skin macrophages.
3. Infected cells rupture, and amastigotes are released to reinfect another cell*.
4. Cellular inflammatory reaction supervenes at the site of inoculation.
5. Inflammatory products to the site of inoculation lead to induration of the skin.

6. Finally, the indurated inflamed skin ulcerates forming a cutaneous leishmanoid ulcer.

* Occasionally, amastigotes are destroyed or lysed by phagocytes and infection is aborted.

Clinical Picture

There are two clinical forms of cutaneous leishmaniasis:

1. **Old-world cutaneous leishmaniasis:**
   It is caused by *L. tropica* complex, transmitted mainly by *Phlebotomus*; it includes:
   - A chronic disease caused by *L. tropica*, with dry lesions in the face, which ulcerate after months. It is called **Oriental sore or Baghdad boil.**
   - An acute disease caused by *L. major*, with moist lesions in the lower limb, which ulcerate early.
2. **New-world cutaneous leishmaniasis:**
   It is caused by *L. mexicana* complex, transmitted by *Lutzomyia,* and it includes:
   - Skin ulcers caused by *L. mexicana*, occur mainly in the ear and in Mexico, these ulcers are called **Chiclero ulcers or Bay sores.**
   - Diffuse cutaneous leishmaniasis caused by *L. mexicana* and *L. aethiopica.*

General characteristics of cutaneous leishmaniasis:

1. The incubation period of cutaneous leishmaniasis ranges from 2 weeks in *L. maj*or to 2 months in *L.* tropica, after a sandfly bite.
2. Appearance of red papule at the bite site.
3. Itching at the bite site.
4. Ulceration.
5. Healing finally supervenes except in anergic or hypersensitive patients.
6. **Leishmania diffusa:** macronodular skin lesions all over the body usually demonstrated in immunocompromised or in anergic hypersensitive persons.
7. **Leishmaniasis recidivans:** recurring cutaneous leishmaniasis (a rare condition that occurs in Iraq and Iran), with significant tissue damage and difficulty in detecting organisms in the lesions.

Diagnostic Procedures

- **Clinical:**
  Diagnosis of cutaneous leishmaniasis in endemic areas usually depends on the clinical picture. Unhealed cutaneous ulcers for more than 2 months should be considered as leishmanoid ulcer till proven otherwise.
- **Sampling Procedures:** There are two sampling methods:
  1. Invasive sampling:
     a. Cutaneous scraping.
     b. Shave and punch biopsy.
     c. Needle aspirate.
  2. Noninvasive sampling:
     a. Filter paper lesion impression (FPLI).
     b. Cytology brush (CerviSoft/histobrush)
- **Direct:**
  1. **Staining** of the aspirate with Leishman or Giemsa stain is done to demonstrate the leishmanoid bodies (amastigotes) in the lesion.
  2. **Culture** of the aspirate for demonstration of the promastigotes in:
     - NNN medium.
     - *Drosophila* medium.
  3. **Animal inoculation**: the aspirate is inoculated subcutaneously in the nose of the hamster and infection is noticed later on the animal.
- **Indirect Immunological:**
  1. **Montenegro skin test**: injecting dead promastigotes in the skin and demonstrating the hypersensitivity reaction in positive cases.
  2. **ELISA, IFA**, and immunochromatography, for detection of the specific antibodies.
  3. **Agglutination tests:** mainly for field and survey studies. It includes a group of rapid tests detecting either specific antigens or antibodies, e.g., DAT.
- **Molecular:**
  Polymerase chain reaction **(PCR)** can be used for diagnosis and typing of *Leishmania.*

Treatment

The following are commonly used drugs for the treatment of both cutaneous and mucocutaneous leishmaniasis:

1. **Pentostam:**
   - 10 mg/kg body weight, once daily for 10 days, by I.V. or I.M.
   - It could be administered up to three courses every 10 days intervals.
2. **Glucantime:**
   - 50 mg/kg once daily for 10 days.
   - May be repeated every 20 days.
3. **Amphotericin B:**
   - 1 mg/kg body weight, by slow I.V. on alternative days.
4. **Ketoconazole:**
   - 400 mg orally for 4–8 weeks.
5. **Miltefosine:**
   - Oral 100–150 mg/day for 28 days.
   - **S.E.:** Vomiting and diarrhea and toxicity.
6. **Local application of:**
   - Topical chlorpromazine (2%).
   - Paromomycin (aminocidine) (15%).

- In addition to the above-mentioned drugs, an intradermal injection of **γ-interferon** could be used.
- **Immunotherapy** using a combined vaccine of BCG and heat-killed promastigotes is now adopted.
- **Heat, cryotherapy, and photodynamic therapies** could be used for the cutaneous lesions.

Prevention and Control

1. Vaccination formed of cultured promastigotes of different strains proved to be effective in the endemic areas.
2. Insect control (insecticides, net, repellants…).
3. Screening and treatment of the infected cases.
4. Combating the reservoir animals, e.g., gerbils and dogs.

- **Parasitology**: Both *L. tropica* and *L. mexicana* infect the skin, causing leishmanoid ulcers that contain amastigotes. Promastigote is the infective stage and is only found in the vector and NNN culture. Sandfly is the vector.
- **Clinical:** Skin ulcer (oriental sore in the old world, chiclero ulcer in the new world). *Leishmania* diffusa in immunocompromised and anergic patients. Pentostam is used in treatment.

B. Mucocutaneous Leishmaniasis

Mucocutaneous leishmaniasis is caused by *L. braziliensis* complex. It is an endemic disease in Latin American countries, where sandfly species belonging to the genera *Lutzomyia and Psychodopygus* exist.

Pathogenesis

The following pathological events are demonstrated during infection with *L. braziliensis* (Fig. 15.1):

1. Sandfly inoculates promastigotes in the host skin, especially at the mucocutaneous junctions, such as nasal and oral mucosa.
2. After inoculation, promastigotes are taken by phagocytic cells and transformed into amastigotes.
3. Amastigotes multiply in infected cells; rupture of these cells releases amastigotes to reinfect other adjacent cells.
4. The skin at the mucocutaneous junctions becomes indurated, followed by multiple ulcerations.
5. Secondary bacterial infection supervenes.
6. The ulceration starts to affect the adjacent mucous membrane and finally causes a mucocutaneous ulcer.

Clinical Picture

The disease pattern of mucocutaneous leishmaniasis is known as **Espundia** in Brazil and **Utah** in Peru. Clinically, it is manifested by the following:

1. Appearance of cutaneous ulcerations mainly at the mucocutaneous junction of the nose or mouth.
2. Ulceration of mucous membrane starts when the adjacent skin ulcer starts to heal or sometimes after many years after leaving the endemic area. The mucous ulceration progresses slowly.
3. If the patient does not receive effective treatment, the entire nasal mucosa, even the soft palate and nasal septum, could be destroyed **(no bone lesion).**

Diagnostic Procedures
The same diagnostic and sampling procedures for cutaneous leishmaniasis are followed in the diagnosis of mucocutaneous leishmaniasis.

Treatment
The same as in cutaneous leishmaniasis.

Prevention and Control
The same as in cutaneous leishmaniasis.

C. Skin Manifestations of *Leishmania donovani*

Usually, *L. donovani* does not affect the skin, but in some cases, especially after treatment, the following skin manifestations of visceral leishmaniasis are demonstrated:

1. Small papule at the site of sandfly bite.
2. **Post Kala-azar Dermal Leishmanoid** (PKDL); depigmented macules-like lesions that could contain parasites and never ulcerate.
3. Lepromatous-like skin lesions.
4. Blackening of the forehead skin in Kala-azar patients.

Dracunculus medinensis (*D. medinensis*)

(Guinea Worm, Fiery Serpent, Dragon Worm, Medina Worm)

D. medinensis is a tissue nematode that inhabits just under the dermis of the lower limb.

Geographical Distribution
D. medinensis is prevalent in areas where standing water sources, like ponds and step wells, are still used for drinking. It is detected in Africa, Iran, Pakistan, India, Saudi Arabia, and Yemen.

Parasite Morphology
The following are morphological features of *D. medinensis* worm and its infective rhabditiform larva (Fig. 15.2)

- Adult worm (♀):
  - It has a long, slender body, filled with uteri.
  - 90–100 cm × 2 mm.
  - Female *D. medinensis* is larviparous.
- Adult worm (♂):
  - It is shorter than the female.
  - 2 cm × 0.4 mm.

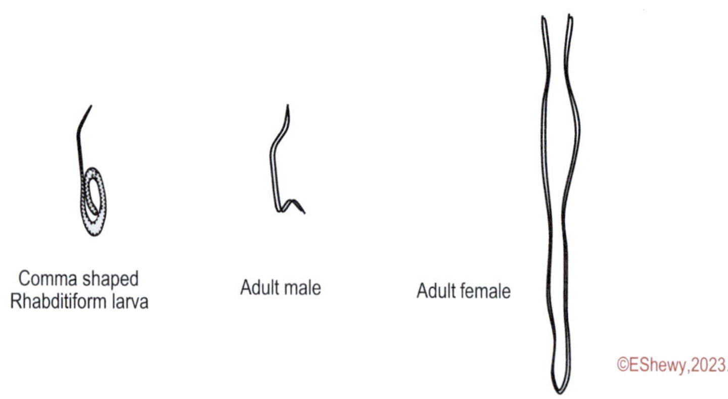

Fig. 15.2 *Dracunculus medinensis*: morphology

- Rhabditiform larva:
  - **Comma shape.**
  - 500–750 × 20 μm.
  - It is never detected in the blood or in human tissues.

Transmission

D. medinensis is transmitted through drinking water contaminated with infected copepods (*Cyclops*).

Habitat

Adult female *D. medinensis* inhabits the subcutaneous tissues of humans, particularly that of the lower limb, while its rhabditiform larva inhabits the freshwater *Cyclops*.

Life Cycle

- After fertilization, the female medina worm becomes gravid and migrates to the subcutaneous tissues of the lower limb.
- A vesicular eruption, which ulcerates through the skin, appears and a loop of worm uterus is prolapsed when the patient touches the water.
- In water, a large number of rhabditiform (comma shape) larvae discharged out from the worm uterus to be ingested by water copepods (*Cyclops*).
- After 2–3 weeks inside the *Cyclops,* larvae reach maturity and become infective.
- *Cyclops* is swallowed with water. Larvae in the *Cyclops* are liberated and then penetrate the digestive tract to enter the loose connective tissues reach the subcutaneous tissues. After around 1 year, larvae reach maturity (Fig. 15.3).

Final hosts:	Humans.
Infective stage:	Comma-shape rhabditiform larvae after 21 days in the *Cyclops*.
Diagnostic stage:	Rhabditiform larvae or the worm in its position.
Intermediate hosts:	*Cyclops*.

Pathogenesis

1. Usually, one or rarely more than one worm could infect the subcutaneous tissues.
2. Migration of the worm results in allergic reactions and localized skin erythema.
3. The localized inflammatory reaction usually occurs around the worm uterine opening, forming a skin blister (ulcer) and possibly causing a secondary bacterial infection.
4. After emptying the uterus, the female worm withdraws onto tissues to be resorbed or expelled.
5. Calcification is formed around the dead worms or the nonerupted worms.

Clinical Picture

1. Fever, urticaria, itching, allergic manifestations, e.g., asthma or periorbital edema in the allergic patients.
2. Worms are detected under the skin with a **serpentine pattern** and with a blister at the uterine opening.
3. A marked relief of symptoms after the blister is ruptured.

Diagnostic Procedures

1. Diagnosis is typically based on clinical manifestations. In endemic areas, the characteristic serpentine track in the lower limb is easily recognizable
2. The worm can be stimulated to discharge larvae by submerging the ulcer with water.
3. A plain X-ray may demonstrate the calcified dead worm in the subcutaneous tissues.

Treatment

1. **Niridazole** (Amblihar):
   - 25 mg/kg/b. w daily for 15 days orally.
   - **S.E.:** nausea, vomiting, diarrhea, and headache.
2. **Metronidazole:**
   - 250 mg/3 times a day for 10 days orally.
3. **Thiabendazole:**
   - 50 mg/kg/b. w daily for 2 days.

N.B.

- Drugs usually relieve the symptoms and make the worm removal easy, but they have no direct effect on the worm itself.
- Usually, the traditional treatment of the worm is done by twisting the worm around a stick.

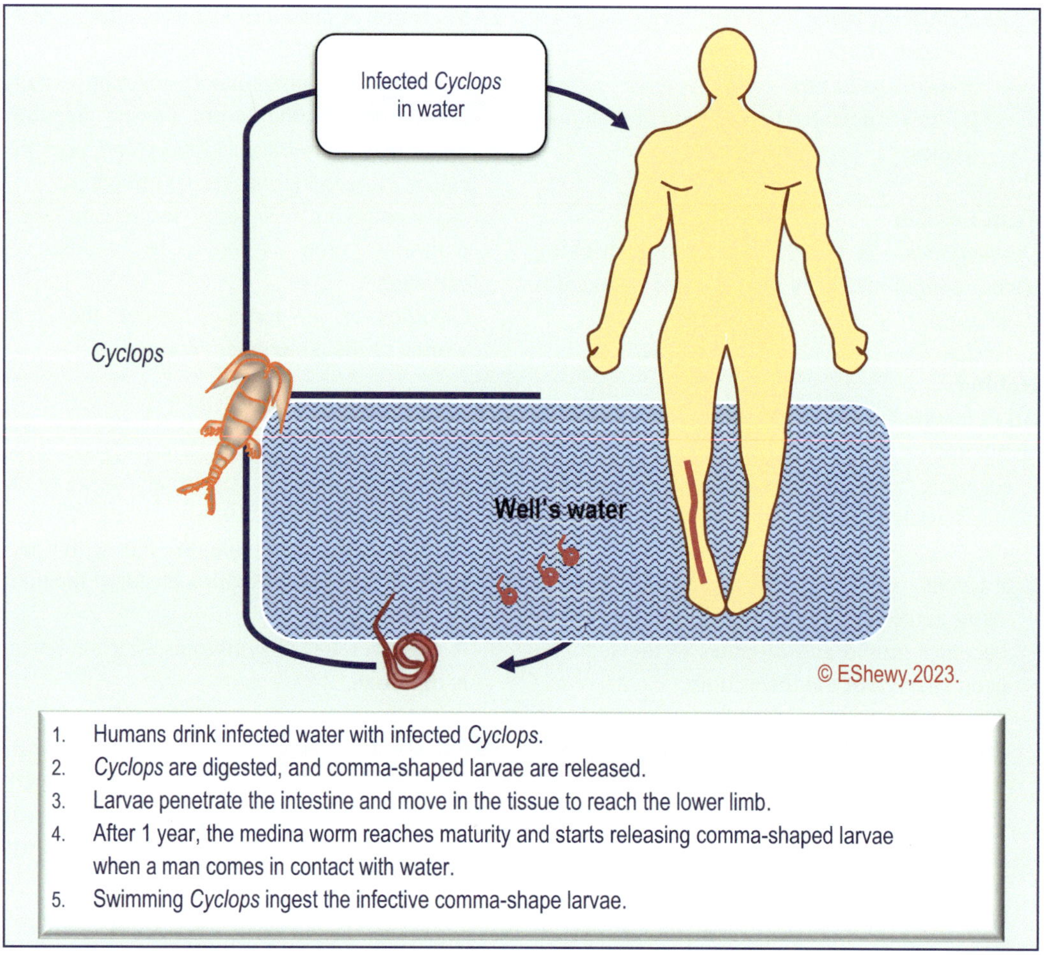

Fig. 15.3 *Dracunculus medinensis*: life cycle.

The worm should be extracted completely as a whole; otherwise, a secondary infection or a severe allergic reaction develops.

Prevention and Control

1. A safe and healthy drinking water supply should be used.
2. Water suspected to contain the *Cyclops* should be boiled, filtered, or chemically treated.
3. Health education of the people.
4. Proper treatment of the infected persons.

Cutaneous Larva Migrans (CLM)

Cutaneous larva migrans (CLM) result from infection of human skin with certain nonhuman and sometimes human helminths. CLM includes:

a. **Creeping eruption:** infection of the skin with larvae of nonhuman species of hookworms, e.g., *Ancylostoma braziliense* and *Ancylostoma caninum*.
b. **Larva currens:** skin infection with *Strongyloides stercoralis* larvae.

A. Creeping Eruption

Ancylostoma caninum and *braziliense* (*A. caninum* and *braziliense*)

(Dog and Cat Hookworms)

A. caninum and *braziliense* are nonhuman species of hookworms. They are dog and cat parasites with many biological and morphological similarities with human hookworms.

Geographical Distribution

A. caninum & braziliense have a worldwide distribution.

Transmission

Human infection takes place through piercing the bared skin by *A. caninum and braziliense* filariform larvae when they come in contact with dog stool in soil, e.g., beach sand.

Habitat

A. caninum & braziliense inhabit the small intestine of dogs and cats.

Life Cycle

The life cycle of both *A. braziliense and caninum* is diagramed in Fig. 15.4.

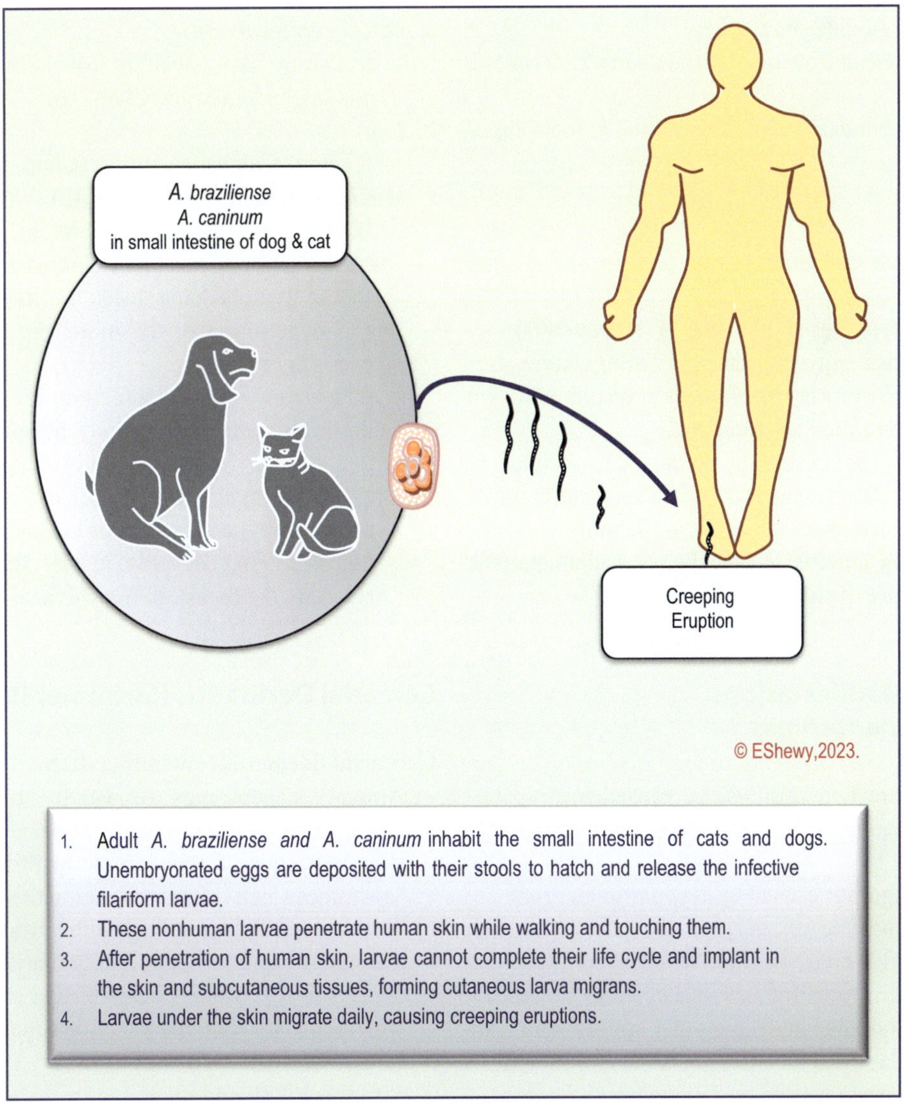

Fig. 15.4 *Ancylostoma braziliense and caninum*: life cycle

Clinical Picture

1. Legs, buttocks, feet, and arms are the main sites for cutaneous larva migrans.
2. The patient suffers from intense itching that lasts for months.
3. Larvae cause zigzag or serpentine tunnels with advancing edges by 1–2 cm/day.

Diagnostic Procedures

Diagnosis of CLM infection depends on the clinical picture and history of dealing with dogs or dog excreta.

Treatment

1. Cryotherapy using ethyl chloride spray or carbon dioxide snow.
2. Thiabendazole oral tablets 25 mg/kg/2 times a day.
3. Thiabendazole 10% suspension as local applicant.

B. Larva currens

Strongyloides stercoralis (*S. stercoralis*)

- Occasionally, migration of *Strongyloides stercoralis* larvae in human skin results in a skin lesion called larva currens.
- It is manifested by skin tracks similar to creeping eruption but with a very rapid rate of advance under the skin (5–10 cm/h).
- Larva currens is not always a characteristic feature of strongyloidiasis.

Skin Manifestations of *Trypanosomes*

There are two skin lesions caused by trypanosomes, commonly at the bite site of the vector:

a. **Chagoma:** caused by *Trypanosoma cruzi:*
   At the site of entry of the parasite by the winged bug, swelling with an erythematous margin turns into a violaceous nodule with a central edematous area that appears and lasts for days and then disappears in 2–3 months.
b. **Chancre:** caused by *Trypanosoma brucei:*
   A painful skin area appears at the site of *Tsetse* fly bite. It includes a circumscribed, indurated red nodule that desquamates after 2–3 weeks and slowly disappears, leaving behind a central scar.

Skin Manifestations of Filarial Worms

Filarial Worms

Infection with the following filarial worms is associated with skin manifestations:

a. ***Onchocerca volvulus:***
   - It causes skin swelling (onchocercoma) and onchodermatitis (Chap. 16).
b. ***Loa*:**
   - It causes Calabar swellings (Chap. 16).
c. ***Wuchereria bancrofti*:** (Chap. 12).
   - *W. bancrofti* infection results in lymphedema in the lower limb; chronic and persistent lymphedema leads to thickening and hyperplasia of the tissue around and overlying skin.
   - After many years of infection, the skin of the limb starts to be woody fibrous indurated with verrucous thickenings like elephant skin. **Elephantiasis** (dermal hypertrophy) and verrucous changes (leopard skin) may develop in any limb, the scrotum, the breast, or the vulva.

Cercarial Dermatitis (Swimmer Itch)

Cercarial dermatitis (swimmer Itch)

- Animal schistosomes (especially of birds and mammals) can penetrate exposed human skin and induce a transient dermatitis called "swimmer's itch" (Cercarial dermatitis).
- This is typically presented with pruritic erythematous papular lesions and urticaria immediately after exposure. Later, vesicular eruptions with secondary bacterial infection supervene.
- It resolves after 7–10 days.
- Treatment is symptomatic.

Arthropods of the Skin and Mucocutaneous Tissues

Arthropods are the most commonly encountered parasites in the skin and subcutaneous tissues. They may be a primary parasite infesting the skin or may affect the skin during inoculation of other pathogens.

A. *Sarcoptes scabiei* (Itch Mite)

N.B. See Chaps. 19 and 20 for morphology, pathogenesis, clinical picture diagnosis, and treatment of arthropods of the skin.

B. Cutaneous myiasis

S. scabiei is the causative arthropod of scabies, an itchy skin lesion (Chap. 20).

Cutaneous myiasis could be:

1. Traumatic (wound) myiasis: Larvae invade open wounds or ulcers, e.g., *Calliphora, Lucilia, Phormia, Musca,* and *Fannia.*
2. Furuncular myiasis (follicular): The larva remains at a single spot, causing a boil-like lesion, e.g., *Dermatobia, Cordylobia, Wohlfahrtia.*
3. Creeping myiasis: Larvae burrow into the skin, e.g., *Gasterophilus and Hypodermia* (Chap. 19).

C. *Tunga penetrans* (Sand flea)

- *Tunga penetrans* (sand flea) is responsible for skin lesions in the human feet and is called **Chigger** foot. It prevails in Latin America.
- The fertilized female of *Tunga* burrows in the epidermis of human feet. Subsequently, a pruritic, painful, and white papular lesion with a central dark discoloration develops.
- It stays 2 weeks with potential complications of gangrene and secondary infection (Chap. 19).

D. Pediculosis

- Pediculosis (**Vagabond's Disease**):
- It is caused by heavy lice infestation among poor hygienic overcrowded slums. It is manifested by cutaneous lesions, with itching and ulcerations of the skin.
- In a heavy infestation with *Phthirus pubis* (crab lice), the patient's skin may be inflamed with some blue spots.
- The typical presentation of heavy lice infestation includes erythematous papules, pruritus, and regional lymphadenopathy. Potential complications include impetigo and infection (Chap. 19).

E. *Demodex folliculorum* (Hair Follicles Mites)

- *Demodex* mites (hair follicle mites) infest the space within the hair follicles.
- It infects the sebaceous glands of humans and animals.
- It gives the acne-like appearance or keratitis with no itching.
- During heavy *Demodex* infestation (demodicosis): there are pustular folliculitis, dermatitis, and papulopustular scalp eruptions (Chap. 20).

F. Skin Allergy, Eruption, and Nodules (*Arachnida*)

Members of arachnid could affect human skin through the bite or sting of his skin; this could be manifested by:

- Skin allergy and contact dermatitis caused by house dust mites.
- Raised papule, pustule, or wheal at the site of the spider stings.
- Skin inflammation with swelling around the sting during a scorpion sting (Chap. 20).

Review Questions

1. **Chiclero ulcer is a skin ulcer caused by *Leishmania***
   a. *tropica*
   b. *mexicana*
   c. *donovani*
   d. *braziliensis*
2. **N.N.N. medium is used for the detection of**
   a. *Babesia*
   b. *Leishmania donovani*
   c. *Toxoplasma*
   d. *Plasmodium*
3. **The treatment of choice for *Leishmania tropica* is**
   a. tetracycline
   b. sulpha compounds
   c. pentostam
   d. metronidazole
4. **Aspiration fluid from *Leishmania* ulcer should be taken from**
   a. the wall of the ulcer
   b. the surface of the ulcer
   c. beneath the surface of the ulcer
   d. anywhere from the ulcer
5. **Sowda is a disease caused by**
   a. *Dracunculus medinensis*
   b. *Onchocerca volvulus*
   c. *Taenia saginata*
   d. *Wuchereria bancrofti*
6. **Medina worm is**
   a. an intestinal nematode
   b. a tissue nematode affecting the subcutaneous tissues
   c. a tissue nematode that causes river blindness
   d. transmitted by eating contaminated meat
7. **Chancre is a skin lesion of**
   a. *Trichinella spiralis*
   b. *Onchocerca volvulus*
   c. *Trypanosoma brucei*
   d. *filarial elephantiasis*
8. **Black fly (*Simulium*) is a vector for**
   a. *Dracunculus medinensis*
   b. *Onchocerca volvulus*
   c. *Ancylostoma duodenale*
   d. *Wuchereria bancrofti*
9. **Ground itch (swimmer's itch) is caused by *Schistosoma***
   a. adult
   b. egg
   c. cercaria
   d. miracidium

16 Parasites Affecting Organs of Special Senses (Eye, Ear, Nose, and Mouth)

Test your learning and check your understanding of this book's contents: use the "Springer Nature Flashcards" app to access questions using ▶ https://sn.pub/oyFLIx. To use the app, please follow the instructions in Chap. 1.

Parasites of the Eye

A. *Onchocerca volvulus* (*O. volvulus*)

(River Blindness Worm)

O. volvulus is one of the filarial tissue nematodes responsible for onchocerciasis characterized by river blindness, onchocercoma, and onchodermatitis.

Geographical Distribution

O. volvulus is prevalent in more than 37 countries all over the world, particularly in Africa, Latin America, Yemen, and Saudi Arabia.

Parasite Morphology

The following are morphological features of *O. volvulus* filarial worm and their microfilariae:

- Adult worm:
  - It is a threadlike filarial worm with tapered on both ends.
  - Female is about 30–50 cm.
  - Male reaches 20–40 cm in length.
- Microfilariae:
  - 200–330 × 6–9 μm.
  - Unsheathed body with striated cuticle.
  - Head and tail are free of nuclei.
  - They have no periodicity.

Transmission

O. volvulus is transmitted through the bite of the black fly (buffalo gnat) (*Simulium spp.*), which breeds in fast-running rivers, streams, and the lands along the riverbanks.

Habitat

- Adult worm of *O. volvulus* inhabits the subcutaneous tissues, forming a nodule, especially at the junction of the skin with bones, e.g., scalp, pelvis, chest, and spine.
- *O. volvulus* microfilaria could be detected in the skin and eye.

Life Cycle

- *O. volvulus* microfilariae are taken from the infected person through a bite of *Simulium* fly.
- Microfilariae undergo cyclic development inside the fly, which transforms them into infective microfilariae (I.S.).
- Infective microfilariae are inoculated into a new host during the fly bite.
- They migrate under the conjunctiva and the skin until they mature into adult worms.

K. ElShewy, *Medical Parasitology*, https://doi.org/10.1007/978-3-031-64736-9_16

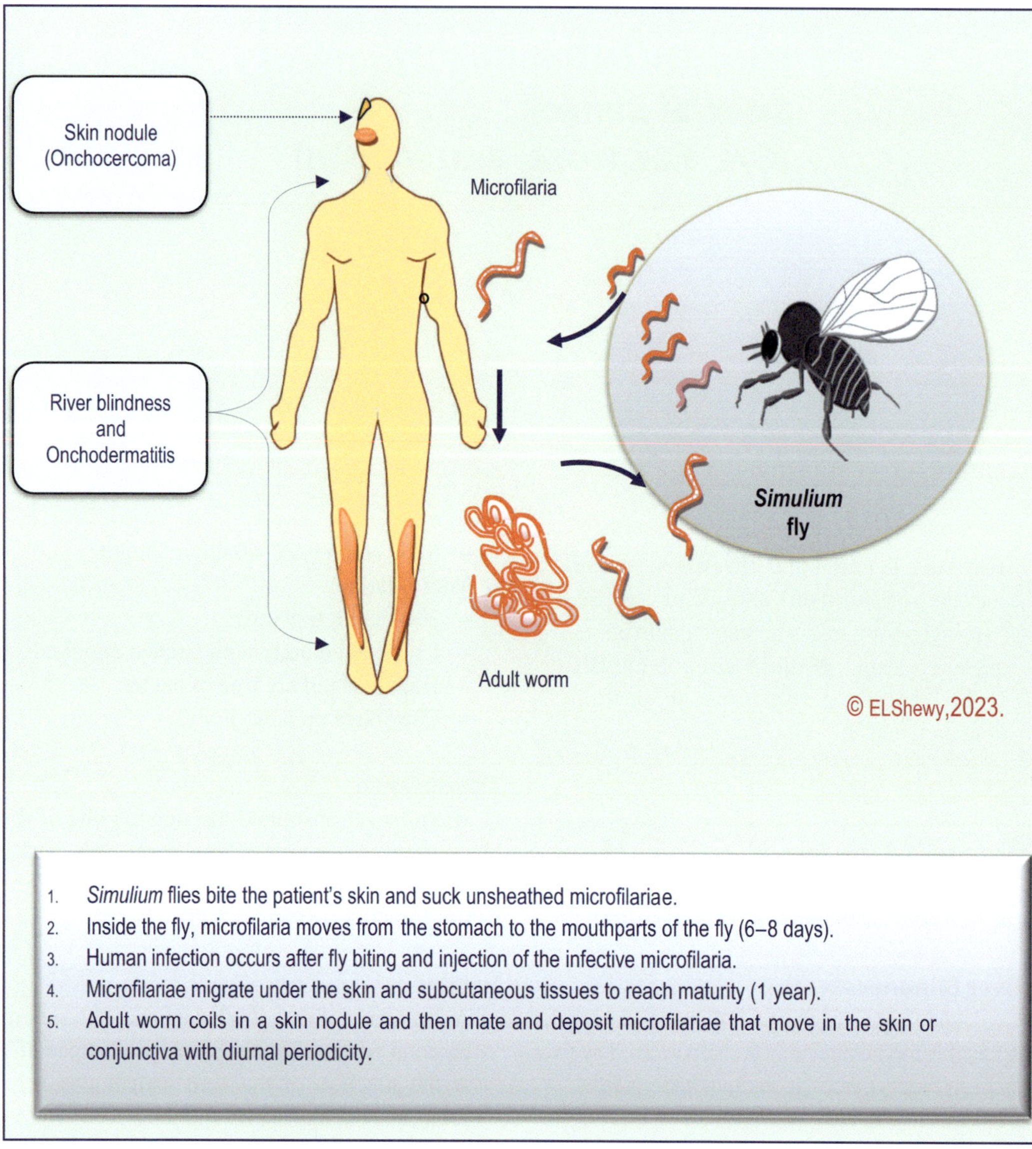

Fig. 16.1 *Onchocerca volvulus*: life cycle

– After many years, most worms coiled inside capsulated nodules (onchocercoma) ranging in size from a few millimeters to several centimeters (Fig. 16.1).

Final hosts:	Humans.
Infective stage:	Infective microfilariae inside the vector.
Diagnostic stage:	Unsheathed microfilariae in skin snip and tissues.
Intermediate hosts:	*Simulium spp.*

Pathogenesis

The eye and skin are the main sites for the pathologic changes during onchocerciasis:

1. Migrating microfilariae in the eye results in punctuate keratitis, corneal opacities that coalesce in the lower portion of the cornea and then affect the entire cornea. Iridocyclitis with displacement of the pupil and finally blindness supervenes **(River blindness).**

2. Migrating microfilariae under the skin results in an acute allergic reaction (the same reaction could be demonstrated after treatment).
3. Deposition of a large amount of immune complex due to continuous antigenic stimulation results in an acute inflammatory reaction in the skin, L.N. Later fibrosis and loss of elasticity of the skin with lymphadenitis **(onchodermatitis)** could occur.
4. There is an individual variation in the frequency of skin lesions, which ranges from localized skin affection to marked generalized affection. This variation depends on the immune status of the individuals.
5. Adult worms finally end in encapsulated painless nodules **(onchocercoma).**

Clinical Picture

1. **River blindness:**
   - It is very common among infected Africans (endemic).
   - The patient suffers from corneal opacities, blurring of vision with a possibility of blindness.
2. **Onchodermatitis:**
   - It starts with hot skin and painful itching on the face, ears, and neck.
   - Later, inflammation subsides, and the skin thickens and assumes a violaceous color.
   - Onchodermatitis assumes different forms in different localities:
     a. In Mexico and Latin America: acute patchy reddish eruptions and chronic atrophic wrinkled skin with thickening.
     b. In Africa: atrophy and thickening of the skin with depigmentation (leopard skin). The affected skin at the groin forms a long sac called the "**Hanging groin.**"
     c. In Yemen and Saudi Arabia: it is confined to a single limb (around the ankle); the affected skin is dark, swollen, and pruritic with eruptions; this is called "**Sowda.**"
3. **Onchocercoma**: painless masses at the scalp, pelvis, and at the elbow.

Diagnostic Procedures

1. Skin needle snip: snip biopsy placed in saline, teased to liberate microfilariae and examined under the microscope.
2. In light infection, the **Mazzotti test** is applied, i.e., oral administration of 50 mg DEC results in local pruritus at the site of microfilariae.
3. Ultrasonography could be used for the detection of deep onchocercal nodules.
4. Detection of the specific antibodies using ELISA and IHA.

Treatment

1. **Ivermectin** (The drug of choice):
   - 100–200 μg/kg/b.w., to be repeated every year.
   - It is superior to DEC for the treatment of eye affection but should be preceded with corticosteroids days before its use.
   - **S.E**: mild pruritus, skin edema, and arthralgia.
2. **Diethylcarbamazine (DEC)** (Hetrazan):
   - 2 mg/kg/b.w./3 times daily, for 2–4 weeks
   - **S.E.:** allergic reactions, malaise, vertigo, headache, severe pruritus, and skin edema, and if given with an eye affection, punctuate keratitis usually happens.

N.B. Surgical removal of the palpable disfiguring nodules could be applied.

Prevention and Control

1. Repeated treatment for the infected persons.
2. Personal protection from black fly bites during travel to endemic areas.
3. Vector control by destroying the fly larvae and the breeding sites.

B. *Loa loa* (*L. loa*)

(African Eye Worm)

Loa loa is a filarial tissue nematode with sheathed microfilaria and diurnal periodicity. It is responsible for extensive eye disease, skin inflammation, and Calabar swellings.

Geographical Distribution

L. loa has an extensive distribution in the rainforest areas of Sudan, Congo, and West Africa.

Parasite Morphology

- Adult worm:
  - Threadlike filarial worm with tapered both ends.
  - The female is about 5–7 cm, and the male reaches 2–3 cm in length.
- Microfilariae:
  - 250–300 μm × 6–9 μm
  - Sheathed body with striated cuticle. The tail is full of nuclei
  - They have diurnal periodicity (between noon and 2.00 p.m.).

Transmission

L. loa is transmitted by inoculation of the infective microfilariae through the bite of *Chrysops* fly (Mango fly).

Life Cycle

L. loa has a similar life cycle as *W. bancrofti*, with *Chrysops* fly as a vector (Fig. 16.2).

- Fly ingests blood from an infected person that contains sheathed microfilariae of *L. loa*.

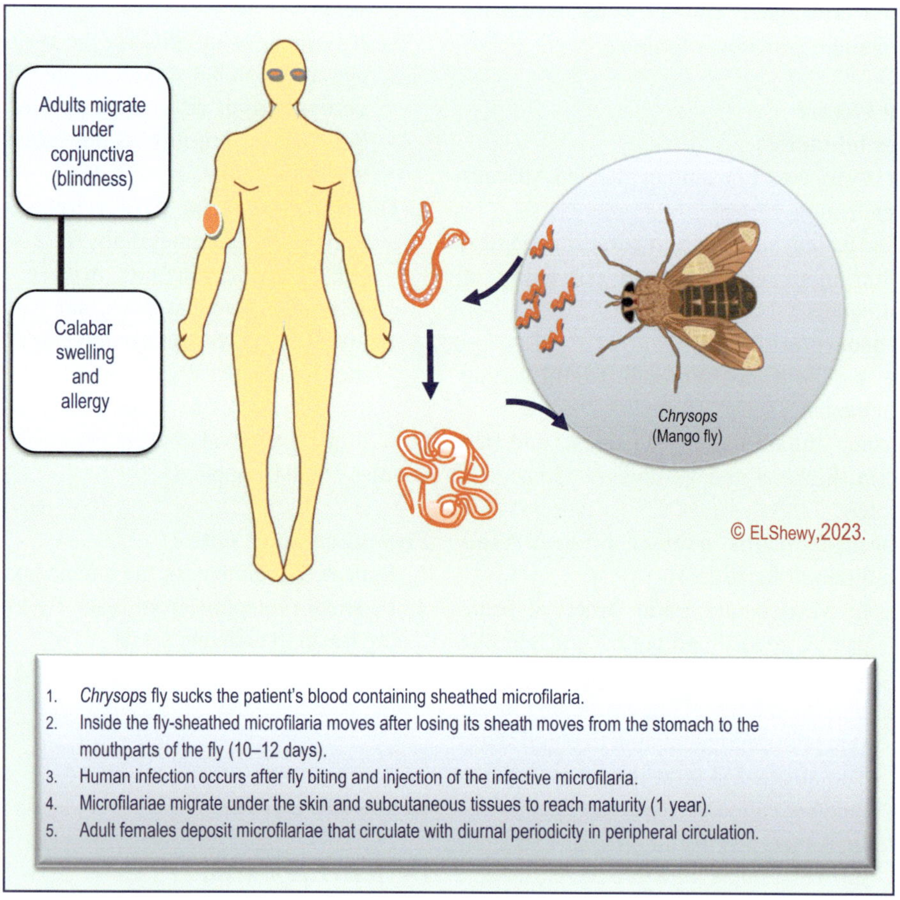

Fig. 16.2 *Loa loa* (eye worm): life cycle

Final hosts:	Humans.
Infective stage:	Infective microfilariae inside the vector.
Diagnostic stage:	Microfilariae in the skin snip and in the affected tissues.
Intermediate hosts:	*Chrysops* fly (Mango fly).

- Microfilariae lose their sheaths, mature within 10–12 days, and migrate to reach the fly's mouthparts.
- Microfilariae are injected into another host to migrate in his blood and mature into adult worms within 1 year. The adult worm then invades subcutaneous tissues, and it may invade the conjunctiva and the nasal bridge. Adult female worm discharges microfilariae into the bloodstream to be taken by *Chrysops*.

Pathogenesis and Clinical Picture

Pathologic changes during loiasis are mainly caused by adult worms and not by microfilariae. The adult worms can live up to 17 years in human.

- Loiasis is mostly asymptomatic.
- Subconjunctival migration of the adult worm in the eyes occurs frequently and at irregular intervals; it takes less than 15 min each.
- *L. loa* crossing under the conjunctiva is associated with inflammatory reactions, repeated conjunctivitis, and eventually blindness.
- Migration of the worm under the skin is not painful. However, it is associated with an allergic reaction causing pruritus and papules, and the skin may be thickened.
- **Calabar swelling (Fugitive swollen lumps of subcutaneous tissues):** It appears anywhere in the body; it lasts for a few days and then disappears. It is due to the worm excretory products and lymphedema.
  a. Calabar swelling is 3–10 cm, non-erythematous and non-pitting mass.
  b. It lasts for 1–3 days, with localized skin eruption and pruritus.
  c. It can form cyst-like enlargement of the connective tissue around the sheaths of muscle tendons. It is very painful when moved.
  d. Dead worms may stimulate an inflammatory reaction with abscess formation.

Diagnostic Procedures

Diagnosis of loiasis depends on the clinical picture and the following investigations:

- **Direct:**
  1. Microscopic demonstration of sheathed microfilariae in peripheral blood:
     - Specimen should be taken between noon and 14:00 (diurnal periodicity). Thin smear and thick smear stained with Giemsa.
     - Concentration methods are helpful in the detection of light infection, e.g., centrifugation of fluid fixed in 2% formalin (**Knott's technique**).
     - Filtration of citrated blood or fluid specimen through Nucleopore.
  2. Extraction of adult worms in skin and eye tissues.
  3. The blood picture demonstrates significant eosinophilia.
- **Indirect Immunological:**
  IHA, IFA, and ELISA are used for the detection of the specific anti-filarial antibodies; however, they are mostly nonspecific.

Treatment

1. **Diethylcarbamazine (DEC)** (Hetrazan):
   - 2 mg/kg/b.w., 3 times daily, for 2–4 weeks.
   - It has a rapid response against microfilaremia.
   - S.D: allergic reactions could cross the blood–brain barrier and cause encephalitis.
2. **Ivermectin:**
   - 100 μg/kg/b.w., as single dose.

N.B. Surgical removal of the migrating adult worms under the conjunctiva or when crossing the nose could be done after local anesthesia.

C. *Acanthamoeba* (Amebic keratitis)

Acanthamoeba species are free-living amebae (FLA) that spread freely in nature. Most of these strains are causes of chronic granulomatous amebic encephalitis (**GAE**) and a severe form of corneal infection (amebic keratitis).

Geographical Distribution

Acanthamoeba has a worldwide distribution; it can be found in soil (moist sands, flowerpot soil), air, brackish water, sewage, humidifiers, heating hospital environments, dental and dialysis units, and contact lenses.

Transmission

Acanthamoeba could affect the human eye through the following portals of entry:

1. A minor corneal lesion.
2. Previous episodes of *Herpes simplex*.
3. Abrasion from hard or soft contact lenses and sharing lenses.
4. Scratch or severe itching of the eye.
5. The use of homemade saline or tap water for contact lens cleaning.

Pathogenesis

1. *Acanthamoeba* trophozoites and cysts affect the superficial then the deeper corneal stroma, causing a moderate granulomatous inflammation.
2. Later, a chronic ulcerative keratitis appears.
3. Finally, a localized corneal ulceration may coalesce and progress to a complete corneal perforation and blindness.

Clinical Picture

- Painful red eye with itching.
- Superficial keratitis with decreased visual acuity.
- Deep keratitis with opacification and blindness.

Diagnostic Procedures

Corneal swab is taken for:

1. Staining with microscopic examination to detect the trophozoites and/or the cysts.
2. IFA.
3. Culture followed by microscopic examination.

Treatment

1. **Itraconazole** (orally).
2. **Otramidryl** (ointment).
3. **Brolene** (drops).
4. **Neomycin** (drops).

Prevention and Control

1. Conducting safety campaigns to educate the public.
2. Contact lens wearers should not use nonsterile homemade saline to clean their contact lenses.
3. Contact lens wearers should not wear their lenses while swimming, with corneal abrasion, or during *Herpes simplex* infection.
4. When there is any skin lesion, do not swim in common pools or deal with soil and peach sand.

Parasites of the Mouth, Nose, and Ear

A. *Entamoeba gingivalis* (*E. gingivalis*)

(Oral Ameba)

E. gingivalis is an oral, potentially pathogenic ameba. Morphologically, it resembles *E. histolytica*. It is recovered from the mouth of individuals with pyorrhea alveolaris.

Geographical Distribution

E. gingivalis has a worldwide distribution, especially among individuals with decreased oral hygiene.

Parasite Morphology

E. gingivalis has only a trophozoite form without a cyst (Fig. 16.3).

- Trophozoite:
  - It has an irregular amebic shape, 15–40 μm.
  - Cytoplasm is granular and contains bacteria, leukocytes, and RBCs.
  - Single nucleus, with fine, evenly distributed peripheral chromatin dots and a small central karyosome.

N.B. *E. gingivalis* is the only species that ingests leukocytes in its cytoplasm.

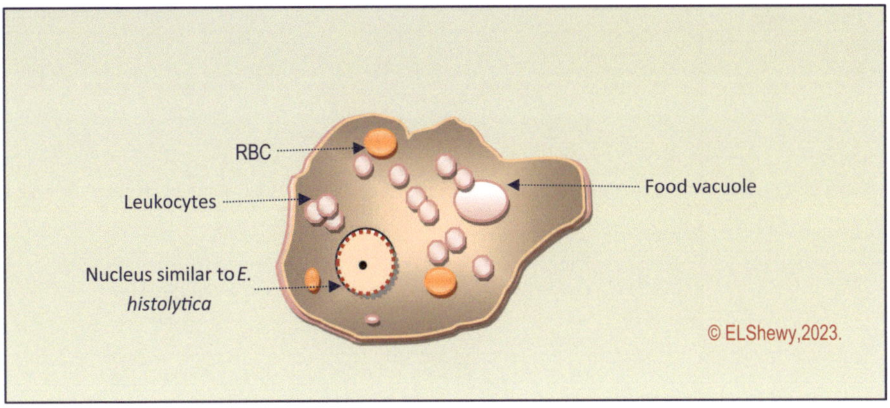

Fig. 16.3 *Entameba gingivalis*: morphology

Transmission

1. Drinking from common glass.
2. Kissing with bad oral hygiene.
3. Contaminated dental procedures and surgical instruments.

Habitat

E. gingivalis inhabits the area between teeth and gums, in the tonsillar crypts, and in bronchial mucus.

Life Cycle

Final host:	Humans.
Infective stage:	Trophozoite.
Diagnostic stage:	Trophozoite.

E. gingivalis trophozoites are found to multiply in the bronchial mucus, and they reside between teeth and gums, in the tonsillar crypts, and the bronchial mucus. They are transmitted from one host to another through the above-mentioned routes.

Pathogenesis and Clinical Picture

Detection of *E. gingivalis* in patients with dental problems could throw light on an association between the presence of *E. gingivalis* and dental diseases. *E. gingivalis* is found to be associated with pyorrhea and gingivitis patients more in normal control individuals.

Diagnostic Procedures

- *E. gingivalis* is diagnosed by staining a swab taken with gingival scraping and examination under the microscope.
- It is very important to distinguish it from the *E. histolytica* that might be found in the sputum.

B. *Trichomonas tenax/buccalis* (*T. tenax* or *T. buccalis*)

(Oral flagellate)

T. tenax is an oral, potentially pathogenic flagellate. Morphologically, it resembles *T. vaginalis* and *T. hominis*. It is recovered from the mouth of individuals with pyorrhea alveolaris.

Geographical Distribution

T. tenax has a worldwide distribution, especially due to decreased oral hygiene.

Parasite Morphology

T. tenax has only a trophozoite form without a cyst. Trophozoite of *T. tenax* is similar to that of *T. vaginalis* with the following criteria (Fig. 16.4):

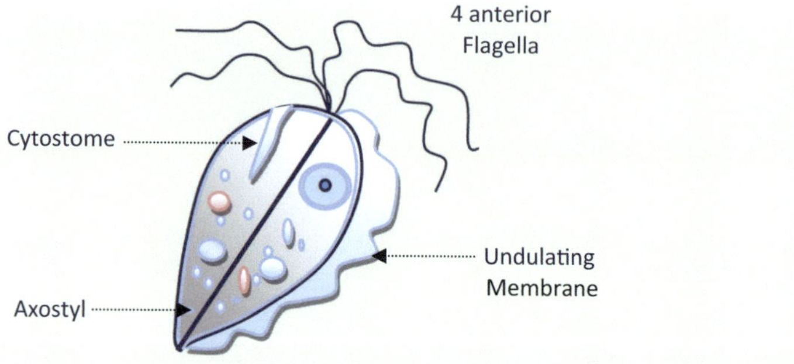

Fig. 16.4 *Trichomonas tenax* (*buccalis*): morphology

Trophozoite:

- Shape: oval, 15–30 μm.
- Cytoplasm is granular, containing bacteria.
- Single nucleus with fine, evenly distributed peripheral chromatin dots and a small central karyosome.
- It has four anterior flagella, and the fifth forms an undulating membrane. Both the undulating membrane and axostyle are not extended beyond the posterior end.

Transmission

T. tenax is transmitted through the oral route, drinking from the same source, and kissing with bad oral hygiene.

Habitat

T. tenax resides in between teeth and gums, in the tonsillar crypts, and in bronchial mucus.

Life Cycle

T. tenax has a direct life cycle from host to host without an intermediate host (Fig. 16.5).

Final host:	Humans.
Infective stage:	Trophozoite.
Diagnostic stage:	Trophozoite.

Pathogenesis

- Detection of *T. tenax* in patients with dental problems could suggest an association between the presence of *T. tenax* and dental diseases.
- *T. tenax* is found to be associated with pyorrhea and gingivitis patients more than in normal control individuals.
- *T. tenax* has been aspirated from the bronchial lavage in some cases of pulmonary and brachial infection.

Clinical Picture

- Infection with *T. tenax* is generally asymptomatic.
- *T. tenax* is associated with many cases of dental infections, e.g., pyorrhea and gingivitis.
- *T. tenax* recovered from tonsillar crypts and cases with pulmonary manifestations. However, the question to be answered is whether this relation is an association or a causation.

Diagnostic Procedures

T. tenax is diagnosed by staining of a swab taken by a gingival scraping. It takes a characteristic shape of a **teardrop** appearance.

Treatment

There is no effective treatment for *T. tenax.*

Prevention and Control

Following the standards of health and oral hygiene is the main line of prevention of such oral parasitic infections.

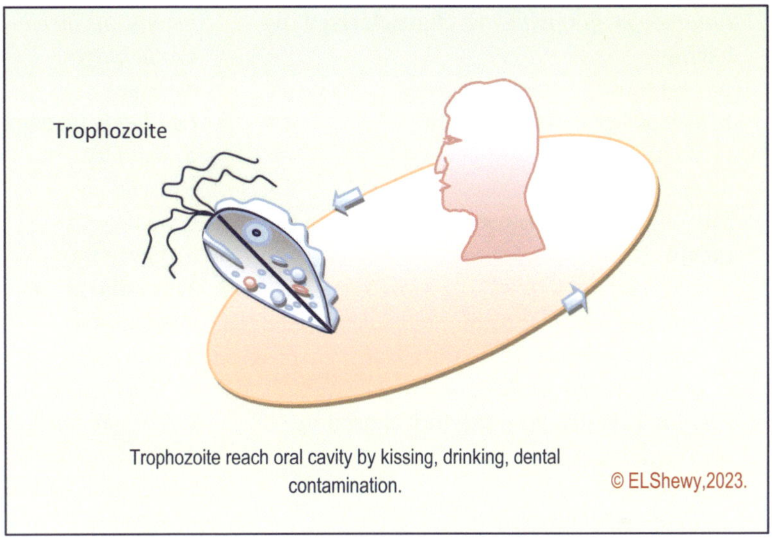

Fig. 16.5 *Trichomonas tenax* (*buccalis*): life cycle

C. Myiasis

1. **Auricular myiasis:** Infection of the ear with larvae (maggots) of insects, e.g., *Wohlfahrtia and Sarcophaga* (Chap. 19).
   - It is a rare clinical condition that usually occurs in patients who have mental or physical disabilities.
   - Auricular myiasis involves infestation of the external ear and/or middle ear with fly larvae.
   - **Clinically, auricular myiasis is manifested by:**
     1. Larvae in the ear.
     2. Itching, tinnitus, furuncle of the external ear, and restlessness.
     3. Otalgia and otorrhea.
     4. Perforation of the eardrum and bleeding.

2. **Nasopharyngeal myiasis:** the nose, sinuses, and pharynx could be infected with larvae of insects, e.g., *Wohlfahrtia, Sarcophaga, Chrysomya, and Oestrus.*
   - Nasal myiasis is mostly accidental.
   - It is more in rural areas and among low socioeconomic groups.
   - Flies are often attracted toward odoriferous suppurative lesions.
   - **Clinically, nasopharyngeal myiasis is manifested by:**
     1. Nasal discharges and nasal block.
     2. Erosion of the nasal bridge.
     3. Sloughing and spread to paranasal sinuses.
     4. Diffuse cellulitis of the face.

- **Treatment of aural and nasal myiasis:**
  1. An immediate hospitalization.
  2. Surgical removal of the maggots.
  3. Topical application of turpentine oil, ether, chloroform, mineral oil, or ethyl chloride.
  4. Systemic drugs, e.g., butazolidin, ivermectin, or thiabendazole.
  5. Broad-spectrum antibiotics, e.g., amoxicillin or ampicillin, when secondarily infected.

Review Questions

1. **One of the following parasites could infect the oral cavity**
   a. *Leishmania donovani*
   b. *Trichomonas tenax*
   c. *Giardia lamblia*
   d. *Onchocerca volvulus*

2. ***Entamoeba gingivalis* is characterized by having**
   a. cyst form
   b. leucocytes in its cytoplasm
   c. 2 nuclei
   d. cyst wall
3. **The following parasites could infect the eye except**
   a. *Loa loa*
   b. *Trichomonas*
   c. *Acanthamoeba*
   d. *Onchocerca volvulus*
4. **Amebic keratitis is commonly caused by**
   a. *Naegleria*
   b. *Entamoeba coli*
   c. *Entamoeba histolytica*
   d. *Acanthamoeba*
5. ***Leishmania braziliensis* could affect the following except**
   a. *skin*
   b. *mucous membrane*
   c. *cartilage*
   d. *bone*
6. ***Loa loa* is transmitted by**
   a. sand fly
   b. tsetse fly
   c. mango fly
   d. house fly
7. **Microfilaria of the following worms is unsheathed**
   a. *Loa loa*
   b. *Wuchereria bancrofti*
   c. *Brugia malayi*
   d. *Onchocerca volvulus*
8. **Congenital retinitis is caused by**
   a. *Loa loa*
   b. *Trypanosoma*
   c. *Toxoplasma*
   d. *Onchocerca volvulus*

17 Uncommon Parasitic Infections

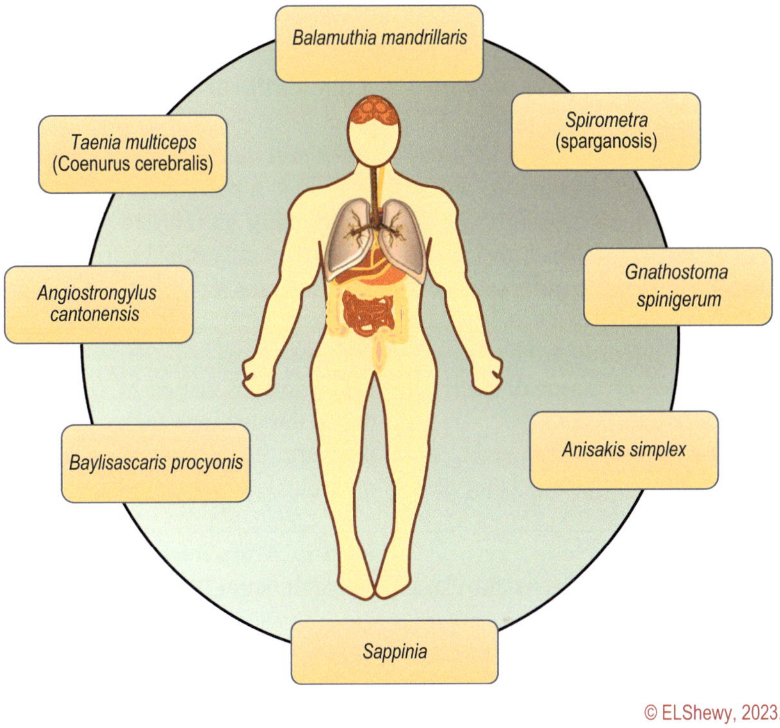

Uncommon parasitic infections

K. ElShewy, *Medical Parasitology*, https://doi.org/10.1007/978-3-031-64736-9_17

Test your learning and check your understanding of this book's contents: use the "Springer Nature Flashcards" app to access questions using ► https://sn.pub/oyFLIx. To use the app, please follow the instructions in Chap. 1.

Introductory
Occasionally, humans could be exposed to rare and uncommon parasitic infections. Although most of these infections are asymptomatic, however, in certain instances, these uncommon parasites result in the development of serious or even lethal diseases. Parasites causing uncommon human infections could be:

a. Nonhuman species of some parasites (Human acts as **paratenic or incidental host**): these parasites fail to complete their normal life cycle in humans. Instead, their early developmental stages migrate in host tissues, leading to serious diseases.
b. Parasites that could infect both humans and animals (humans act as **natural hosts** or **dead-end hosts**).
c. Free-living parasites in the environment are usually nonpathogenic except when they accidentally infect humans.

The above-mentioned parasites could cause the following clinical conditions:

1. **Cutaneous larva migrans (creeping eruption):** caused by human infection with animal hookworms, e.g., *Ancylostoma braziliense, Ancylostoma caninum,* and *Gnathostoma spinigerum.*
2. **Visceral larva migrans:** caused by *Toxocara cati, Toxocara canis* (Chap. 15), *Gnathostoma spinigerum, Baylisascaris procyonis,* and *Angiostrongylus cantonensis.*
3. **Ocular larva migrans**: caused by *Toxocara canis* and *Spirometra spp.*
4. **Sparganosis:** caused by nonhuman tapeworm of the genus *Spirometra* resulting in cutaneous and/or visceral lesions.
5. **Brain cyst and mass***:* caused by *Taenia multiceps* (Chap. 10) and *Sappinia spp***.**
6. **Granulomatous and eosinophilic meningitis:** mainly caused by *Gnathostoma spinigerum, Baylisascaris procyonis, Angiostrongylus cantonensis, and Balamuthia mandrillaris* (Chap. 10).

Some of these parasites are going to be discussed in this chapter in detail, and some other parasites have been mentioned in other chapters.

Gnathostoma spinigerum (*G. spinigerum*)

(Larva migrans profundus)

G. spinigerum is a nonhuman *Phasmid* nematode. It is a parasite of dogs and cats. Humans could rarely be infected with its developmental larval stages, causing both cutaneous and visceral larva migrans (larva migrans profundus).

Transmission
Larva migrans caused by *G. spinigerum* is transmitted through ingestion of the third stage larva of the worm in undercooked fish, frogs, and snakes or through using their tissues in medication.

Parasite Morphology
- Adult worm (♀):
  - Stout, reddish-colored worm.
  - It has a bulbous head surrounded by rings of hooks and a pair of fleshy lips.
  - The body is surrounded by leaf-like spines around its anterior half.
  - Its length is around 30–50 mm (Fig. 17.1).
- Adult worm (♂):
  - It is similar to the female but with a smaller body size, 10–20 mm.
- Egg:
  - It is oval with a plug at one end.
  - It has a superficial outer pitted coat.
  - 70–40 μm in size.

Habitat
G. spinigerum lives in a tumor-like mass in the stomach wall of dogs and cats.

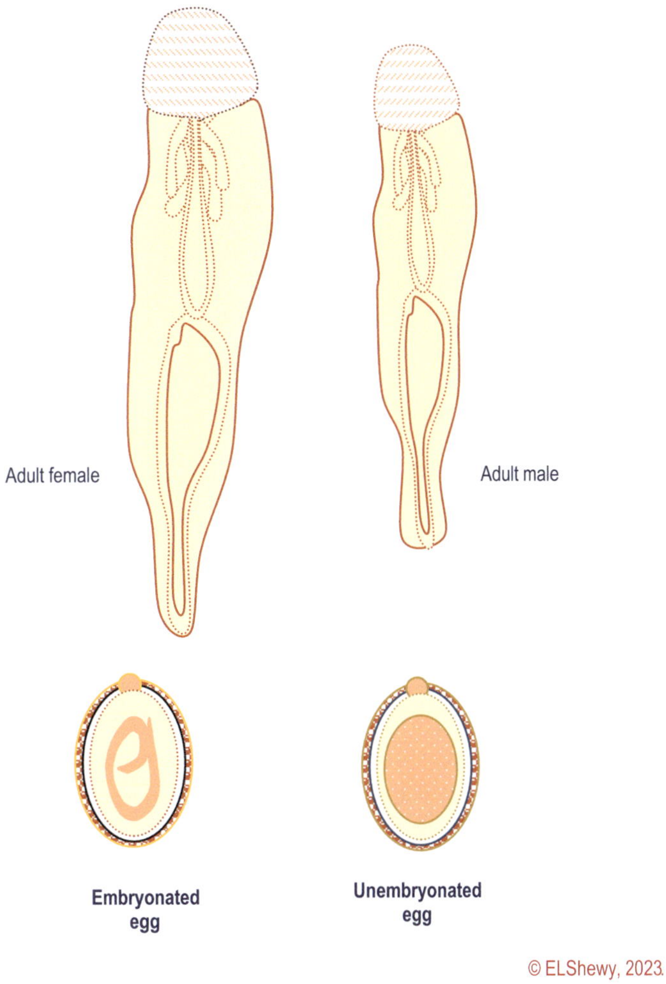

Fig. 17.1 *Gnathostoma spinigerum*: adult and egg morphology

Life Cycle

Adult *G. spinigerum* inhabits the stomachs of cats and dogs, with unembryonated eggs deposited in their stool. Egg hatches and releases 1st larval stage, to be ingested by *Cyclops* (in which 2nd larval stage is formed), which in turn ingested by snakes, fish, or frogs to develop into the 3rd larval stage. Occasionally, man can get an infection with 3rd larval stage by eating undercooked fish or frogs. The 3rd larval stage could not reach maturity, and they migrate, causing cutaneous larva migrans in the superficial tissues (skin and subcutaneous tissue). Sometimes, larvae could reach deeper tissues, causing visceral larva migrans, particularly in the brain (Fig. 17.2).

Pathogenesis

The presence of the 3rd stage larva of *G. spinigerum* in the superficial or deep human tissues stimulates an inflammatory reaction with granulomatous tissues and abscess formation. If it reaches the brain, it causes severe eosinophilic meningitis.

Diagnostic Procedures

Diagnosis of *G. spinigerum* infection depends on:

1. Clinical picture.
2. History of eating raw or undercooked fresh or brackish water fish.

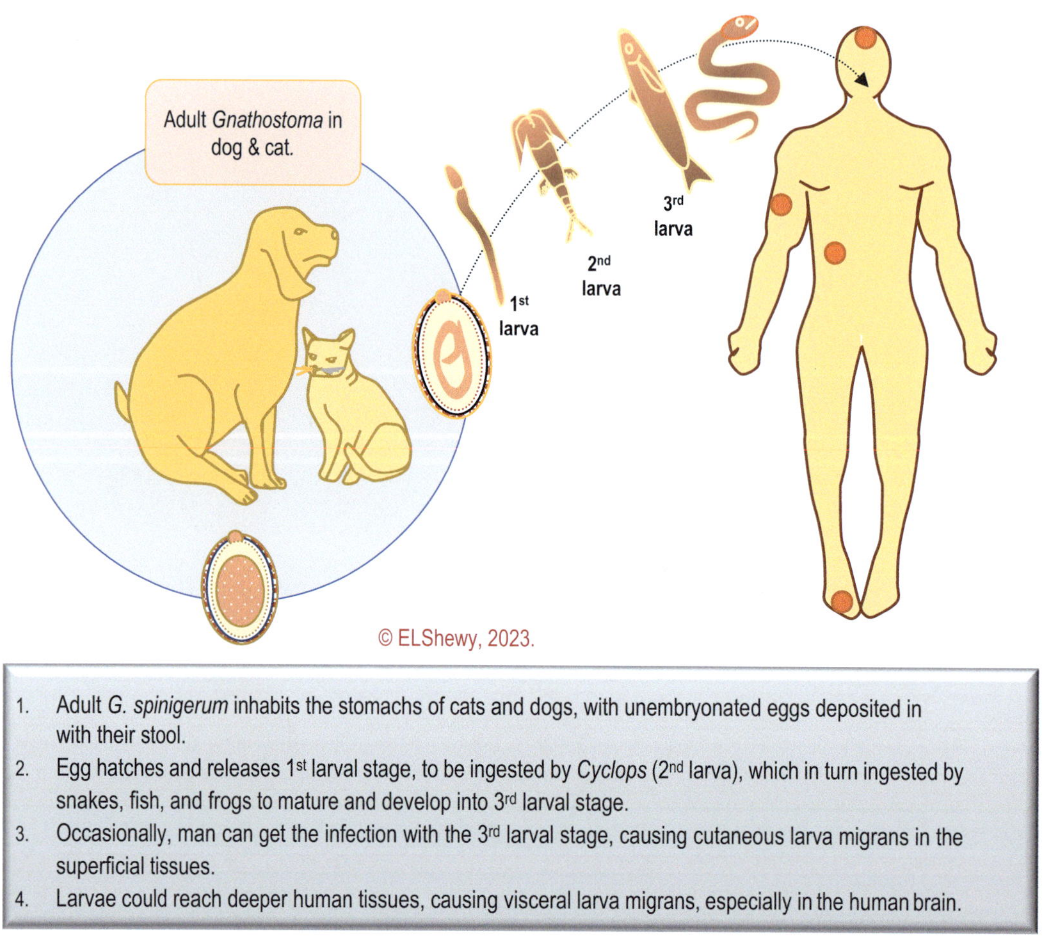

Fig. 17.2 *Gnathostoma spinigerum*: life cycle

3. Serological tests for detection of the specific antibodies.
4. Histologic examination of excised lesions.

Spirometra spp.

Spirometra spp. are nonhuman tapeworms that occasionally infect humans. Inside the human body, larvae of these nematodes (Sparganum) fail to complete their normal life cycle and implant in the superficial and sometimes in deeper tissues, forming sparganosis.

Geographical Distribution

Sparganosis caused by *Spirometra* is a rare and uncommon parasitic infection. It could be detected in the Far East and sporadically elsewhere. *Spirometra mansoni* is the most common species causing sparganosis.

Transmission

Sparganosis is transmitted through:

1. Drinking infected *Cyclops* containing procercoid larvae.
2. Using frog or snake tissues as foments on wounds.

Parasite Morphology

Sparganum may be procercoid (1st larval stage) or plerocercoid of *Spirometra*. Both are similar to *Diphyllobothrium latum* larval stages.

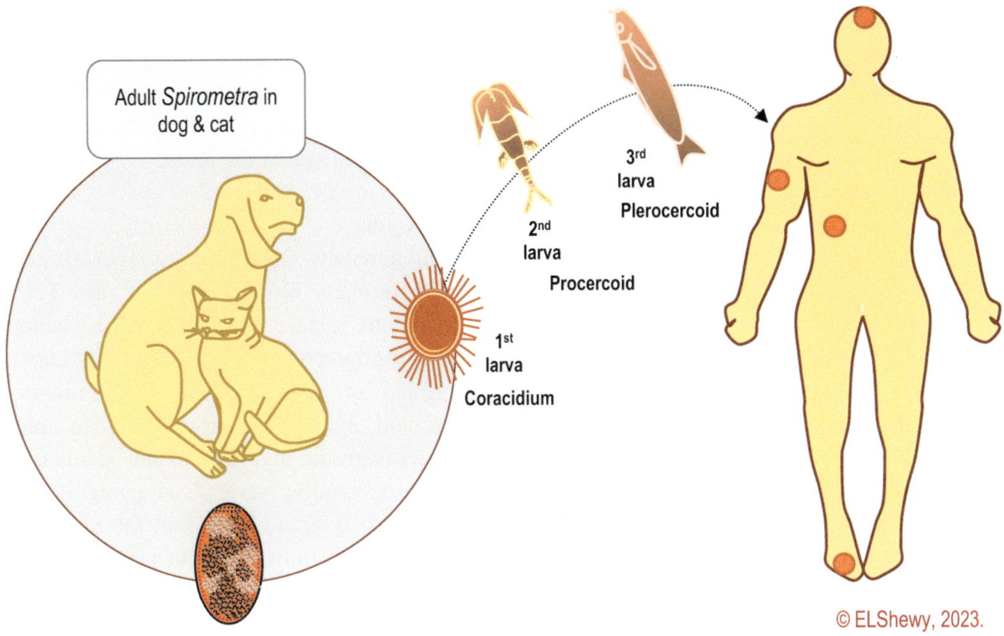

1. Adult *Spirometra* inhabits the small intestine of cats and dogs. Unembryonated eggs are deposited with their stool.
2. Egg hatches and releases coracidium, to be ingested by *Cyclops* (1st intermediate host), to be developed into procercoid, which in turn ingested by fish or frogs (2nd intermediate host) to develop into plerocercoid.
3. Occasionally, man can ingest raw or undercooked intermediate hosts and get infection with procercoid or plerocercoid larvae,
4. Larvae migrate in human tissues, forming larva migrans in the superficial and deeper tissues.

Fig. 17.3 *Spirometra:* life cycle

Habitat

Adult worms of *Spirometra* inhabit the small intestine of dogs and cats.

Life Cycle

Adult *Spirometra* worms inhabit the small intestine of cats and dogs. They deposit unembryonated eggs with their stool. Eggs hatch to release coracidium, which is ingested by *Cyclops* (1st intermediate host). In *Cyclops*, coracidium developed into procercoid, which in turn was ingested by fish or frogs (2nd intermediate host) to be developed into plerocercoid larvae. Occasionally, man can get an infection with procercoid or plerocercoid larvae, triggering larva migrans in his superficial and/or deep tissues (Fig. 17.3).

Pathogenesis

1. Living and dead larvae cause marked inflammatory reactions at the tissues they infect, with marked monocellular infiltration and many eosinophils.
2. Sparganosis could affect deeper tissues, e.g., the eye or even the brain, with granulomatous reactions in these tissues.
3. In rare conditions, *Sparganum proliferum*, which is a rare parasitic infection, could proliferate by lateral branches.

Clinical Picture
Painful edematous swelling, abscess formation, eye and visual problems, or sometimes nervous manifestations.

Diagnostic Procedures
Diagnosis of sparganosis infection depends on:

1. Clinical picture.
2. History of drinking water contaminated with *Cyclops* or eating undercooked fish.
3. Histologic examination of the biopsy taken.
4. Serological test for detection of the specific antibodies.
5. Examination of the soil is done to detect the characteristics of the eggs.

Angiostrongylus cantonensis (*A. cantonensis*)

(Rat Lungworm)

A. cantonensis is a roundworm of rats. Humans are incidental hosts for this worm. Infection with the worm is called angiostrongyliasis; it is characterized by eosinophilic meningitis (neural angiostrongyliasis), which may lead to death or serious brain and nerve damage.

Geographical Distribution
Angiostrongyliasis is a rare disease confined to the southeast of Asia and the Western Pacific Islands, with sporadic cases reported all over the world.

Parasite Morphology
A. cantonensis is a roundworm with a tough outer cuticle and unsegmented body. It is 2 cm long, with a slender thread-like body.

Transmission
A. cantonensis is transmitted through ingestion of its developing 3rd stage larvae with raw snails, crabs, frogs, shrimps, or any foods and drinks that contain the infective larvae.

Habitat
Adult worm of *A. cantonensis* resides in the pulmonary blood vessels of rats. The developing larvae of the worm could be detected in human tissues, particularly, the brain.

Life Cycle
Adult female worms of *A. cantonensis* lay eggs in the pulmonary blood vessels of rats. The eggs hatch and release 1st stage larvae, which are swallowed to pass with the rat's stool. Larvae are ingested by intermediate snails and mature into 2nd and 3rd stages larvae. Human infection results from the ingestion of raw snails containing the developing larvae. Larvae migrate through the blood to reside mainly in the human brain, where they mature more but never develop into adult worms (Fig. 17.4).

Pathogenesis and Clinical Picture
The presence of the worm larvae in the brain causes an acute inflammatory reaction with a high level of eosinophils (eosinophilic meningitis).

1. The incubation period lasts from 1 to 3 weeks.
2. Nausea, vomiting, severe headache, neck rigidity, and sensory changes.
3. High level of eosinophils in the CSF.
4. Sometimes, infection affects the eyes with inflammation, blurring, and/or loss of vision.
5. Complications can rarely occur, leading to neurologic dysfunction or death.

Diagnostic Procedures
Diagnosis of eosinophilic meningitis infection depends on:

1. Clinical picture.
2. History of eating raw or undercooked snails, frogs, shrimps, or crabs.
3. High eosinophils in the CSF and peripheral blood.
4. Serological tests for the detection of specific antibodies.

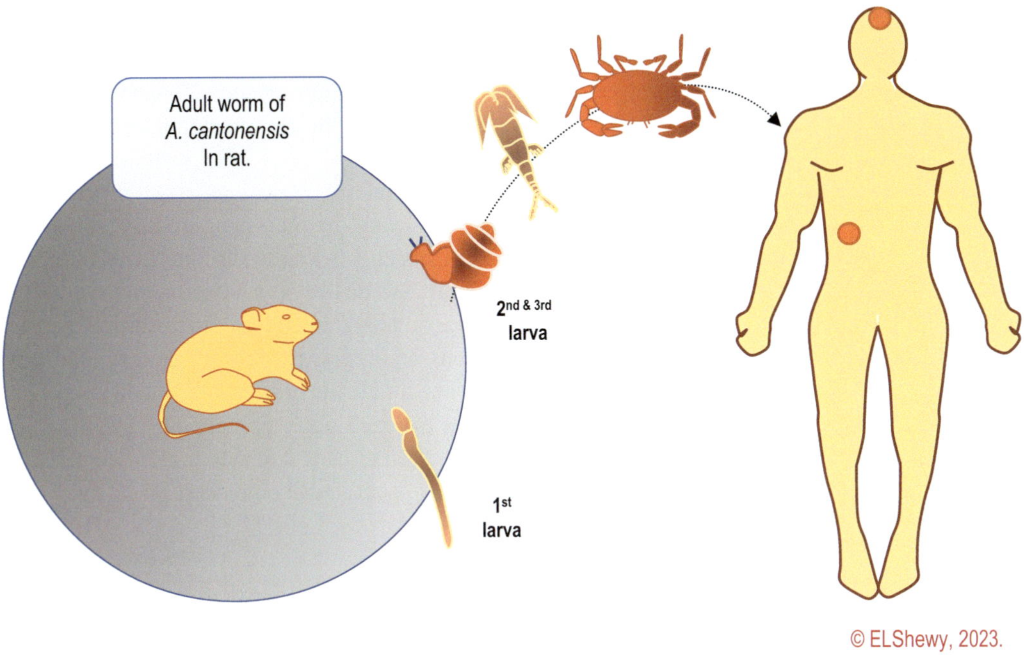

1. Adult female *A. cantonensis* inhabits the pulmonary blood vessels of rats. It deposits eggs in pulmonary tissues.
2. Eggs swallowed to reach the rat's intestine, hatch and release the 1st stage larva in the rat's stool.
3. Snails ingest the 1st larval stage, which develops into the 2nd and 3rd larval stages. Larvae are released.
4. Released larvae are taken by shrimps, crabs, or contaminated foods and drinks of humans who can get an infection with the 3rd larva.
5. Larvae reach the human circulation to reach different tissues, particularly the brain.

Fig. 17.4 *Angiostrongylus cantonensis:* life cycle

Treatments

Most infections of *A. cantonensis* resolve over time without specific treatment because the parasite cannot survive in the human body.

Prevention and Control

1. Avoid eating raw or undercooked snails, frogs, shrimps, or crabs.
2. Sanitary and healthy foods and drinks and combat rats.

Baylisascaris procyonis (*B. procyonis*)

(Raccoon Roundworm)

B. procyonis is a roundworm of raccoon. It could infect other animals and on very rare occasions, it can infect humans. Human bayliscariasis is severe and life-threatening; it could cause severe eosinophilic meningitis and ocular and visceral diseases.

Geographical Distribution

B. procyonis is a parasite of raccoon. It could infect humans in areas close to raccoons. e.g., North America and Canada.

Parasite Morphology

B. procyonis is a roundworm, almost similar to the *Ascaris* worm; however, its larvae are bigger, reaching 1.8 mm.

Transmission

B. procyonis is transmitted through accidental ingestion of infective eggs of the worm in foods and drinks contaminated with raccoon's stool.

Habitat

B. procyonis inhabits the small intestine of the final hosts mainly raccoons, sometimes it could infect dogs as the second definitive host.

Life Cycle

B. procyonis in the small intestine of raccoon deposits eggs with the raccoon's stool. Eggs last about 2 weeks to be mature (containing larvae). Occasionally, humans can ingest these eggs with foods or drinks. Eggs hatch and release larvae that penetrate the human intestine to stomach tissues to reach other tissues, e.g., the brain, eye, and other viscera. Larvae cannot complete their development inside humans (Fig. 17.5).

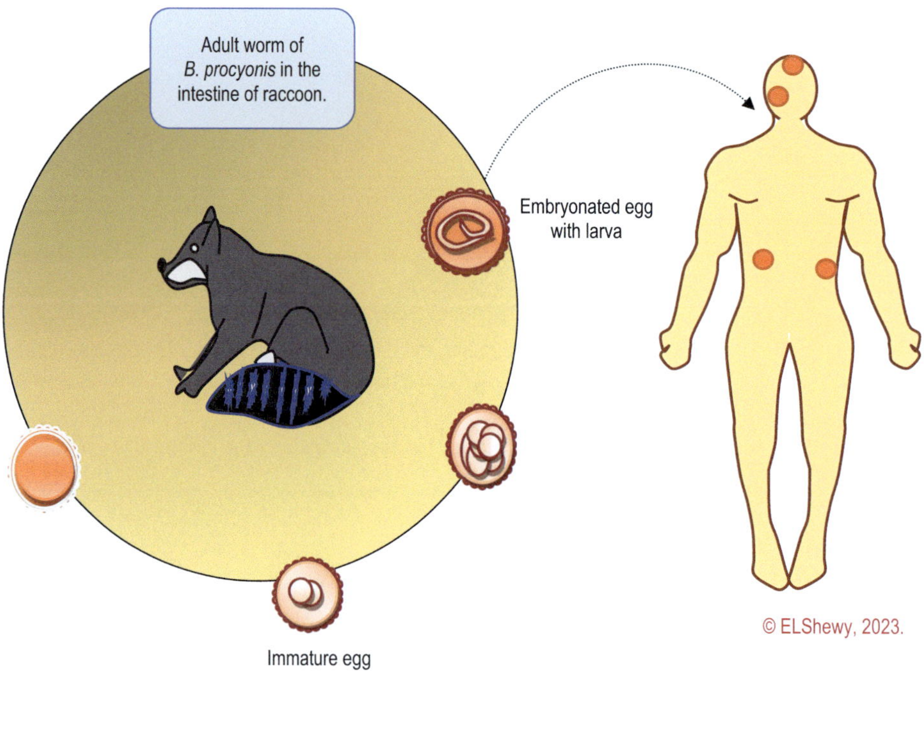

1. Adult females *B. procyonis* inhabit the small intestine of raccoons. They deposit eggs in the animal's stool.
2. Eggs embryonated in the external environment in 2–3 weeks to be infective (containing larva) .
3. Accidentally, humans and other animals ingest the infective eggs with food and drinks.
4. Larvae are released in the intestine and pass to reach the brain, eye, and other viscera.

Fig. 17.5 *Baylisascaris procyonis:* life cycle

Pathogenesis and Clinical Picture

The presence of the worm larvae in the brain causes an acute inflammatory reaction with a high level of eosinophils (eosinophilic meningitis). Infected tissues could be mechanically destroyed by the large-sized larvae. The infection could also affect the eye and other viscera, forming ocular larva migrans (OLM) and visceral larva migrans (VLM). Bayliscariasis is associated with the following clinical manifestations:

1. Nausea, vomiting, severe headache, neck rigidity, and sensory changes.
2. High level of eosinophils in the CSF and peripheral blood.
3. Sometimes, the eyes can be affected by diffuse unilateral subacute neuroretinitis (DSNR), with blurring of vision and even loss of vision.
4. Most infections of *B. procyonis* resolve over time without specific treatment because the parasite cannot survive in the human body.
5. Complications can rarely occur, leading to neurologic dysfunctions or blindness.

Diagnosis

Diagnosis of bayliscariasis is difficult; it depends on:

- History of contact with raccoon.
- Clinical picture.
- Imaging.
- Serology for detection of specific antibodies.

Prevention and Control

1. Avoid coming in close contact with raccoons.
2. Sanitary and healthy foods and drinks.

Anisakis simplex (A. simplex)

(Herring Worm)

A. simplex is a zoonotic nematode worm of marine animals. Humans could be infected with its larvae inside raw or undercooked fish. Anisakiasis, in humans, is mainly a gastrointestinal or ectopic disease, which may lead to severe and life-threatening complications.

Geographical Distribution

A. simplex is a parasite of marine mammals. Human infection is detected in regions where raw or undercooked fish is commonly consumed, e.g., Japan, the North Sea, and Spain. It is rarely infecting humans in North America and Canada.

Parasite Morphology

A. simplex is a roundworm with three to four stages of large-sized developing larvae:

- Early larva detected in the stool of the final hosts is around 5 mm in length, with a tedious tooth.
- 3rd stage larvae reach up to 20–30 mm in the body cavity of the host, cylindrical pink to white in color.
- 4th stage female larvae begin to develop a reproductive system in the final hosts.

Transmission

A. simplex is transmitted by eating raw or incompletely salted fish containing the larval stages of the worm, e.g., smoked herring, sushi, sashimi, and pickled anchovies.

Life Cycle

The life cycle of *A. simplex* can be postulated as follows:

- *A. simplex* worm deposits immature eggs in the stool of the final hosts.
- Eggs are released in the seawater. They start to mature, forming 1st and 2nd stages of larvae. Eggs hatch and release larvae.
- Released larvae stay alive for up to 1 month in the seawater to be taken by the 1st intermediate host (crustacean), where they mature to 3rd stage larvae.

- Crustaceans with 3rd stage larvae are ingested by paratenic hosts (salmon, mackerel, and herring fish).
- Final hosts (whales, squid, and dolphins) eat the paratenic hosts, and larvae are released, penetrate the stomach wall, and mature more into adult worms.
- Humans can be infected by 3rd stage larvae when eating the paratenic hosts raw, undercooked, or incompletely salted.
- Larvae are released in the human stomach, intestine, or other viscera, penetrating and localized deeper into these organs.

Pathogenesis and Clinical Picture

Once the worm reaches the stomach, it uses its protruding lips to penetrate the gastric mucosa and releases tissue-dissolving enzymes to pass into the submucosa. It starts an acute allergic inflammatory and allergic reaction, and if it reaches the submucosa, it stays there with the initiation of a chronic inflammatory response.

- Anisakiasis is characterized by acute abdominal pain shortly after eating contaminated seafood.
- Allergic manifestations in the form of urticaria to anaphylaxis within a few hours from ingestion.
- Nausea, fever, diarrhea, and bloating within a few hours.
- The clinical picture is confused with food poisoning or peptic ulceration.
- Later, there is dull aching gastric pain with chronic indigestion.
- In intestinal anisakiasis, the symptoms appear after some days with abdominal pain, diarrhea, and indigestion lasting for months.

Diagnosis

Diagnosis of anisakiasis depends on:

1. History of eating seafood.
2. Clinical picture.
3. Gastroscopy with histologic examination of tissue biopsy.

Treatment

1. Surgical removal of the worm out of the tissues using endoscopy.
2. Anti-histaminic drugs.
3. Anti-inflammatory drugs.

Prevention and Control

1. Avoid eating raw or undercooked seafood.
2. Seafoods should be sufficiently cooked or frozen before consumption.

Sappinia diploidea and pedata (*S. diploidea and S. pedata*)

Sappinia spp. are free-living amebae. They could be found in soil, freshwater, and animal manure. They could cause disease in both humans and animals. They are newly discovered human pathogens of the central nervous system (CNS) causing *Sappinia* Amoebic Encephalitis (SAE).

Geographical Distribution

They have been described from Europe, North America, Egypt, the Middle East, the West Indies, and Japan. They are found in soil, freshwater, forest litter, and in places where farm animals are known to eat soil containing rotting plants.

There are two species of Sappinia: *S. diploidea* and *S. pedata*.

Parasite Morphology

Sappinia diploidea has two forms: a trophozoite and a cyst form.

- **Trophozoite**:
  - Ameboid in shape, 40–70 μm
  - Two adjacent nuclei with two nucleoli.
  - Granular cytoplasm.
- **Cyst:**
  - Rounded with double cyst wall, 20–35 μm
  - Two adjacent nuclei with filaments.

Clinical Picture

The incubation period and the route of infection are unknown but thought to be by inhalation through the nasopharynx or by hematogenous spread to the brain. Symptoms of *Sappinia* Amebic encephalitis (SAE) include:

1. Headache and photophobia.
2. Nausea or upset stomach and vomiting.
3. Blurred vision.
4. Loss of consciousness.
5. A tumor-like mass on the back left section of the brain was reported in one case.

Introductory to Medical Entomology

18

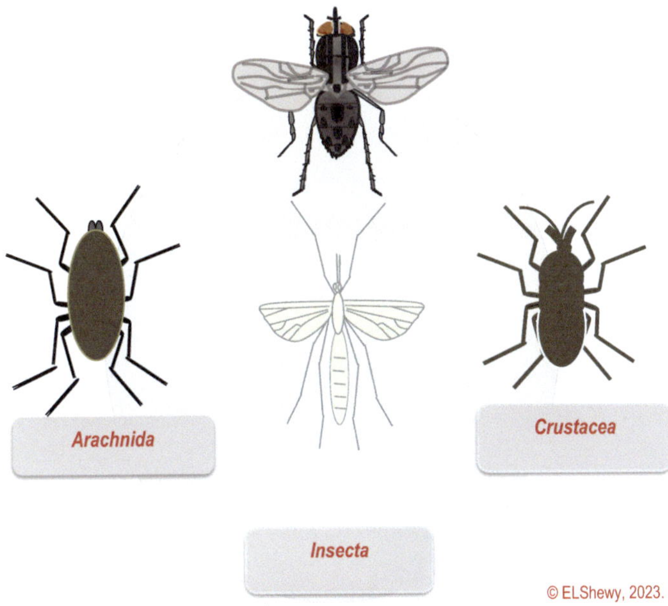

Medical entomology

Test your learning and check your understanding of this book's contents: use the "Springer Nature Flashcards" app to access questions using ► https://sn.pub/oyFLIx. To use the app, please follow the instructions in Chap. 1.

In this chapter, we discuss medical entomology in a very simplified way that suits medical students without unnecessary details, which is not crucial for the practice of medicine.

K. ElShewy, *Medical Parasitology*, https://doi.org/10.1007/978-3-031-64736-9_18

What Is an Arthropod?

An arthropod is an invertebrate animal with an exoskeleton, a segmented body, and rigid cuticle, and it is characterized by its jointed limbs. The rigid cuticle inhibits growth, so the arthropod replaces it periodically by molting. Arthropods are bilaterally symmetrical. Some species are known to spread severe disease to humans, livestock, and crops.

Classification of Arthropods

There are three classes of arthropods with medical significance to humans:

1. **Insecta**: includes arthropods with a head, thorax, abdomen, three pairs of legs, wings, and one pair of antennae, e.g., mosquitoes, flies, lice, bugs, and fleas.
2. **Arachnida:** arthropods with a head and thorax attached (cephalothorax), four pairs of legs, with no wings, and no antennae, e.g., ticks, mites, scorpions, and spiders.
3. **Crustacea:** arthropods with a head and thorax attached (cephalothorax), with an abdomen and four pairs of legs with no wings, and two pairs of antennae, e.g., crabs and *cyclops*.

Each class of the above-mentioned arthropods has its morphological criteria that could be used for the identification of their members (Fig. 18.1).

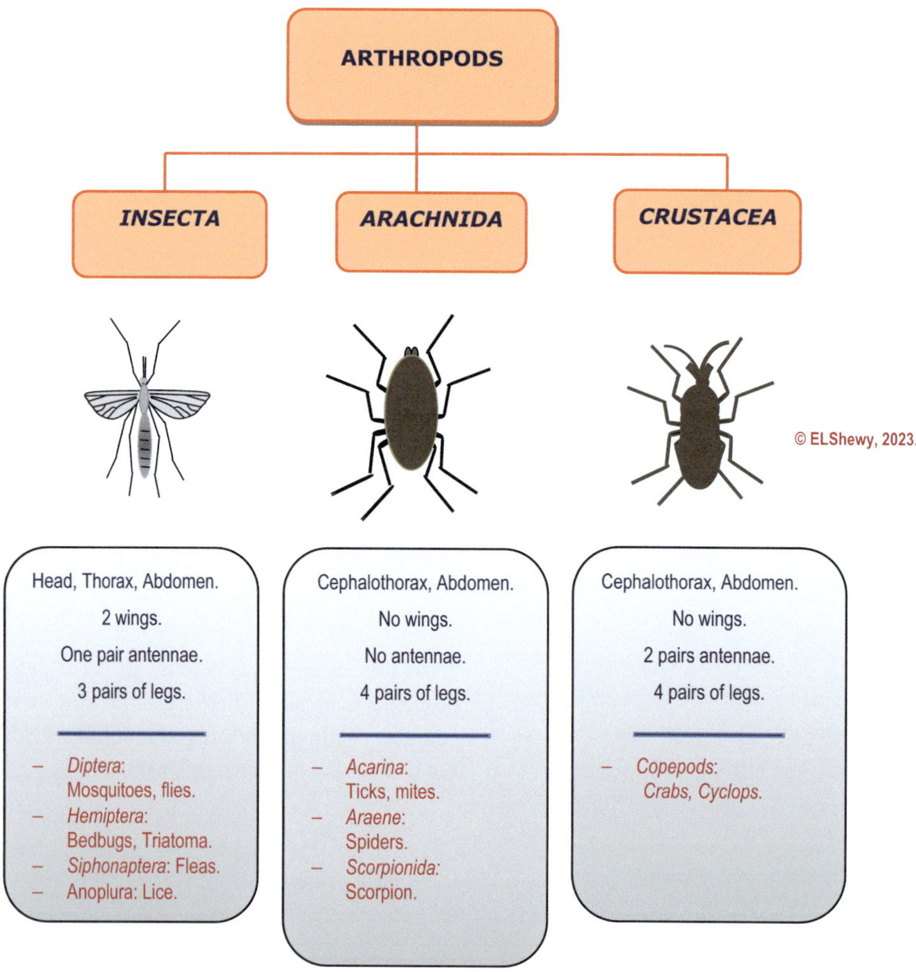

Fig. 18.1 Classification and general characteristics of medically important arthropods

Arthropod Metamorphosis (Development)

Arthropods develop in two ways:

1. **Complete metamorphosis (egg–larva–pupa–adult):** young stages (larva, pupa) differ morphologically from the adult stages, e.g., mosquitoes, flies and fleas.
2. **Incomplete metamorphosis (egg–larva–nymph–adult):** young stages (nymph) morphologically resemble the adult stages, e.g., ticks and mites.

Effects of Arthropods on Human Health

Arthropods could affect human health through:

A. **Direct effects:**
   1. **Injury:** invasion of human and animal living bodies, e.g., myiasis and scabies.
   2. **Pest of man:** ectoparasites, e.g., lice, ticks, and bugs.
   3. **Envenomation**: scorpion sting and spider.
   4. **Psychological disturbance:** fear of insects (entomophobia).

B. **Indirect effects:**
   Arthropods could act as vectors transmitting pathogens (parasites, bacteria, and viruses) from one host to another or from animal to man. Pathogens are transmitted through the following methods:
   1. **Mechanical Transmission:**
      Arthropods act as mechanical vehicles transporting the pathogen or its infective stage to many other hosts through contamination of their foods or drinks, e.g., house flies (*Musca domestica*).
   2. **Biological transmission (Appendix D):**
      The arthropod acts as an obligatory vector or a natural host for the pathogen, and it is considered an essential part of the pathogen's life cycle. Biological transmission of the pathogens by an arthropod could be:
      a. **Propagative:** multiplication of the pathogen in the arthropod without developmental changes, e.g., plague in fleas.
      b. **Developmental***:* developmental changes without multiplication of the pathogens, e.g., filariasis in mosquito and *Diphyllobothrium latum* in *Cyclops*.
      c. **Cyclopropagative***:* multiplication and developmental changes of the pathogens in the arthropod, e.g., *Plasmodium* in *Anopheles* mosquito, *Trypanosoma* in *Tsetse* fly, and *Leishmania* in Sandfly.
      d. **Transovarian:** the pathogen is transmitted through the ovaries of the adult arthropod to the nymph, e.g., scrub typhus and *Babesia* in the tick's ovaries.
      e. **Transstadial:** pathogen is transmitted from the larva to nymph and then to adult stages of the arthropod, e.g., *Borrelia* in ticks.

Control of Arthropods

The following are general outlines for control of arthropods:

1. **Ecological and Mechanical Control:**
   - Screening windows, openings, and air vents, e.g., mosquito nets and fly nets with suitable mesh size.
   - Air current barriers at the entrance of public buildings.
   - Ultraviolet light traps with an electric grid.
   - Environmental sanitation, proper sewage, and garbage disposal.
   - Filling or drainage of the breeding places.
   - Alteration of the habitat of arthropods, e.g., changes in pH, water current, and depth.
   - Controlling the light in the location.
2. **Biological Control:**
   a. **Natural predators:** the use of natural enemies to insects, e.g., fish, frogs, and birds.

b. **Transgenic insects:** release of genetically modified insects proved promising although costly and still under development:
   - Sterile Insect Technique **(SIT):** release of irradiated sterile male insects to mate with females, aiming at decreasing the insect population over time.
   - Release of Insects carrying Dominant Lethal **(RIDL):** releasing insects in the environment with lethal genes.
   - Incompatible Insect Technique **(IIT):** infection of insects with *Wolbachia* bacteria renders them sterile, decreasing their ability to transmit pathogens and/or destroy the insect.

3. **Chemical Control:**
   Unfortunately, this is the commonest control method; it includes:
   a. **Repellents:** substances that keep arthropods away from contact with the human body; repellents could be:
      - **Synthetic:** e.g., DEET (N, N-diethyl-m-toluamide), ethyl hexanediol, indalone, permethrin, and SS220; a more recent repellent.
      - **Natural**: e.g., citronella oil, neem oil, lemon grass, tea tree oil, and tobacco.

   b. **Insect attractant bait and traps:** Pheromones and volatile compounds similar to food sources could be used as attractants to trap and destroy insects.

   c. **Insecticides:**
      - **Synthetic insecticides:**
        Organic and inorganic metal insecticides, which are working by:
        - Direct contact with the insect, e.g., Pyrethrum.
        - Residual effect, e.g., Hydrocarbons (DDT, Gammaxene, Chlordane, and Dieldrin); Organochlorides (Malathion, dipterx); Neonicotinoids (Acetamiprid, Clothianidin).
      - **Bioinsecticides** (Biopesticides):
        Biopesticides are biological agents that achieve pest management in an environmentally friendly way:
        - Microbial pesticides consist of bacteria, entomopathogenic fungi, or viruses.
        - Bioderived chemicals: pyrethrum, rotenone, neem oil, and essential oils.
        - Plant-incorporated protectants such as nicotine, pyrethrum, and neem extracts.

4. **Transmission and Infection Blocking Vaccines:**
   - Vaccines against infective stages transmitted by arthropods, e.g., anti-gametocyte vaccines given to the human host.
   - Vaccine against insect salivary proteins to block infections during insect biting.

19 Medically Important Insects

> Test your learning and check your understanding of this book's contents: use the "Springer Nature Flashcards" app to access questions using ▶ https://sn.pub/oyFLIx. To use the app, please follow the instructions in Chap. 1.

Mosquitoes (*Anopheles, Culex, Aedes*)

General Characteristics of Mosquitoes

- Mosquitoes are dipterous insects widely distributed in the continent according to climate and ecological variation.
- There are approximately 4000 species of mosquitoes in the world.
- Three most significant genera, namely: *Anopheles*, *Aedes,* and *Culex*, are responsible for the transmission of various fatal diseases to humankind.
- These three genera could transmit malaria, various forms of filariasis, and numerous arboviruses.
- The life span of mosquitoes ranges from 1 to 2 weeks and rarely reaches up to 3 weeks.
- Only the female mosquito is blood-sucking (female needs blood to lay eggs).
- The male mosquito feeds on the nectar or honeydew of plants to get sugar. It does not transmit diseases.
- Mosquito has a slender segmented body, a pair of wings, three pairs of long hair-like legs, feathery antennae, and elongated mouthparts (proboscis).
- The head and its mouthparts are important for differentiation between female and male mosquitoes (Fig. 19.1).
- Mosquitoes undergo complete metamorphosis. Typically, mosquitoes lay their eggs in standing water. Larvae hatch from the eggs and remain in the water to feed on tiny organisms. They change to pupa and then to adult mosquitoes (Fig. 19.2).

The following are general characteristics, bionomics, common species, and the three medically important genera of mosquitoes.

A. *Anopheles*

Characteristics

- There are more than 400 species of *Anopheles* in the world.
- It is 5–10 mm in length, yellowish, with spotted wings.
- *Anopheles* rests at an angle on surfaces.
- Palps of the mouth are as long as proboscis in females and males but with swelling at their ends.
- Only females are bloodsuckers.
- The female lays eggs on the surface of standing water.
- *Anopheles* is an indoor nocturnal mosquito (blood feeding and oviposition at night).

K. ElShewy, *Medical Parasitology*, https://doi.org/10.1007/978-3-031-64736-9_19

Fig. 19.1 Head of mosquito

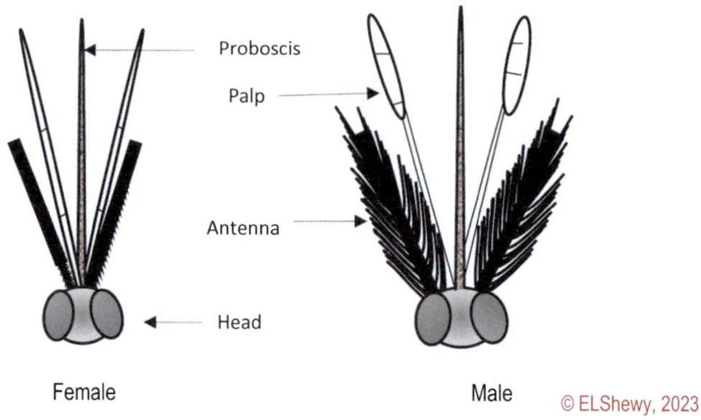

Fig. 19.2 General life cycle of mosquitoes

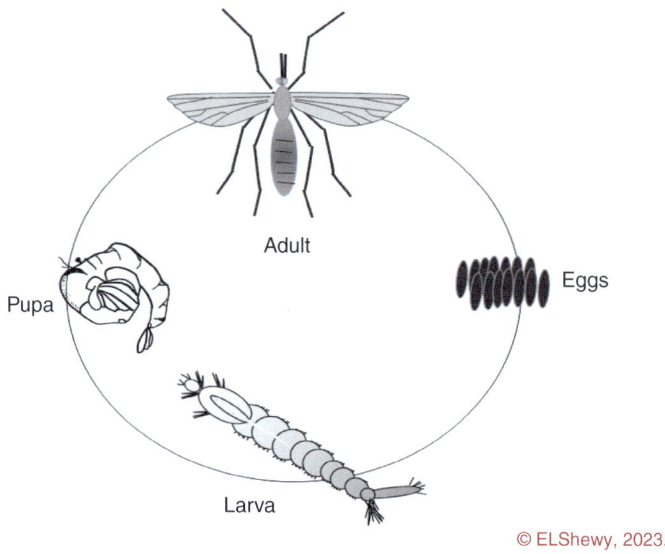

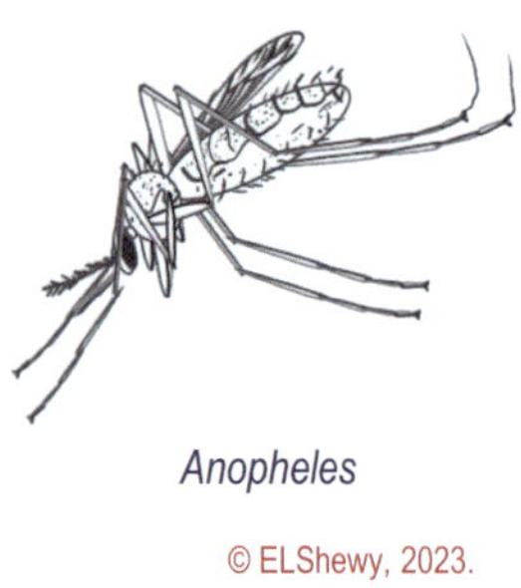

Common Species

- *A. pharoensis* and *A. sergentii* are common in the Middle East.
- *A. funestus* and A. *gambiae* are the most efficient vectors of malaria in Africa.

Medical Importance

1. Vectors of malaria *(Plasmodia)*.
2. Vectors of lymphatic filariasis (*W. bancrofti, B. malayi, and B. timori*).
3. Allergic skin lesions.

B. *Culex*

Culex pipiens (House Mosquito)

Characteristics

- It is called the house mosquito.
- Yellowish brown with no spots on the wings.
- Most active at dawn and dusk, night biter.
- Resting position parallel to the surface.
- Palps are much shorter than proboscis in females.

- Palps are longer than proboscis in males but with no swelling at the ends.
- They prefer nonhuman blood for feeding.

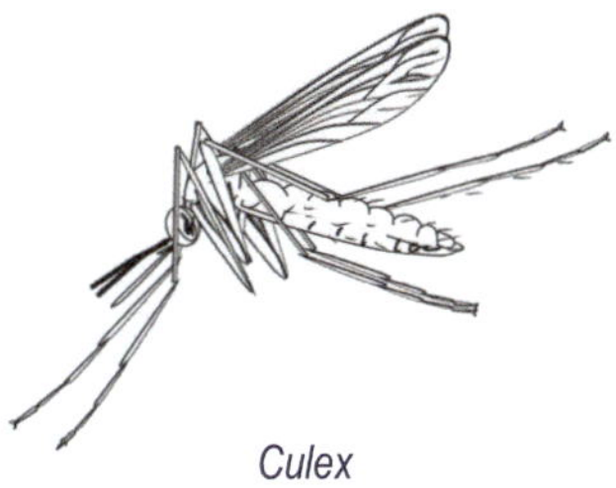

Culex

Common Species
- *Culex pipiens* prevails all over the world.
- Least dangerous species of mosquitoes.

Medical Importance
1. Vectors for viral infections: e.g., West Nile fever, Japanese encephalitis, and St. Louis encephalitis.
2. Vector of lymphatic filariasis (*B. malayi* and *B. timori*).
3. Allergic skin lesions.

C. *Aedes*

Aedes aegypti and *Aedes albopictus* (Asian Tiger Mosquito)

Characteristics
- It is the most common and dangerous species of mosquitoes.
- It has a black background with silver spots on the thorax, abdomen, and legs.
- No spots on the wings.
- Resting position parallel to the surface.
- Palps are much shorter than proboscis in females.
- Palps are longer than proboscis in males but with no swelling at the ends.
- *Aedes albopictus* adult females bite aggressively, usually during the day and preferably outdoors. It is called the "Asian tiger mosquito."
- It feeds on human and animal blood.

Aedes albopictus.

Common Species
- *Aedes aegypti prevails* in many regions of the world.
- *Aedes albopictus* originated in Southeast Asia; recently, it has spread to the Middle East, Europe, South America, and many other areas.

Medical Importance
1. Vectors for viral infections, e.g., Dengue fever, Yellow fever, West Nile fever, Zika virus, and Chikungunya fever.
2. Vector of lymphatic filariasis (*B. malayi* and *B. timori*).
3. Allergic skin lesions.

Flies

General Characteristics of Flies
- There are over 120,000 species of flies.
- Fully-grown flies are usually about 0.5–2.5 cm in length.
- In addition to the general characteristics of dipterous arthropods, flies have two pairs of wings. One pair functioning, and the other is rudimentary (halters) used for balance.
- Flies lay up to about 3,000 eggs in their lifetime (around one month)
- Flies undergo complete metamorphosis. Their life cycle consists of eggs, larvae (maggots), pupae, and adults. Flies lay eggs mainly on larval food sources (Fig. 19.3).
- Flies with medical importance to humans are found in some families in the class *Insecta*:
  1. ***Psychodidae*:** *Phlebotomus* (sandfly).

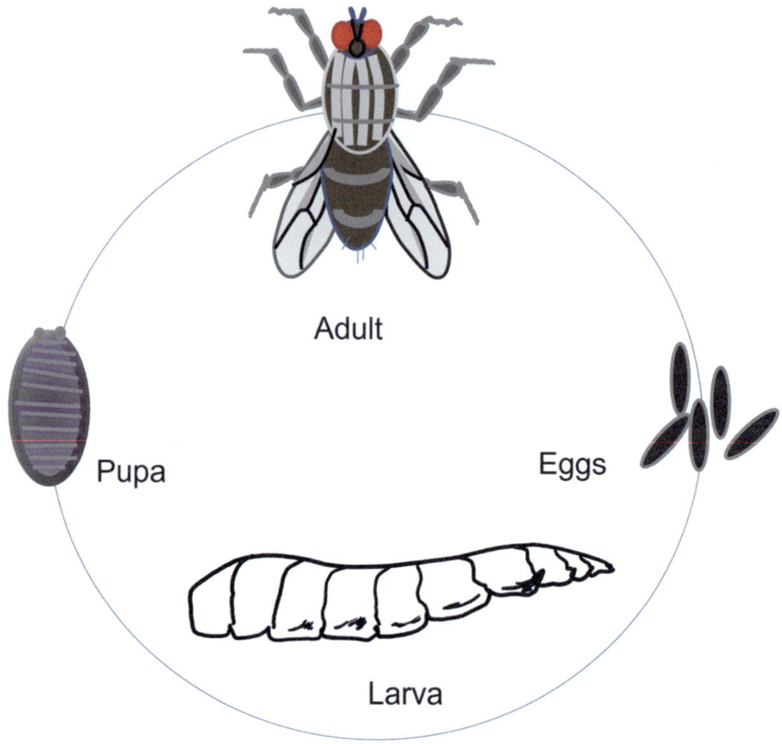

Fig. 19.3 General life cycle of flies

2. ***Muscidae:*** *Musca domestica, Stomoxys, and Glossina* (tsetse fly).
3. ***Calliphoridae***: *Calliphora and Lucilia* (metallic flies/blow flies).
4. ***Sarcophagidae:*** *Sarcophaga and Wohlfahrtia* (Flesh flies).
5. ***Simuliidae:*** *Simulium* (black fly).
6. ***Oestridae:*** *Dermatobia*, *Hypoderma bovis, and Oestrus ovis*.
7. ***Tabanidae***: *Tabanus*, *Chrysops* (Deer fly/Mango fly).

A. *Psychodidae*

Phlebotomus (Sandfly)

Characteristics

- Sandfly is 2–3 mm long, pale sandy to golden in color, covered with hairs.
- Its thorax is humped with two lancet shape hairy wings.
- At the resting position, the hairy wings are at a right angle on the body.
- The female is a night biter; its bite is very painful.
- Female lays eggs in racks in the wall and ground.
- Sandfly undergoes complete metamorphosis, it takes 2 months to complete its life cycle.

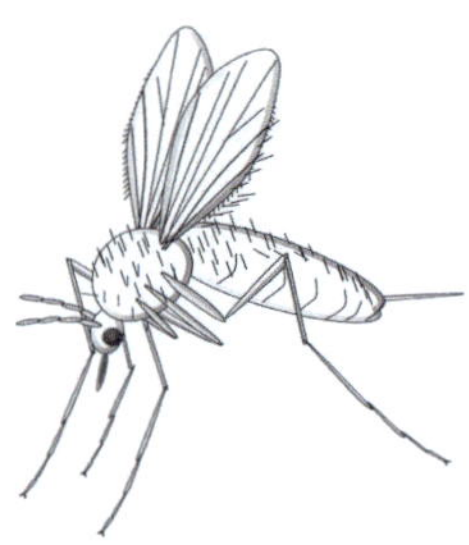

Sandfly

Common Species

Sandflies are worldwide in distribution.

- New world countries: *Lutzomyia* (*L. longipalpis*).
- Old world countries: *Phlebotomus* (*P. sargentii* and *P. papatasi*).

Medical Importance
Sandflies are vectors of:

1. Leishmaniasis (*L. tropica, L. braziliensis,* and *L. donovani*).
2. Sandfly fever (3 days fever).
3. Oroya fever (Carrion fever) is caused by *Bartonella*.

B. *Muscidae*

a. *Musca domestica* (House Flies)

Characteristics

- Housefly measures 0.7–1 cm long, greyish.
- The head has two red eyes; the thorax of the housefly has four longitudinal strips. The abdomen has four segments with one longitudinal strip.
- It is sucking, not biting fly.
- Life span ranges from 15 to 25 days.
- Housefly lays eggs (3000 eggs during its life).
- Larva (maggot) of house fly measures 5–9 mm, creamy in color, tapered anterior end and broad posterior one, D-shape posterior spiracle with three sinusoid slits.

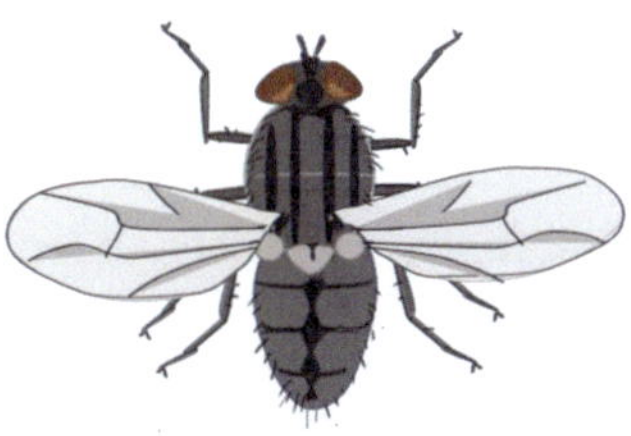

Musca domestica
(Housefly)
© ELShewy, 2023.

Common Species

- *Musca domestica* with worldwide distribution.

Medical Importance

1. Mechanical transmission of many pathogens, e.g., *Salmonella, Shigella, Campylobacter, Escherichia, Enterococcus, Chlamydia, Giardia, Balantidium, Giardia, Entamoeba, Ascaris, Hymenolepis*, and other.
2. It is involved in outbreaks of food poisoning and diarrheal illness.
3. Accidental myiasis.
4. Experimentally, houseflies are used in research on aging and sex determination.

b. *Stomoxys calcitrans* (Stable Fly)

Characteristics

- Stomoxys (stable fly) measures 0.7–1 cm long. It looks like a house fly; however, it is lighter in color and has a slightly wider abdomen with four spotted segments.
- Its mouthparts have biting structures (piercing and sucking blood).
- The fly breeds in and around where cattle and horses are kept.
- Its maggot is found in the rotting manure of these animals.
- During the day, it attacks the ankles of humans and the limbs of animals.
- Larvae (maggots) of stable flies are similar to house flies but with triangular posterior spiracles with three S-shape slits.

Stomoxys
(Stable fly)
© ELShewy, 2023.

Common Species

- *Stomoxys calcitrans* has a worldwide distribution.

Medical Importance

1. Myiasis.
2. *Trypanosoma evansi*, brucellosis, Equine infectious anemia, African horse sickness.

c. *Glossina* (Tsetse Fly)

Characteristics

- *Glossina* prevails in Africa; there are two species:

G. palpalis: it is 10–15 mm long, black and *G. morsitans:* it is brown and the same size.

- The abdomen has many segments with a median pale strip in *G. palpalis* and brown with yellow transverse bands in *G. morsitans.*
- The distal cell in the wing has a **cleaver shape**.
- In the resting position, both wings are crossed together.
- Both flies have long proboscis.
- Both males and females are biting and sucking; females are larviparous.
- They have diurnal activity.

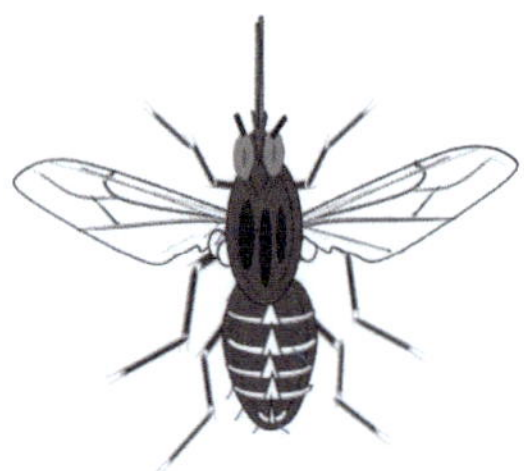

Glossina palpalis
© ELShewy, 2023.

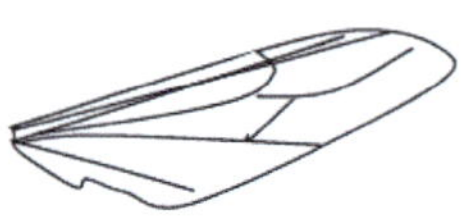

Glossina wing
© ELShewy, 2023.

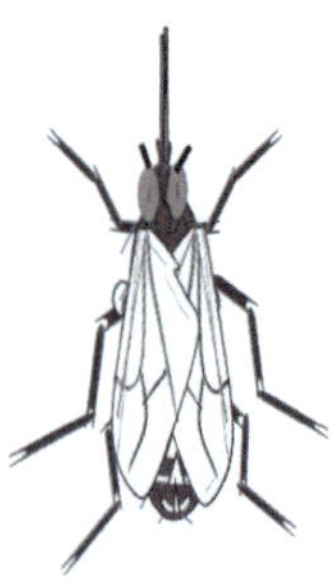

Crossed wing in rest
© ELShewy,2023.

Common Species

- *G. palpalis* prevails in West Africa
- *G. morsitans* prevails in East Africa.

Medical Importance

Both are vectors of human African trypanosomiasis (African sleeping sickness):

1. *G. palpalis* transmits *Trypanosoma gambiense.*
2. *G. morsitans* transmits *Trypanosoma rhodesiense.*
3. They both transmit Nagana in animals.

C. *Calliphoridae*

Metallic Flies

a. *Calliphora* (Bluebottle Fly, Blowfly)

Characteristics

- *Calliphora* (bluebottle) fly is 5–12 mm long, metallic blue, with pearly gray spots on its abdomen and stripes on its thorax.
- The fly has large compound eyes. Stiff transparent wings.
- Flies are sucking; feed on any meat product, decaying matter, excrement and manure.
- Life span is 30–35 days.
- Bluebottle fly lays eggs in dead manure or fresh meat.
- The larva is similar to that of *Musca* but longer (18 mm long).
- Larva feeds on decaying animals (obligatory necrobiots) or invade skin and wounds (facultative sarcobiots).

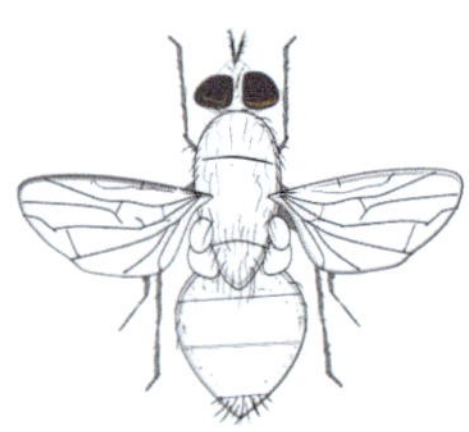

Calliphora
© ELShewy, 2023.

Medical Importance

1. Semi-specific myiasis and it is used in the healing of wounds.
2. They are called blowflies as detected inside dead blown bodies (Blowflies).

b. *Lucilia* (Green Bottle Fly, Blow Fly)

Characteristics

- *Lucilia* is similar to *Calliphora* in everything; both are metallic flies of the same size, wings, and eyes. However, *Lucilia* fly has a brilliant, metallic, **green** coloration with black markings.
- Lucilia flies are sucking; they feed on any meat product, decaying matter, and manure.
- Life span is 30–35 days.
- Larvae of green bottle fly are similar to blue-bottle fly.
- Larva feeds on decaying animals (obligatory necrobiots) or invade skin and wounds (facultative sarcobiots).

Medical Importance

Semi-specific (facultative) myiasis can sometimes be used in wound healing.

D. *Sarcophagidae*

Flesh Flies

a. *Sarcophaga*

Characteristics

- *Sarcophaga* (flesh fly) is 5–12 mm long, black bristly body, with three gray longitudinal stripes on the thorax.
- The abdomen has dark markings in a chess-board pattern.
- Large compound eyes. Stiff transparent wings.
- *Sarcophaga* is larviparous and blood-sucking.
- Larvae are similar to *Musca* but larger; they burrow into food material.
- Flesh flies prefer decaying matter, carrion, and any meat product.
- *Sarcophaga* fly is commonly found during summer in waste areas.
- They seldom enter buildings but can be found in meat processing plants with poor waste disposal.

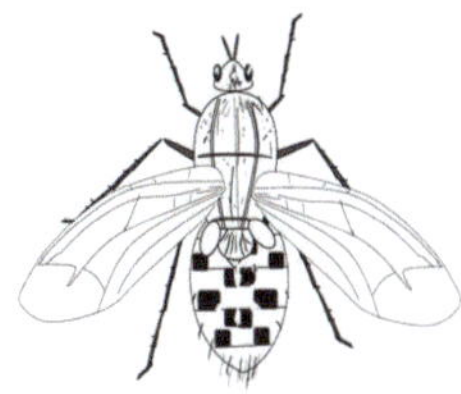

Sarcophaga

Medical Importance

1. Semi-specific myiasis (could be used in the healing of wounds).
2. The larvae infest wounds, carrion, or excrement.
3. The larvae feed on eggs, nymphs, or larvae of more harmful insects.

b. *Wohlfahrtia*

Characteristics

- *Wohlfahrtia* (flesh fly) is hairy greyish, around 5–12 mm.
- It has three strips on the chest and dark markings as round lateral spots with a median strip on the abdomen.
- Female is larviparous. Larvae are laid in scratches, wounds, and sores of man and animals.
- Larvae may be laid on the nose, ear, or eyes.
- Larvae of *Wohlfahrtia* are similar to that of *Sarcophaga*.

Wohlfahrtia

Medical Importance

1. Semi-specific myiasis.
2. Nasal, eye, and wound myiasis.

E. *Simuliidae*

Simulium (Black Fly)

Characteristics

- *Simulium* or black fly is a small humped blackish body.
- It is 1.5–4 mm long with short stout legs.
- The antenna is very short and stout, and the wings are not spotted.
- The only female is a diurnal blood-sucking, and its bite is very painful.
- The female secrets anticoagulant and histamine during biting.
- Females lay eggs on **running water** (rivers, water canals), and larvae attach themselves to rocks.

Simulium
(Black fly)

Medical Importance

1. *Simulium* is a vector of *Onchocerca volvulus* responsible for river blindness.
2. *Simulium* bite may cause a systemic illness and allergic reactions mediated by histamine. This reaction is known as "black fly fever" and is characterized by headache, fever, nausea, adenitis, generalized dermatitis, and allergic asthma.

F. *Tabanidae*

Chrysops (Deer Fly, Mango Flies)

Characteristics

- *Chrysops* prevails mainly in Africa.
- It is up to 10 mm long, has a yellow to black body, striped abdomen, and large head.
- Their wings are clear membranous with dark patches.
- They live in forested and muddy habitats like swamps and streams.
- Female deer flies require blood during the day for egg deposition.

Chrysops
(Deer fly)

Medical Importance

1. The bite of the *Chrysops* is very painful, as it lacerates skin to bleed.
2. They are vectors of filarial worm *Loa loa* in humans and anthrax among cattle and sheep.

Myiasis

Definition

Myiasis means invasion of the dead or living human or animal tissues by fly larvae (maggots). Flies commonly involved in myiasis are among three main groups:

1. *Calliphoridae* (blowflies).
2. *Oestridae* (botflies).
3. *Sarcophagidae* (flesh flies).

Predisposing Factors of Myiasis

The following are predisposing conditions to myiasis:

1. Open untreated wounds, contaminated war, and surgical wounds.
2. Diving and swimming in stagnant water.
3. Workers in contact with goats, sheep, and cattle are more susceptible to myiasis.
4. Extreme ages in geriatric units and hospitals (nosocomial myiasis).
5. Low economic status and poor hygienic conditions.
6. Mental illness.
7. Uncontrolled diabetes with open wounds.
8. Vascular diseases obstructing blood flow.

Classification of Myiasis (Table 19.1)

There are several methods for the classification of myiasis depending on the species of the fly, food habits, habitat of larvae, organs infested by myiasis larvae, and ecological factors. However, flies causing myiasis could share the same habits and habitats, so there are several classifications, but none are clear-cut.

The following are some of these classifications:

A. **According to the habit of the fly:**
   1. **Specific(obligatory) Myiasis:** Caused by flies that need living hosts for the development of their larvae **(Obligatory Sarcobiots),** e.g., *Dermatobia, Oestrus, Hypodermia, Gasterophilus, and Cordylobia.*
   2. **Semi-specific (facultative) Myiasis:** Caused by flies that could lay eggs in decaying animal or vegetable matter (**Obligatory Necrobiots**), also could lay eggs in open wounds or sores (**Facultative Sarcobiots**), e.g., *Calliphora, Lucilia,* and *Sarcophaga.*
   3. **Accidental Myiasis (pseudomyiasis):** This includes free-living larvae that may accidentally come in contact with the host and cause pathological reactions, e.g., larvae of *Musca, Stomoxys,* and *Fannia.*

B. **According to the tissue infected by larvae:**
   1. **Cutaneous myiasis:**
      – **Traumatic (wound) myiasis:** Larvae invade open wounds or ulcers, e.g., lar-

Table 19.1 Common flies causing human myiasis

Fly	Common name	Site of myiasis
A. Specific Myiasis		
Dermatobia hominis	Human botfly	Skin
Cordylobia anthropophaga	Tumbu fly	Skin
Oestrus ovis	Sheep botfly	Nose
Hypoderma spp	Cattle botflies	Skin, wounds
Gasterophilus spp	Horse botfly	Skin, wounds
Cochliomyia hominivorax.	New world screwworm fly	Skin, wounds
Chrysomya	Old world screwworm fly	Nose, sinus, pharynx
Auchmeromyia	Congo-floor maggot	Hematophagous myiasis
B. Semi-specific Myiasis		
Lucilia spp.	Green-bottle fly (blowfly)	Decaying animals or vegetable matter, open wounds, or sores
Cochliomyia spp	Screwworm fly	Skin, wounds
Phormia spp	Black-bottle fly	Wounds
Calliphora spp	Bluebottle fly (blowfly)	Decaying animals or vegetable matter, open wounds, or sores
Sarcophaga spp	Flesh fly	Decaying animals or vegetable matter, open wounds, or sores
C. Accidental Myiasis		
Musca domestica	Housefly	Gastrointestinal tract, wounds
Fannia spp	Latrine flies	Urogenital tract, wounds
Eristalis tenax	Rat-tailed maggots	Gastrointestinal tract, wounds

vae of *Calliphora, Lucilia, Phormia, Musca,* and *Fannia.*
   - **Furuncular myiasis (follicular):** The larva remains at a single spot, causing a boil-like lesion, e.g., larvae of *Dermatobia, Cordylobia,* and *Wohlfahrtia.*
   - **Creeping myiasis:** Larvae burrow into the skin, e.g., larvae of *Gasterophilus* and *Hypodermia.*
2. **Gastrointestinal myiasis:** Intestinal, gastric, and rectal, e.g., larvae of *Musca, Calliphora, Lucilia, Sarcophaga, Fannia,* and *Eristalis.*
3. **Nasopharyngeal myiasis:** Larvae infecting the nose, sinuses, and pharynx, e.g., larvae of *Wohlfahrtia, Sarcophaga, Chrysomya,* and *Oestrus.*
4. **Ocular myiasis:** The eyes, orbits, and periorbital tissues are affected by larvae of Wohlfahrtia, *Sarcophaga,* and *Oestrus.*
5. **Urogenital myiasis:** It involves the vagina and penile opening, e.g., larvae of *Fannia.*
6. **Auricular myiasis:** Infection of the ear, e.g., larvae of *Wohlfahrtia* and *Sarcophaga.*
7. **Hematophagus myiasis:** This is a very rare disease commonly found in infants younger than one year where the larvae of Congo floor maggot (*Auchmeromyia*) suck blood from the patient skin.

Clinical Picture of Myiasis

The clinical picture of myiasis depends on the site affected:

- Cutaneous myiasis: Painful, slow-developing ulcers or furuncles (boils) that can last for a prolonged period.
- Nasal myiasis: Obstruction of nasal passages and severe irritation. In some cases, facial edema and fever can develop. Some cases could end fatly.
- Aural myiasis: Crawling sensations and buzzing noises. Smelly discharge is sometimes present. If located in the middle ear, larvae may get to the brain.
- Ophthalmomyiasis: Severe irritation, redness, and swelling of the eye.
- Urogenital myiasis: Frequency, dysuria, passage of larvae in urine or vaginal discharge.

Diagnosis of Myiasis

- History of traveling to infested areas and socioeconomic and hygienic status of the patient.
- Unhealed skin lesions, sense of moving subject under skin, itching.
- Physical examination of the lesion with detection of visible larvae.
- Internal myiasis needs US, CAT, or endoscopic visualization.
- Identification of posterior spiracles of the larvae.
- Growing larvae to flies on blood agar or a dish of raw meat with moistened sand.

Treatment of Myiasis

- Covering breathing holes of visible larvae with paraffin, beeswax, or oil.
- Larvae are moved spontaneously otherwise, surgical removal should be done
- Larvae should be removed intact since the remaining parts may cause further inflammation.
- Antiseptics application.
- Infected wounds require debridement, removal of maggots, and antiseptic dressings, all under local anesthesia.
- Internal myiasis requires systemic antiparasitic administration of ivermectin.

Prevention and Control of Myiasis

- General improvement of sanitation.
- Personal hygiene.
- Control of the flies by insecticides.
- Clothes are washed thoroughly in hot water, dried away from flies, and ironed.

Fleas

Characteristics

- There are more than 3000 species of fleas, and only very few of them are medically important.
- The body of the flea is 2–3 mm in length, bilaterally compressed, and has no wings. It is divided into the head, thorax, and segmented abdomen.
- Fleas have three pairs of very strong legs that help in jumping.
- Both males and females have piercing and sucking mouthparts.
- Both males and females feed on blood.
- Fleas undergo complete metamorphosis.

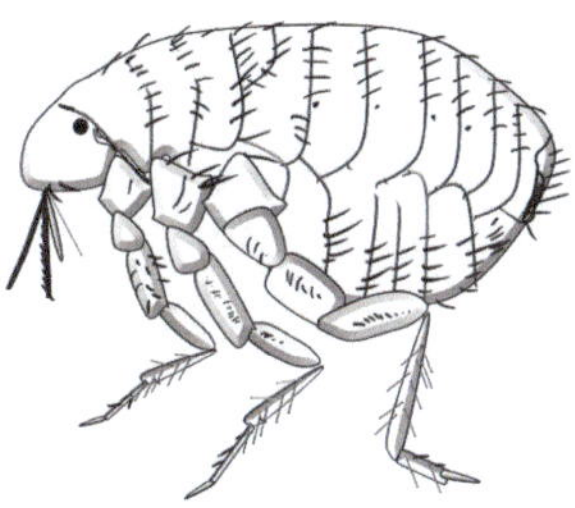

Pulex
© ELShewy,2023.

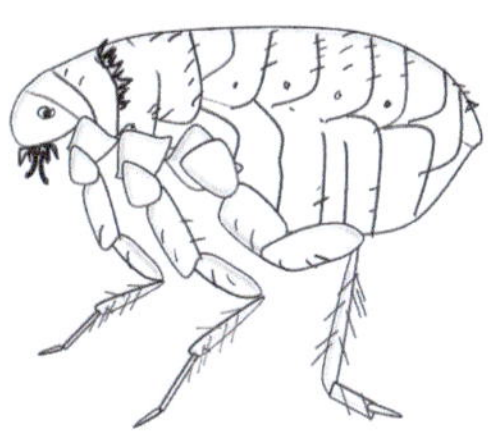

Ctenocephalus
© ELShewy,2023.

Common Species

- *Pulex irritans*: human flea.
- *Xenopsylla cheopis:* rat flea.
- *Ctenocephalus canis* (dog flea); *Ctenocephalus felis* (cat flea).
- *Tunga penetrans:* Sand flea (Jigger or chigger flea).

Medical Importance

1. Flea **nuisance**: caused by flea bites, especially at the ankle and legs.
2. *Xenopsylla cheopis* (rat fleas) are vectors for:
   - **Plague** caused by *Pasteurella pestis,* which is transmitted by a bite of the flea.
   - **Murine typhus** is caused by *Rickettsia mooseri,* transmitted by fecal matter of the flea, contaminating skin abrasions or mucous membranes.
3. *C. canis* and *C. felis* are intermediate hosts of *Dipylidium caninum* and *Hymenolepis diminuta.*
4. **Chigger** foot caused by *Tunga penetrans*:

   It prevails in Latin America, where fertilized female burrows in the skin of human feet cause painful skin nodular swellings that ulcerate and secondarily infect or even become gangrenous.

Lice

a. *Pediculus humanus*

Characteristics

- The body of the louse is 2–5 mm, compressed dorsoventrally, divided into a head, thorax, and segmented abdomen.
- They have three pairs of short legs ending in claws.
- Mouthparts are retractile used for piercing and sucking blood.
- Lice are permanent ectoparasites and die if taken from their hosts.
- They leave the host during fever and after its death.
- The female lays 10 eggs daily cemented to hair or clothes.
- They undergo incomplete metamorphosis.

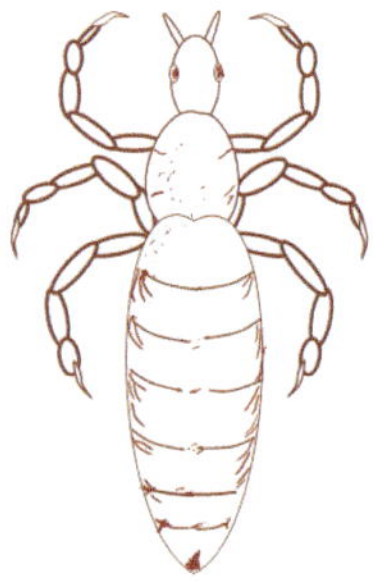

Pediculus humanus capitis
(Head louse)

Common Species

There are two species of *Pediculus humanus*:

1. *Pediculus humanus capitis* (Head Louse): Smaller than the body lice and darker in color. It infects the head, sometimes, it could affect the body.
2. *Pediculus humanus corporis* (Body Louse): Larger than head lice, and it is restricted to the human body only.

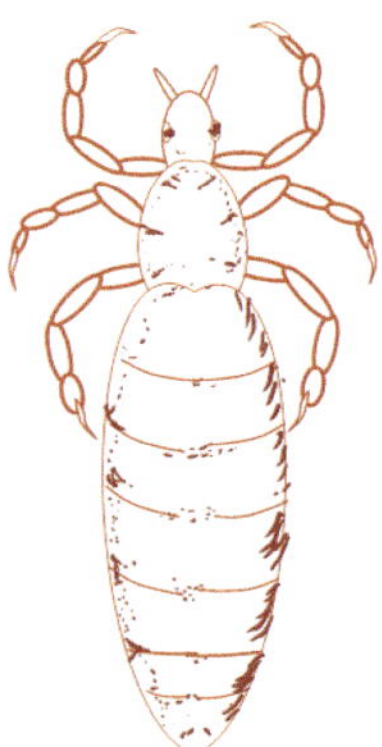

Pediculus humanus corporis
(Body louse)

Medical Importance

- Head louse does not transmit any diseases.
- Pediculosis (**Vagabond's disease**):
  It is caused by heavy lice infestations in poor, hygienic slums and overcrowded areas. There are cutaneous lesions with itching and ulcerations.
- Body louse could transmit the following:
  1. **Epidemic typhus** is caused by *Rickettsia prowazekii*: transmitted by fecal matter of the louse contaminating skin abrasions or mucous membranes.
  2. **Trench fever** is caused by *Rickettsia quintana:* transmitted as in epidemic typhus.
  3. **Epidemic relapsing fever** is caused by *Borrelia recurrentis:* infection occurs by crushing the lice and contamination of bite wounds with the lice body fluid.

b. *Phthirus pubis* (Pubic Louse, Crab)

Characteristics

- It is smaller than the *Pediculus*; and it is around 1–2 mm in size .
- It has an almost rounded body with a compressed thorax and abdomen together and a small head.
- It has three pairs of legs; the first pair is not well-developed.
- It inhabits the pubic area, axilla, and eyelashes.
- It is not known to transmit any disease to humans.
- In a heavy infestation, the patient's skin may be inflamed with blue spots.
- It causes itchy eyes with inflammation of the lashes.
- Pediculosis ciliaris is an infestation of the eyelid margin with conjunctivitis and edema, which may lead to keratitis.

Phthirus pubis
(Pubic louse)

Management of Lice

- Personal cleaning, frequent washing, boiling, and ironing of clothes.
- Cutting hair short (head and pubic).
- Blowing heated air into the scalp could kill more than 97% of lice eggs daily over more than a week.
- Frequent combing.
- Delousing using the following preparations:
  1. 5–10% DDT, Gammexane as body dust or head cream.
  2. Ivermectin lotion (0.5%).
  3. Pyrethrum and Permethrin lotion (1%).
  4. Benzyl alcohol lotion (5%).
  5. Malathion lotion (0.5%).

Bugs

a. *Cimex lectularius* (Bed Bug)

Characteristics

- *C. lectularius* (bed bug) is found all over the world in almost every area with humans.
- The bed bug is about 5 mm long and dark brown.
- The body is compressed dorsoventrally, divided into a head, a three-segmented thorax, and an eight-segmented oval abdomen.
- Its mouthparts are piercing and sucking in both males and females.
- They are obligate ectoparasites and feed on blood at night when the victim sleeps.
- They undergo incomplete metamorphosis.
- They inhabit cracks and bed mattresses and produce characteristic odors.
- They withstand starvation for a long time.

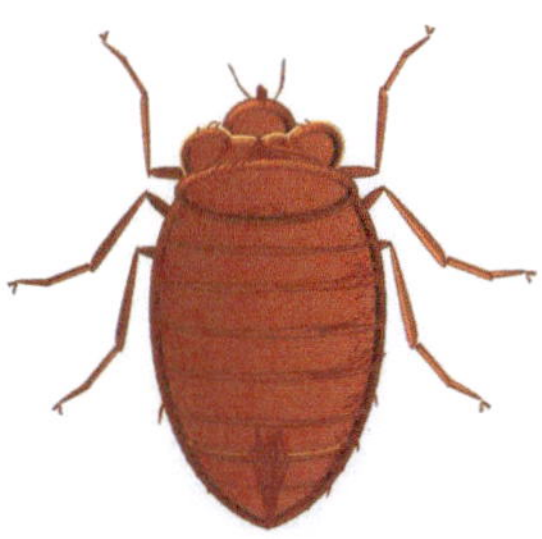

Cimex lectularius
(Bed bug)
© ELShewy,2023.

Medical Importance

1. *C. lectularius* is not known to transmit disease to humans.
2. Bed bug bite is irritating, causing insomnia.

b. *Triatomine* Bugs (Kissing or Winged Bug)

Characteristics

- Reduviid bugs, assassin bugs, and cone-nose bugs, all are names of *Triatomine* bugs.
- They prevail in South America.
- They have elongated bodies ranging from 0.5 to 4.5 cm, with three pairs of legs.
- Two pairs of wings: the proximal parts of the wings are thickened and hard, and the distal parts are membranous.
- They bite at the face during the night and hide in the day in cracks.

Triatoma (*Reduviid*)
(Kissing bug)
© ELShewy,2023.

Common Species

- *Triatoma infestans; Panstrongylus megistus; Rhodnius prolixus*

Medical Importance

Vector for *Trypanosoma cruzi*, which is transmitted through contamination of the bite site by the insect fecal matter (posterior station).

Review Questions

1. **Crustaceans are characterized by having the following except**
   a. No wings
   b. 3 pairs of legs
   c. 2 pairs of antennae
   d. cephalothorax
2. **The following flies are responsible for specific myiasis except**
   a. *Dermatobia hominis*
   b. *Cordylobia*
   c. *Fannia*
   d. *Hypoderma spp.*
3. ***Aedes albopictus***
   a. is a house mosquito
   b. is an aggressive day biter?
   c. has spots on its wings
   d. rest at right angle with surfaces
4. **Sandfly could transmit**
   a. Cholera
   b. Oroya fever
   c. *Trypanosoma*
   d. *Babesia*
5. ***Glossina* is characterized by**
   a. short proboscis
   b. only female is sucking blood
   c. crossed wings at rest
   d. vector of *Trypanosoma cruzi*
6. ***Sarcophaga* is**
   a. a metallic fly
   b. vector of viral meningitis
   c. larviparous
   d. indoor fly
7. ***Simulium* (black fly) transmits**
   a. *Leishmania*
   b. *Onchocerca*
   c. *loa loa*
   d. *Babesia*
8. **Creeping myiasis is caused by**
   a. *Lucilia*
   b. *Hypodermia*
   c. *Dermatobia*
   d. *Calliphora*

20 Medically Important *Arachnida* and *Crustacea*

Test your learning and check your understanding of this book's contents: use the "Springer Nature Flashcards" app to access questions using ▶ https://sn.pub/oyFLIx. To use the app, please follow the instructions in Chap. 1.

Mites

A. *Sarcoptes scabiei* (*S. scabiei*), (Itch Mite)

Sarcoptes scabiei (itch mite) is the causative arthropod of **scabies**.

Characteristics

- *S. scabiei* is a minute arthropod belonging to the class *Arachnida*.
- It has an oval unsegmented body.
- Male (0.2 mm) is smaller than the female (0.4 mm).
- The body is convex dorsally and flat ventrally.
- The body is covered with a cuticle with many hairs.
- They have four pairs of legs.
- The posterior pairs of legs end with long bristles in females and suckers in males.
- *S. scabiei* inhabits human skin; the female burrows tunnel in the superficial skin layers, moving slowly at night.

Life Cycle

1. Mating between male and female occurs on the skin.
2. The female burrows a tunnel in the superficial layers of the skin to lay eggs.
3. Larvae hatch within 1 week and molt into nymphs.
4. Nymphs mature to adults within 2 weeks.

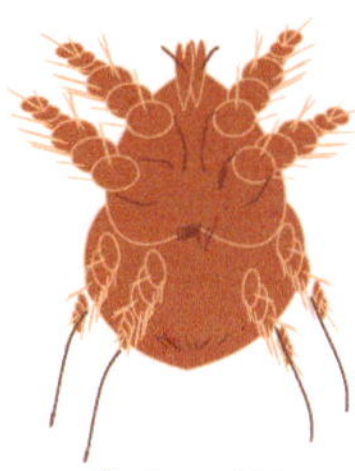

♀ *S. scabiei*
© ELShewy,2023.

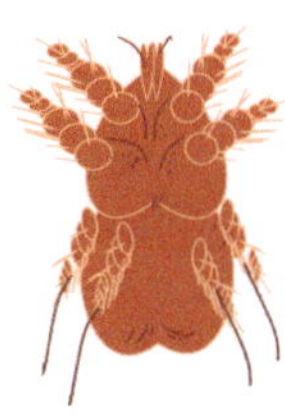

♂ *S. scabiei*
© ELShewy,2023.

Pathogenesis and Clinical Picture

- *S. scabiei* preferentially affects the skin at the interdigital spaces, flexor, surfaces of the

K. ElShewy, *Medical Parasitology*, https://doi.org/10.1007/978-3-031-64736-9_20

wrists and forearms, axillae, groin, and genitalia
- Because the skin is thick and compact, the palm and plantar surfaces of the feet are not affected.
- The burrowing of the female is associated with an itching sensation.
- Burrowing is activated by warmth, so itching increases by night.
- Itching leads to reddish tracts (scratching) with minute vesicles.
- Secondary bacterial infection leads to vesicular eruptions and pustules.
- Infection is transmitted by close contact with an infected person, especially in overcrowding and poor hygiene (in slums and jails).

Norwegian Scabies

This is a clinical condition that appears among immunocompromised individuals, e.g., individuals receiving corticosteroids, antidepressants, anticancer drugs, or individuals with HIV. Norwegian scabies is characterized by:

1. Increased skin lesions with an increased number of parasites in these lesions.
2. Thick keratolytic crusts over hands and feet.
3. Scaling eruption all over the body with decreased intensity of itching.
4. Norwegian scabies are highly contagious.

Nodular Scabies

Itchy raised bumps or lumps commonly found around genitalia, armpits, and groin.

Diagnosis of Scabies

1. Itching that increases by night.
2. Demonstration of the characteristic scratching and tunnels in the infected skin.
3. Snip biopsy by fine needle and examine it under the microscope to demonstrate the parasites.

Treatment of Scabies

A. **Noncrusted scabies:**
   1. **Permethrin cream 5%** (topical cream):
      - Drug of choice for scabies.
      - 2 or more applications a week apart is enough.
      - Safe for children and pregnant women.
   2. **Sulfur (5–10%) ointment:**
      - It is safe for topical use in children, including infants less than 2 months of age.
      - The odor and cosmetic quality may make it unpleasant to use.
   3. **Crotamiton 10%,** (Eurax).
      - This medication is available as a cream or a lotion.
      - It is applied once a day for two days.

B. **Crusted scabies**:
   1. **Ivermectin:**
      - Ivermectin should be administered together with a topical agent.
      - Oral ivermectin (200μg/kg/dose) should be taken with food.
      - It should be taken in three doses (approximately days 1, 2, and 8).
   2. **Keratolytic cream:**
      - It is used to reduce the crusting of the skin.
   3. **Benzyl benzoate lotion (25%).**

B. House Dust Mites (HDM)

Dermatophagoides spp. (Allergy Mites)

The term "house dust mites" has been applied to a large number of mites found in association with dust in the environment. Their significance is due to the powerful allergens in their cast skins, fecal material, and secretions.

Characteristics

- House dust mites are arthropods that are not capable of biting or stinging humans.
- Both male and female adult house dust mites are globular in shape.
- They are creamy white and have striated cuticles.
- They have four pairs of hairy legs, no eyes, no antennae, and a mouth region at the anterior end that resembles the head.

- They have long setae (bristle-like hairs) that extend from the outer margins of the body.
- The female measures approximately 0.4 mm in length and 0.3 mm in width.
- The male is approximately 0.4 mm long and 0.2 mm wide.
- House dust mites could affect humans through inhalation of the whole organism or its secretion and fecal matter. They feed on the shed skin of humans.
- They undergo incomplete metamorphosis.
- Mated females live for about 2 months.

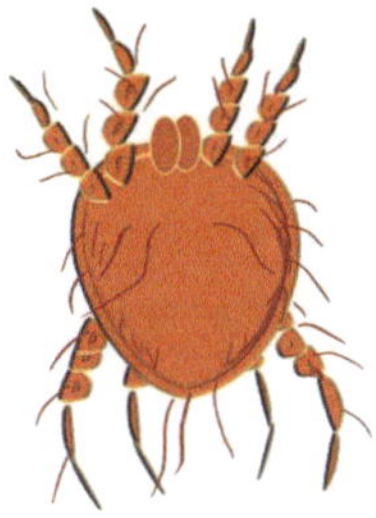

Dermatophagoides (House dust mite)

Common Species
- House dust mites are nearly cosmopolitan in distribution.
- *Dermatophagoides farinae* represents more than 60% of mites, followed by *D. pteronyssinus* more than 20%.

Medical Importance
- Bronchial asthma.
- Contact dermatitis.
- Allergic rhinitis.

C. *Demodex folliculorum* (Hair Follicles Mites)

Characteristics
- It is a worm-like arthropod with a size of 0.3–0.4 mm.
- Its body is elongated with four pairs of stubby protruding legs.
- Legs end in blunt claws.
- Mouthparts are short and blunt.
- The abdomen has many transverse striations.
- It inhabits the hair follicles of humans and animals.

Demodex folliculorum

Medical Importance
1. *Demodex* mites infest the space within the hair follicles.
2. It affects the sebaceous glands of humans and animals.
3. It gives an acne-like appearance or keratitis with no itching.

D. *Trombiculid*

Trombicula akamushi (Harvest, Red Bugs Mites)

Characteristics
- Adult *Trombiculid* is a free-living mite, red in color.
- Adult *Trombiculid* has four pairs of legs with no wings.
- Its body is covered with hair.
- Larva of *Trombiculid* is parasitic to humans and animals; it has three pairs of legs.
- Larvae feed on body fluid and not blood.

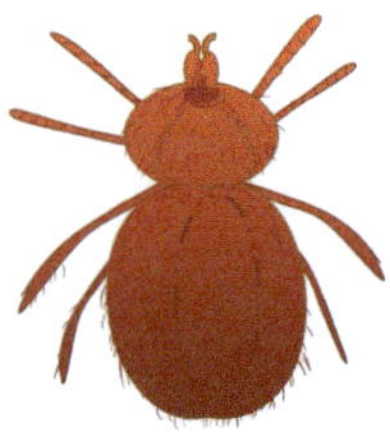

Trombicula
(Adult)
© ELShewy,2023.

Common Species

- *Trombicula akamushi and Trombicula autumnalis.*

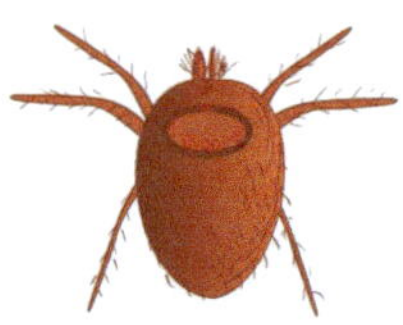

Trombicula
(Larva)
© ELShewy,2023.

Medical Importance

1. The bite of the *Trombicula* causes itching and dermatitis (scrub itch).
2. Vector of Scrub typhus caused by *R. tsutsugamushi* through transovarian transmission.

Ticks

Ticks are minute arthropods that could infest both humans and animals. They are blood-sucking ectoparasites transmitting a wide variety of disease-causing organisms. Ticks are second only to mosquitoes in terms of public health significance.

- They have neither wings nor antennae.
- Capitulum (false head) carrying mouthparts used for sucking blood.
- The body may be covered with a chitinous plate (scutum).
- They have four pairs of legs; the male is smaller than the female.
- They undergo incomplete metamorphosis, with a very long life cycle (1–2 years).

Ticks with medical importance to humans are grouped into two families:

1. *Ixodidae* (hard ticks).
2. *Argasidae* (soft ticks).

A. *Ixodidae* (Hard Ticks)

Characteristics

- **Permanent** ectoparasites.
- Capitulum project from the body.
- The body is covered with scutum in males and only the anterior part in females.
- The posterior end of the body is festooned.
- They are short-lived; males die after fertilization, and females Die after egg deposition.
- They undergo incomplete metamorphosis.

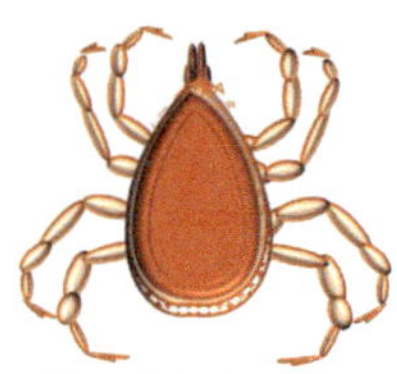

♂ *Ixodidae* (hard tick)
Capitulum & scutum
© ELShewy,2023.

Common Species

- *Dermacentor andersoni.*
- *Boophilus annulatus.*

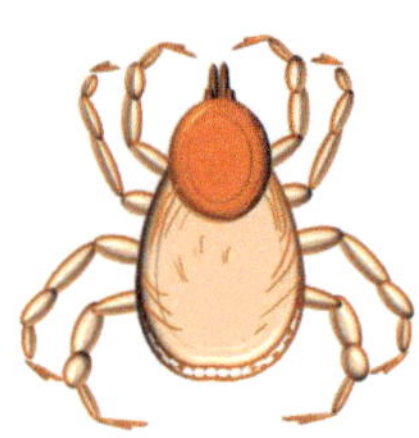

♀ *Ixodidae* (hard tick)
© ELShewy,2023.

Table 20.1 Diseases transmitted by ticks

Ticks	Disease	Causative organism
Hard ticks	Rocky Mountain spotted fever.	*Rickettsia rickettsii.*
	Q fever.	*Rickettsia burneti.*
	Anaplasmosis.	*Anaplasma.*
	Texas cattle fever.	*Babesia.*
	Tularemia.	*Pasteurella.*
	Lyme disease.	*Borrelia.*
	Crimean-Congo hemorrhagic fever.	*Nairo virus.*
Soft Ticks	Endemic relapsing fever (Tick-borne relapsing fever).	*Borrelia duttoni.*
	Q fever.	*Rickettsia burneti.*

Medical Importance

1. Skin lesions due to tick bites.
2. Tick paralysis: ticks release toxins during biting, causing flaccid progressive ascending motor paralysis. Ticks should be removed by gentle traction under local anesthetic.
3. Vectors of diseases (Table 20.1).

B. *Argasidae* (Soft Ticks)

Characteristics

- **Temporary** ectoparasites: visit the host only for a blood meal.
- The capitulum **does not** project from the body.
- The body is **not covered** with scutum in both males and females.
- The posterior end of the body is festooned
- **Long-lived** ticks.
- They undergo incomplete metamorphosis.

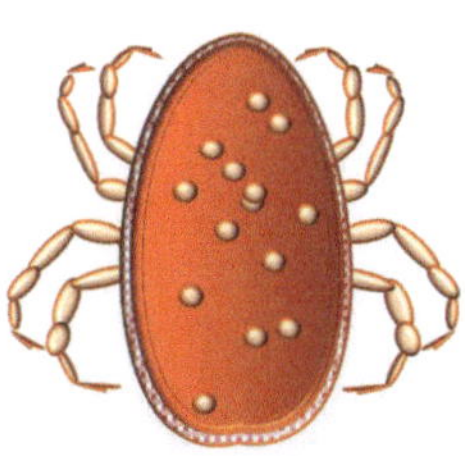

Argas (soft tick)
No apparent capitulum.
No scutum
© ELShewy,2023.

Common Species

- *Argas persicus:* the body is flat and triangular; it is not a human species.
- *Ornithodoros moubata:* body oval with mamillated surface.

Medical Importance

- Vectors of diseases (Table 20.1)
- Skin lesions due to tick bites.
- Tick paralysis: ticks release toxins during biting, causing flaccid progressive ascending motor paralysis. Ticks should be removed by gentle traction with a local anesthetic.

Scorpions

Characteristics

- The body of the scorpion is composed of a cephalothorax with four pairs of legs and two strong pedipalps ending in pincer-like structures.
- Abdomen segmented with posterior thin part forming a tail ending in a stinger with a pair of poison glands.
- Scorpion feeds on insects at night.
- It hides during the day and stings humans only when attacked.
- It is viviparous.

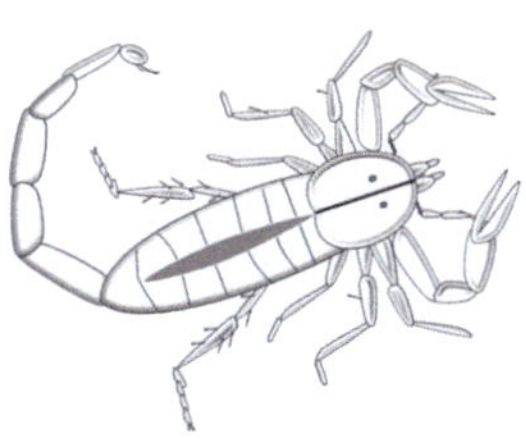

Scorpion
© ELShewy,2023.

Medical Importance

Scorpion sting is characterized by:

- Intense pain.

- Tingling and numbness around the sting.
- Swelling around the sting.
- Profound sweating, nausea and vomiting, and high blood pressure.
- Severe muscular spasm and convulsion, tachycardia, or irregular heart.
- The condition may end fatally from respiratory failure, edema, and shock.

Treatment

- A tourniquet distal to the sting and the wound should be washed and sterilized.
- Pain relieved by ice pack, ethyl chloride spray, and analgesics.
- Injection of specific antidote.
- Corticoids.
- Measures to combat shock (I.V. fluids).

Spiders

Characteristics

- Most spiders do not pose a threat to humans, and their bites are harmless or mildly irritating, like a bee sting.
- Dangerous spiders include:
  Brown recluse, black widow, hobo spider, funnel web spider (Australia), wandering spider (South America), tarantula, and wolf spider.

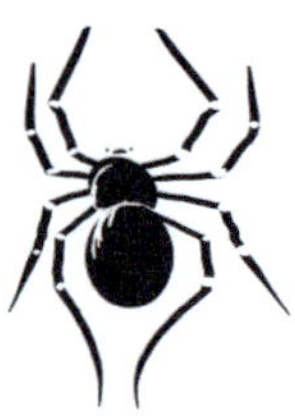

Spider

Medical Importance

Spider bite is characterized by:

- It appears to be two small puncture marks.
- Pain around the area of the bite (less painful than scorpion).
- A single raised papule, pustule, or wheal at the site of the bite.
- Muscle pain or cramping.
- Fever, sweating, difficulty breathing, headache, nausea, and vomiting.

Treatment

The same as in scorpion sting.

Cyclops

Characteristics

- It has a pear-shaped, broad anteriorly, 1–3 mm in length.
- It has two pairs of antennae with no wings.
- Cephalothorax with four pairs of legs.
- Its abdomen has five segments and ends with bristles.
- It has two bags of eggs on the side of the abdomen.

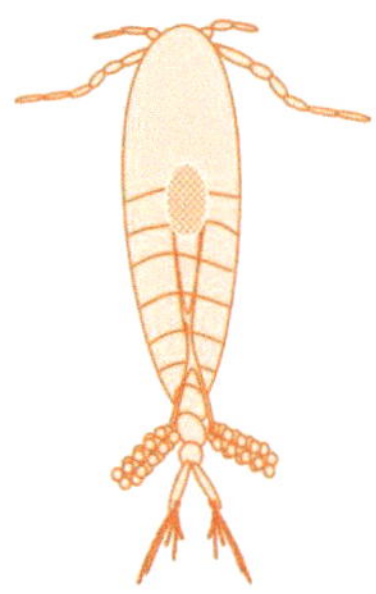

Cyclops

Medical Importance

Cyclops is an intermediate host of:

1. *Diphyllobothrium latum* and *mansoni*.
2. *Dracunculus medinensis*.
3. *Gnathostoma spinigerum*.

Review Questions

1. **Fleas could transmit the following parasites except**
   a. *Dipylidium caninum*.
   b. *Enterobius vermicularis*.
   c. *Hymenolepis diminuta*.
   d. *Hymenolepis nana*.

2. **Chigger is a skin lesion caused by**
   a. *Xenopsylla cheopis.*
   b. *Ctenocephalus canis.*
   c. *Pulex irritans.*
   d. *Tunga penetrans.*
3. **Vagabond's disease is caused by**
   a. Fleas.
   b. Lice.
   c. Flies.
   d. Cimex.
4. **Pediculosis ciliaris is caused by**
   a. *Pediculus capitis.*
   b. *Phthirus pubis.*
   c. *Pediculus hominis.*
   d. *Cimex lectularius.*
5. **Murine typhus is transmitted by**
   a. *Ctenocephalus canis.*
   b. *Xenopsylla cheopis.*
   c. *Pulex irritans.*
   d. *Tunga penetrans.*
6. **Norwegian scabies is**
   a. A regular scabies.
   b. Scabies of Norwegians.
   c. Crusted keratolytic scabies.
   d. An animal scabies.
7. **Scrub itch is caused by**
   a. *Pediculus hominis.*
   b. *Cimex lectularius.*
   c. *Trombicula akamushi.*
   d. *Sarcoptes scabiei.*
8. **Hard ticks transmit the following except**
   a. Texas cattle fever.
   b. Endemic relapsing fever
   c. Tularemia.
   d. Lyme disease.
9. **Plague is transmitted by**
   a. Fleas.
   b. Lice.
   c. Flies.
   d. Bugs.
10. **Skin acne could be caused by infection with**
    a. *Cimex lectularius.*
    b. *Demodex folliculorum.*
    c. *Xenopsylla cheopis.*
    d. *Sarcoptes scabiei.*

Appendices

Appendix A: Common Symptoms and Signs in Parasitology

Parasites Causing Fever

Parasite	Diseases	Fever pattern	Main route of infection
1. *Plasmodium*	Malaria	Early irregular then cyclic patterns (36, 48, 72 h), hyperpyrexia in cerebral malaria	*Anopheles,* blood
2. *Leishmania donovani*	Kala azar	Double or triple peaks on the same day	Sandfly bite
3. *Wuchereria bancrofti*	Lymphatic filariasis, lymphangitis, and adenitis	Early high sudden fever for few hours last few days then subsides	*Anopheles, Culex, Aedes*
4. *Entamoeba histolytica*	Amebic hepatitis	Irregular spiking	Feco-oral, flies
5. *Fasciola*	Fascioliasis, cholangitis	Early irregular and spiking	Freshwater vegetation, having metacercariae
6. *Schistosoma*	Bilharziasis	Early irregular and spiking	Penetration of skin by cercaria
7. *Trichinella spiralis*	Trichinosis myositis	Remittent	Undercooked ham
8. *Trypanosoma cruzi*	Chagas disease	High remittent for weeks	Winged bugs, blood
9. *Trypanosoma brucei*	Human African Trypanosomiasis (HAT)	Irregular with evening peaks and night sweat	*Glossina* bite
10. *Naegleria fowleri*	Primary amebic meningoencephalitis (PAM)	High, sudden fever	Nasal inhalation of trophozoites with water
11. *Acanthamoeba*	Granulomatous amebic encephalitis (GAE)	Irregular, low fever	Skin, GU, respiratory, brain via blood

K. ElShewy, *Medical Parasitology*, https://doi.org/10.1007/978-3-031-64736-9

Parasites Causing Diarrhea and Malabsorption

Parasite	Diseases	Pattern of diarrhea	Main route of infection
1. *Giardia lamblia*	Enteritis and malabsorption	Steatorrhea	Feco-oral, autoinfection, waterborne
2. *Cryptosporidium*	Enteritis and malabsorption	Cholera-like	Feco-oral, autoinfection, waterborne
3. *Strongyloides stercoralis*	Enteritis	Mild with constipation, or severe, prostrating, with dysentery (heavy infection)	Skin filariform larva, internal autoinfection
4. *Capillaria philippinensis*	Protein enteropathy, malabsorption	Profuse diarrhea	Eating infected undercooked freshwater fish
5. *Heterophyes*	Duodenitis	Mild diarrhea or steatorrhea	Eating undercooked or raw freshwater fish
6. *Cystoisospora*	Enteritis	Watery diarrhea in immunocompromised	Feco-oral
7. *Cyclospora*	Enteritis	Watery diarrhea in immunocompromised	Feco-oral
8. *Trichinella*	Duodenitis	Mild diarrhea from 1 to 5 days.	Eating undercooked ham
9. *Taenia*	Enteritis	Mild diarrhea	Undercooked meat
10. *Diphyllobothrium*	Enteritis	Mild to an acute diarrhea for long periods	Undercooked, raw salmon, and trout fish
11. *T. trichiura*	Colitis	Mucoid diarrhea	Feco-oral

Parasites Causing Dysentery

Parasite	Diseases	Pathogenesis	Main route of infection
1. *Entamoeba histolytica*	Amebic dysentery	Mucosal necrosis, ulceration, bleeding	Feco-oral, waterborne
2. *Balantidium coli*	Balantidiasis	Ulceration, secondary bacterial infection	Feco-oral, waterborne
3. *S. mansoni*	Bilharziasis	Mucosal ulceration and necrosis	Skin penetration by cercaria
4. *L. donovani*	Kala azar	Mucosal ischemia	Sandfly bite
5. *P. falciparum*	Malignant malaria	Mucosal ischemia	*Anopheles*, blood
6. *T. trichiura*	Trichuriasis	Mucosal irritation and necrosis	Feco-oral
7. *S. stercoralis*	Disseminated infection	Mucosal ulceration and necrosis	Skin filariform larva, internal, autoinfection

Parasites Causing Hepatosplenomegaly (HSM)

Parasite	Diseases	Pathogenesis	Main route of infection
1. *Schistosoma mansoni*	Bilharziasis (HSM)	Portal hypertension, congestion, and granuloma	Skin penetration by cercaria
2. *Leishmania donovani*	Kala azar (HSM)	Proliferation of parasites in liver & spleen	Sandfly bite
3. *Plasmodium*	Malaria (HSM)	Proliferation of parasites in liver and spleen	*Anopheles*, blood
4. *Toxoplasma*	Toxoplasmosis (HM)	Proliferation of the parasites in liver cells	Oocysts, undercooked meat, blood, congenital

Parasite	Diseases	Pathogenesis	Main route of infection
5. *Trypanosoma*	Sleeping sickness (HSM)	Proliferation of the parasites in liver cells	*Glossina* and winged bugs
6. *Entamoeba histolytica*	Amebic liver (HM)	Hepatitis and abscess	Feco-oral, water borne, flies
7. *Echinococcus granulosus*	Hydatidosis (HM)	Space occupying hydatid cysts and hepatitis	Ingestion of *Echinococcus* eggs
8. *Toxocara cati* and *canis*	Visceral larva migrans (HM)	Granulomas formation around larvae	Ingestion of *Toxocara* eggs

Parasites Causing Ascites

Parasite	Diseases	Pathogenesis	Main route of infection
1. *Schistosoma mansoni*	Bilharziasis	Periportal fibrosis, cirrhosis, portal hypertension	Penetration of skin by cercaria
2. *Leishmania donovani*	Kala azar	Chronicity, with nutritional cirrhosis and portal hypertension	Sandfly bite
3. *Capillaria hepatica*	Capillariasis	Liver granulomas around eggs, with later fibrosis	Ingestion of eggs

Parasites Causing Edema

Parasite	Organs affected	Pathogenesis	Main route of infection
1. *Loa loa*	Unilateral circumorbital edema	Inflammatory reaction due to passage of worm under eyelid it is transient	*Chrysops* bite
2. *Loa loa*	Calabar swellings	Vasculitis due to passage of worm under skin. It is transient	*Chrysops* bite
3. *Trypanosoma cruzi*	Unilateral eye lid (Romana sign)	Due to vasculitis	Winged bugs, blood
4. Hookworm	Generalized.	Chronic, severe infection with nutritional deficiency	Filariform larvae, piercing human skin
5. *Wuchereria bancrofti*	Early intermittent leg or breast edema.	Lymphedema due to lymphangitis	*Anopheles, Culex, Aedes*
6. *Schistosoma mansoni*	Bilateral leg edema	Portal hypertension, liver cell failure, nutritional edema, and cor pulmonale	Penetration of the skin by cercaria

Parasites Causing Lymphadenopathy

Parasite	Disease	Pathogenesis	Main route of infection
1. *W. bancrofti*	Lymphatic filariasis	Lymphadenitis inguinal, femoral epitrochlear, during acute attack	*Anopheles, Culex, Aedes*
2. *Toxoplasma*	Toxoplasmosis	Generalized, lymphadenopathy	Oocysts, undercooked meat, blood, congenital
3. *Trypanosoma brucei*	HAT	Early stage of HAT localized Winterbottom's sign. Later generalized lymphadenitis and adenopathy	*Glossina* bite, blood
4. *Trypanosoma cruzi*	Chagas disease	Generalized lymphadenitis and adenopathy, in acute stage	Winged bugs, blood

Appendix B: Blood Changes During Parasitic Diseases

Parasites Causing Anemia

Parasite	Type of anemia	Pathogenesis	Main route of infection
1. Hookworms	Hypochromic microcytic anemia	Chronic blood loss and hyperplasia of bone marrow	Penetration of skin by filariform larvae
2. *Plasmodium*	Hemolytic anemia	Rupture of RBCs and hypersplenism	*Anopheles*, blood
3. *Diphyllobothrium latum*	Macrocytic hyperchromic (pernicious) anemia	Vat. B12 deficiency	Eating undercooked or raw salmon and trout fish
4. *Fasciola*	Normochromic, normocytic anemia	Nutritional deficiency	Eating freshwater vegetation
5. *Trichuris trichiura*	Hypochromic microcytic anemia	Mucosal ulceration and chronic blood loss	Feco-oral
6. *L. donovani*	Normocytic normochromic anemia	Proliferation of parasites in the bone marrow and hypersplenism	Sandfly bite, blood
7. *Trypanosoma*	Normocytic normochromic anemia	Proliferation of parasites in the bone marrow and hypersplenism	*Glossina* and Winged bugs, blood
8. *Schistosoma*	Iron deficiency anemia	Nutritional deficiency	Penetration human skin by cercaria

Parasites Causing Leucopenia

Parasite	Level of leucopenia	Pathogenesis	Main route of infection
1. *L. donovani*	≤4000/mm^3	Bone marrow depression, monocytosis	Sandfly bite, blood
2. *Plasmodium*	3000–6000/mm^3	Bone marrow depression	*Anopheles* bite, blood

Parasites Causing Eosinophilia

Parasite	Level of eosinophilia	Pathogenesis	Main route of infection
1. *S. stercoralis*	High eosinophilia up to 70%	– Most helminths with tissue phase – Tissue invasion with marked inflammatory response – Marked eosinophilic infiltration	Skin filariform larva, internal autoinfection
2. *Trichinella spiralis*	High eosinophilia up to 70%		Eating undercooked ham
3. Hookworm	High eosinophilia up to 70%		Filariform larvae penetrate human skin
4. *Toxocara*	High eosinophilia up to 70%		Ingestion of *Toxocara* eggs
5. *Fasciola*	Moderate eosinophilia up to 25%		Eating fresh water vegetation with metacercaria
6. *W. bancrofti*	Moderate eosinophilia up to 25%		*Anopheles, Culex, Aedes*
7. *Paragonimus*	Moderate eosinophilia up to 25%		Ingestion of raw or undercooked crabs
8. *Schistosoma*	Low eosinophilia		Penetration of human skin by cercaria
9. *A. cantonensis*	≥10 eosinophils/mm^3 in the cerebrospinal fluid (CSF) and/or eosinophils more than 10% of CSF leucocytes		Ingestion of larvae in raw snails, crabs, frogs, shrimps
10. *B. procyonis*			Ingestion of eggs in Raccoon' stool
11. *G. spinigerum*			Ingestion of larvae in undercooked fish, frogs or using their tissues in medication

Appendix C: Opportunistic Parasites

Parasite	Infection in immunocompetent	Infection in immunocompromised	Main route of infection
1. *Cryptosporidium*	Asymptomatic or self-limited diarrhea	Severe life-threatening diarrhea with dehydration and extraintestinal spread	Feco-oral, waterborne
2. *Cystoisospora*	Self-limited diarrhea	Severe diarrhea with extraintestinal spread	Feco-oral, food, drinks
3. *Cyclospora*	Self-limited diarrhea	Severe diarrhea with extraintestinal spread	Feco-oral, food, drinks
4. *Toxoplasma*	Asymptomatic or mild diseases	Reactivation or new fulminate infection, hepatomegaly, LN, brain, retinal, muscular infections	Cat oocysts, undercooked meat, blood, congenital
5. *Microsporidia*	Asymptomatic or mild diseases	Enteritis, persistent diarrhea, hepatitis, keratitis, muscular infection with disseminated infection	Feco-oral, food, drinks
6. *Blastocystis hominis*	Asymptomatic	Enteritis, arthritis, disseminated infection	Feco-oral, waterborne. food
7. *Leishmania tropica*	Localized skin ulcer	*Leishmania diffusa*: macro-nodular skin lesions all over the body	Sandfly bite
8. *Acanthamoeba*	Chronic GAE	Acute GAE, with disseminated infection to skin and muscle	Respiratory, skin, and urogenital
9. *Strongyloides stercoralis*	Asymptomatic or enteritis with diarrhea	Hyperinfection syndrome, chest infiltrate, intractable mucoid diarrhea. Fatal disseminated infection	Skin penetration by filariform larva, internal autoinfection
10. *Hymenolepis nana*	Asymptomatic or mild diarrhea	Hyperinfection syndrome with tissue affection	Feco-oral, foods, drinks
11. *Capillaria philippinensis*	Protein enteropathy, malabsorption	Profuse diarrhea with hyperinfection syndrome	Eating undercooked freshwater fish
12. *Sarcoptes scabiei*	Regular scabies, itchy localized treatable and controlled	Crusted keratolytic extended all over the body with low itching (Norwegian scabies)	Contact

Appendix D: Transmission of Parasites

Animals Transmitted Parasites

Parasite	Animal	Disease	Main route of infection
1. *Cryptosporidium*	Cattles	Diarrheal illness	Feco-oral, water borne
2. *Taenia saginata*	Cattles	Intestinal illness	Undercooked cow meat
3. *Trichostrongylus colubriformis*	Cattle and sheep	Enteritis	Ingestion of larvae
4. *Fasciola*	Sheep	Liver and intestinal disease	Fresh water vegetation
5. *Toxoplasma*	Cats	Multisystem disease	Oocysts, undercooked meat, blood, congenital
6. *Toxocara*	Cats and dogs	VLM, OLM	Ingestion of eggs from cats and dogs
7. *Echinococcus*	Dogs	Hydatid disease	Ingestion of eggs from dogs
8. *Dipylidium caninum*	Dogs	Dipylidiasis	Ingestion of foods or fleas with worm's eggs
9. *Multiceps multiceps*	Dogs	Coenurosis	Ingestion of eggs from dogs
10. *L donovani*	Dogs	Kala azar	Sandfly bite, blood

Parasite	Animal	Disease	Main route of infection
11. *Taenia solium*	Pigs	Cysticercosis	Ingestion of eggs, internal autoinfection
12. *Taenia solium*	Pigs	Intestinal illness.	Undercooked ham
13. *Trichinella spiralis*	Pigs	Trichinosis (encysted larvae)	Undercooked ham
14. *Balantidium coli*	Pigs	Balantidiasis (dysenteric)	Ingestion of cyst
15. *Giardia lamblia*	Beaver, mice	Enteritis	Ingestion of cyst
16. *Capillaria philippinensis*	Sea birds.	Enteritis	Undercooked fresh and brackish water fish
17. *Hymenolepis diminuta*	Rodents	Enteritis	Ingestion of eggs in dogs' fleas
18. *B. procyonis*	Raccoon	VLM, Eosinophilic meningitis	Ingestion of eggs in food with raccoon' stool
19. *Angiostrongylus cantonensis*	Rats	VLM, eosinophilic meningitis	Eating shrimps, crabs, containing larvae.
20. *Gnathostoma spinigerum*	Dogs & Cats	CLM, eosinophilic meningitis	Ingestion of larvae in snakes, frogs, fish.
21. *Spirometra*	Dogs & Cats	Sparganosis	Ingestion of pro- or plero- cercoids in frogs, fish.

Fish Transmitted Parasites

Parasite	Fish type	Habitat	Infective stage
1. *H. heterophyes*	Fresh water fish Mugil and Tilapia	Intestine	Encysted metacercaria
2. *Metagonimus yokogawai*	Fresh water fish	Intestine	Encysted metacercaria
3. *Paragonimus westermani*	Crabs, crayfish	Lung	Encysted metacercaria
4. *Opisthorchis viverrini*	Fresh water fish	Liver	Encysted metacercaria
5. *Clonorchis sinensis*	Fresh and brackish water fish	Liver	Encysted metacercaria
6. *Anisakis simplex*	Herring and Mackerel (Sea water)	Stomach	Larva in fish
7. *Diphyllobothrium latum*	Salmon and lake trout fish	Intestine	Plerocercoid
8. *Capillaria philippinensis*	Fresh and brackish water fish	Intestine	Larval stages in fish

Arthropods Transmitted Parasites

Arthropod	Parasite transmitted	Mode of infection	Disease
Mosquitoes:			
1. *Anopheles*	*Plasmodia*	Biting	Malaria
2. *Culex and Aedes*	*Wuchereria bancrofti*	Biting	Filariasis
3. *Mansonia*	*B. malayi*	Biting	Filariasis
Flies:			
1. *Simulium* (black fly)	*Onchocerca volvulus*	Biting	River blindness
2. *Phlebotomus* (sand fly)	*Leishmania*	Biting	Leishmaniasis
3. *Glossina* (tsetse fly)	*Trypanosoma brucei*	Biting	HAT
4. Winged bugs	*Trypanosoma cruzi*	Fecal contamination	Chagas
5. *Chrysops* (deer fly)	*Loa loa*	Biting	Loiaisis
Fleas: *Ctenocephalides canis and cati*	*Dipylidium caninum* *Hymenolepis diminuta.* *Hymenolepis nana*	Ingestion of flea containing the infective stages	Dipylidiasis Hymenolepiasis
Ticks:	*Babesia*	Transovarian	Babesiosis
Cyclops:	*Dracunculus medinensis* *Diphyllobothrium latum*	Ingestion of the *Cyclops* containing the infective stages	Dracunculiasis Sparganosis

Appendix E: Immunodiagnostic Techniques of Parasitic Diseases

Technique	Principles	Reading	Applications
Direct Agglutination Test (DAT)/Dipstick, card agglutination	Binding between an antibody to a particulate or soluble antigens, e.g., enzymes., histidine rich proteins (HRP1,2,3); aldolase; Dehydrogenase	Visible agglutination/ clumping/colored stick	Chagas, filariasis, leishmaniasis, malaria
Indirect Hemagglutination Test (IHA)	Known antigen coated on erythrocytes, reacted with the patient serum (unknown antibodies)	Visible agglutination/ clumping	Leishmaniasis, cysticercosis, amebiasis, toxoplasmosis
Counter immune-electrophoresis (CIE)	It is an active diffusion test that involves the movement of antigen and antibody toward each other through a gel medium when an electric current is passed through a buffer solution	Visible mark or precipitate of an antigen- antibody precipitate	Amebiasis, leishmaniasis and cryptosporidiosis
Immunoblotting (IB), Western blot (WB)	Unknown antigens are separated by electrophoresis and blotted on nitrocellulose paper. Blotted antigens are subsequently identified by staining with labeled antibodies	Visible stained precipitate of antigen–antibody on the paper	Amebiasis, fascioliasis cryptosporidium, leishmaniasis, research works
Direct Immunofluorescence (DIF)	A fluorescent dye (fluorescein isocyanate) conjugated to serum antibody (or antigen) reacted together and tested under ultraviolet light	A fluorescent tag suggests a positive antigen antibody complex	Malaria, giardiasis, leishmaniasis, trichinosis, cryptosporidiosis
Indirect-fluorescent Antibody (IFA)	A known antigen on a slide with a patient serum i. After binding between antigen and antibody, the slide is washed and covered with a fluorescent dye conjugated to anti-human globulin and tested under ultraviolet light	The observation of fluorescent tag suggests a positive antibody in the patient serum	Malaria, giardiasis, leishmaniasis, trichinosis, cryptosporidiosis, amebiasis, filariasis, schistosomiasis
Enzyme-linked immunosorbent assay (ELISA)	Binding between enzyme labeled antigen or antibody, then a colorimetric substrate is added. The optical density of the color is proportional to the concentration of the antigen–antibody reaction	A color change resulting from enzyme–substrate interaction indicates a positive reaction	Almost all parasitic infections
Intradermal skin test (IDT)	It is a delayed hypersensitivity skin reaction, involving injection of purified parasitic antigen at the skin and after sometimes (up to days) skin is tested	If the patient has specific antibodies, an inflammatory reaction detected at the site of injection (erythema, nodules, itching or even an ulcer)	Toxoplasm in skin test, Montenegro skin test, and Bachman's test

Further Reading

Abdulrazzaq H. Parasites: introduction, classification, lifecycle and relationship with their hosts. IJRASB. 2021;8:141–6.

Capron M, Trottein F, editors. Parasites and allergy. 1st ed. Karger Publishers; 2006.

Cardoso AEC, Cardoso AEO, Talhari C, Santos M. Update on parasitic dermatoses. An Bras Dermatol. 2020;95(1):1–14. https://doi.org/10.1016/j.abd.2019.12.001.

Carpio A, Romo ML, Parkhouse RM, Short B, Dua T. Parasitic diseases of the central nervous system: lessons for clinicians and policy makers. Expert Rev Neurother. 2016;16(4):401–14. https://doi.org/10.1586/14737175.2016.1155454.

Chiodini PL, Moody AH, Manser DW, editors. Atlas of medical helminthology & protozoology. 4th ed. Toronto: Churchill Livingstone; 2003.

Claudia M, Marcos LA. Carcinogenesis associated with parasites other than Schistosoma, Opisthorchis and Clonorchis: a systematic review. Int J Cancer. 2016;138(12):2915–21. https://doi.org/10.1002/ijc.30028.

Cox FEG, editor. Modern parasitology: a textbook of parasitology. 2nd ed. Wiley-Blackwell; 2009.

Crompton DW, Savioli L. Handbook of helminthiasis for public health. Boca Raton, FL: CRC Press; 2007.

Daher EDF, da Silva Junior GB, Trivedi M, et al. Kidney complications of parasitic diseases. Nat Rev Nephrol. 2022;18:396–406.

Despommier DD, Griffin DO, Gwadz RW, Hotez PJ, Knirsch CA. Parasitic diseases. 7th ed. Parasites Without Borders; 2019.

Farrar J, Hotez PJ, Junghanss T, Kang G, Lalloo D, White NJ, Patricia J, Garcia PJ, editors. Manson's tropical diseases. 24th ed. Elsevier; 2023.

Fernald HT. Applied entomology: an introductory textbook of insects in their relation to man. Kessinger Publishing (Amazon); 2008.

Frischknecht F. Malaria: deadly parasites, exciting research and no vaccination. Springer Spektrum Wiesba; 2023.

Garcia LS. Diagnostic medical parasitology. 6th ed. Wiley, ASM Press; 2020.

Goddard J. Physician's guide to arthropods of medical importance. 5th ed. Boca Raton, FL: CRC Press; 2007.

Heelen J, Ingersoll F. Essential of human parasitology. Albany, NY: Delmar Thomson Learning; 2002.

John DT, Petri WA, Martin G, editors. Markell & Voge's medical parasitology. 10th SEA ed. Elsevier (Singapore); 2020.

Khemasuwan D, Farver C, Mehta AC. Parasitic diseases of the lung. In: Mehta A, Jain P, Gildea T, editors. Diseases of the Central Airways, Respiratory medicine. Cham: Humana Press; 2016.

Lynn Peters L, Burkert S, Grüner B. Parasites of the liver – epidemiology, diagnosis and clinical management in the European context. J Hepatol. 2021; https://doi.org/10.1016/j.jhep.2021.02.015.

Magill AJ, Hill DR, Solomon T, Ryan ET. Hunter's tropical medicine and emerging infectious diseases. 9th ed. New York: Saunders; 2013.

Mamaghani AJ, Fathollahi A, Arab-Mazar Z, Kohansal K, Fathollahi M, Spotin A, Bashiri H, Bozorgomid A. Toxoplasma gondii vaccine candidates: a concise review. Ir J Med Sci. 2023;192(1):231–61.

Maule AG, Marks NJ, editors. Parasitic flatworms: molecular biology, biochemistry, immunology and physiology. 1st ed. Oxford, UK: CABI Publishing; 2006.

Mike Service. Medical entomology for students. 4th ed. Cambridge: Cambridge University Press; 2008.

Muller R, Wakelin D, editors. Worms and human diseases. 2nd ed. Oxford, UK: CABI Publishing; 2002.

Murphy GS, Johnson S. Clinical aspects of eosinophilic meningitis and meningoencephalitis caused by Angiostrongylus cantonensis, the rat lungworm. Hawaii J Med Public Health. 2013;2:35–40.

O'Gorman RG, Donnenberg AD, editors. Handbook of human immunology. 1st ed. Boca Raton, FL: CRC Press; 2008.

Pitt SJ, Gunn A. Parasitology: an integrated approach. 2nd ed. Wiley; 2022.

Richard L, Brigitte L, Richard PL, Robert P, Craig R, Richard KG, Ron S, Renate F. The biology of parasites. 1st ed. Wiley-Blackwell; 2017.

Ridley JW. Parasitology for medical and clinical laboratory professionals. Cengage Learning; 2012.

K. ElShewy, *Medical Parasitology*, https://doi.org/10.1007/978-3-031-64736-9

Spineanu E. Protozoal parasites in the gastrointestinal tract: challenges and solutions (Medical care and health). Kindle Edition Publisher, independently published; 2023.

Sripa B, Kaewkes S, Intapan PM, Maleewong W, Brindley PJ. Food-borne trematodiases in Southeast Asia: epidemiology, pathology, clinical manifestation and control. Adv Parasitol. 2010;72:305–50. https://doi.org/10.1016/S0065-308X(10)72011-X.

Vijayan VK. Diagnosis of pulmonary parasitic diseases. In: Barrios R, Haque A, editors. Parasitic diseases of the lungs. Berlin: Springer; 2013.

World Health Organization. Guidelines for the treatment of malaria. 3rd ed. 2015. Archived from the original on 25 April 2015.

Yves C, Carine T, Philippe D, François P. Congenital parasitic infections: a review. Acta Trop. 2011;121:55–70.

MIX
Papier aus verantwortungsvollen Quellen
Paper from responsible sources
FSC® C105338

If you have any concerns about our products,
you can contact us on
ProductSafety@springernature.com

In case Publisher is established outside the EU,
the EU authorized representative is:
Springer Nature Customer Service Center GmbH
Europaplatz 3, 69115 Heidelberg, Germany

Printed by Libri Plureos GmbH
in Hamburg, Germany